Instant *Notes* *in*

MICROBIOLOGY

The INSTANT NOTES series

Series editor
B.D. Hames
School of Biochemistry and Molecular Biology, University of Leeds, Leeds, UK

Biochemistry
Animal Biology
Molecular Biology
Ecology
Genetics
Microbiology

Forthcoming titles
Chemistry for Biologists
Psychology
Immunology
Neuroscience

Instant Notes *in*

MICROBIOLOGY

J. Nicklin
Department of Biology,
Birkbeck College, London, UK

K. Graeme-Cook
Department of Biology,
Birkbeck College, London, UK

T. Paget
Department of Biological Sciences,
University of Hull, Hull, UK

&

R.A. Killington
Department of Microbiology,
University of Leeds, Leeds, UK

*β*IOS
SCIENTIFIC
PUBLISHERS

© BIOS Scientific Publishers Limited, 1999

First published 1999

A CIP catalogue record for this book is available from the British Library.

ISBN 1 85996 156 8

BIOS Scientific Publishers Ltd
9 Newtec Place, Magdalen Road, Oxford OX4 1RE, UK.
Tel. +44 (0) 1865 726286. Fax +44 (0) 1865 246823
World Wide Web home page: http://www.bios.co.uk/

Published in the United States of America, its dependent territories and Canada by
Springer-Verlag New York Inc., 175 Fifth Avenue, New York, NY 10010-7858,
in association with BIOS Scientific Publishers Ltd.

Published in Hong Kong, Taiwan, Singapore, Thailand, Cambodia, Korea, The Philippines, Indonesia,
The People's Republic of China, Brunei, Laos, Malaysia, Macau and Vietnam
by Springer-Verlag Singapore Pte Ltd, 1 Tannery Road, Singapore 347719,
in association with BIOS Scientific Publishers Ltd.

Production Editor: Fran Kingston.
Typeset and illustrated by The Florence Group, Stoodleigh, UK.
Printed by Biddles Ltd, Guildford, UK.

CONTENTS

ABBREVIATIONS

A	adenine	HSV	herpes simplex virus
AcCoA	acetyl coenzyme A	I	inosine
ACP	acyl carrier protein	ICNV	International Committee on Nomenclature of Viruses
ADP	adenosine 5′-diphosphate		
Ala	alanine	Ig	immunoglobulin
AMP	adenosine 5′-monophosphate	IHF	integration host factor
A-site	amino-acyl site (ribosome)	Inc group	incompatible group (of plasmids)
ATP	adenosine 5′-triphosphate		
ATPase	ATP synthase	IS	insertion sequence
BHK	baby hamster kidney	Kb	kilobase
bp	base pair	KDO	2-keto-2-deoxyoctonate
C	cytosine	KDPE	2-keto-2-deoxy-6-phospho-gluconate
C-phase	chromosome replication phase (bacterial cell cycle)		
		Lac	lactose
CAP	catabolite activator protein	LBP	luciferin-binding protein
CAT	chloramphenicol acetyl transferase	LPS	lipopolysaccharide
		MAC	membrane-attack complex
CFU	colony-forming unit	MCP	methyl-accepting chemotaxis protein
CMV	cytomegalovirus		
CNS	central nervous system	MEM	minimal essential medium
CPE	cytopathic effect	MHC	major histocompatibility complex
CRP	cAMP receptor protein		
CTL	cytotoxic T lymphocyte	m.o.i.	multiplicity of infection
Da	Dalton	mRNA	messenger ribonucleic acid
D-Ala	D-alanine	MTOC	microtubule organizing centre
DAP	*meso*-diaminopimelic acid	NAD⁺	nicotinamide adenine dinucleotide (oxidized form)
D-Glu	D-glutamic acid		
DNA	deoxyribonucleic acid	NADH + H⁺	nicotinamide adenine dinucleotide (reduced form)
dNTP	deoxyribonucleoside triphos-phate		
		NADP	nicotinamide adenine dinucleotide phosphate (oxidized form)
DOM	dissolved organic matter		
D-phase	division phase (bacterial cell cycle)		
		NADPH + H⁺	nicotinamide adenine dinucleotide phosphate (reduced form)
ds	double-stranded		
EF	elongation factor		
EM	electron microscopy	NAG	*N*-acetylglucosamine
ER	endoplasmic reticulum	NAM	*N*-acetylmuramic acid
FAD	flavin adenine dinucleotide (oxidized)	NB	nutrient broth
		NTP	ribonucleoside triphosphate
FADH₂	flavin adenine dinucleotide (reduced)	O	operator
		OD	optical density
G	guanine	omp	outer membrane protein
G-phase	gap phase (bacterial cell cycle)	P	promoter
GTP	guanosine 5′-triphosphate	PCR	polymerase chain reaction
HA	hemagglutination	pfu	plaque-forming unit
Hfr	high frequency recombination	PHB	poly-β-hydroxybutyrate

Phe	phenylalanine	S	Svedberg coefficient
P_i	inorganic phosphate	snRNA	small nuclear ribonucleic acid
PMF	proton motive force	SPB	spindle pole bodies
PMN	polymorphonucleocyte	ss	single-stranded
PP_i	inorganic pyrophosphate	T	thymine
PS	photosystem	TCA	tricarboxylic acid
PSI and II	photosystems I and II	TCID	tissue culture infective dose
P-site	peptidyl site (ribosome)	tRNA	transfer RNA
R	resistance (plasmid)	Trp	tryptophan
RBC	red blood cell	TSB	tryptone soya broth
redox	reduction–oxidation	U	uracil
RER	rough endoplasmic reticulum	U_L, U_S	unique long, unique short
rubisco	ribulose bisphosphate carboxylase	UDP	uridine diphosphate
		UDPG	uridine diphosphate glucose
RNA	ribonucleic acid	UV	ultraviolet light
rRNA	ribosomal RNA		

PREFACE

Microbiological matters have always been in the public eye, but recent occurrences of food poisoning, antibiotic-resistant bacteria, HIV and BSE have heightened awareness of the existence of microbes and emphasized the need for continuing research into microbiology. We need to understand and control the activities of microbes when they are detrimental to mankind, and to enhance and manage their effects when they are beneficial to us.

Very little microbiology is taught in schools within the A-level syllabus, and foundation text books in the subject must deal effectively with the basics of the topic, but must also provide sufficient information to bring the reader's knowledge up to university first- or second-year standards. Many of the available texts are very large and expensive and dwell heavily on the medical, symptom-based aspects of microbiology. This textbook aims to provide the reader with basic information about microbes without this bias.

We have considerable experience in teaching microbiology, especially to mature, part-time undergraduate and postgraduate students at Birkbeck College, University of London. These students, who attend lectures in the evening, have busy life-styles and many commitments. They therefore require information to be presented in a concise, understandable format without excessive frills or diversions. We have written this book using our understanding of their problems.

Instant Notes in Microbiology has been written in a way that gives students easy access to the important key features of microbes. The book has been divided into eleven main sections which cover all areas of microbiology. Each topic starts with a key notes section, which is a revision check list of the topic, and then expands on the subject. Diagrams are kept to the simple outline drawings that a student might produce under examination conditions. Further reading lists are provided for each topic.

In Section A you are introduced to the diverse range of microbes and the roles they can play in the environment. Section B covers the biochemistry of metabolism in microbes and Section C describes salient features of DNA replication, transcription and translation in both prokaryotes and eukaryotes. In section D, bacterial taxonomy, structure, function and growth are discussed and topics on handling bacteria in the laboratory are included. The bacterial genetics section, Section E, is devoted to the means by which bacteria can alter their genetic make-up, including mutagenesis, DNA repair, conjugation, transduction and transformation. Section F, the last on bacteria, discusses their relationship with their environment. This section is heavily slanted to bacterial interactions with a human host as this is still the region of microbiology which is studied the most, and is probably of the greatest interest to the average reader.

Section G introduces the reader to the basic structure of eukaryotic cells, and the taxonomic divisions within the eukaryotic protists. Nuclear (mitotic and meiotic) division and cell division are also covered in this section. Subsequent sections deal with the different taxonomic groups in more detail. Section H describes the fungi, their structure, biology and impact on the environment. Within this section other related phyla are considered, organisms which by parallel evolution have come to share many characteristics of fungi. The important features of the algae are covered in Section I, and in Section J the biology of the protozoa and their disease-causing capacity is detailed. The final subject of virology is dealt with in Section K.

This book has been designed to allow students instant access to subjects that are written as free-standing topics. Cross-referencing to other topics allows the reader to follow-up different lines of interest, and the reading list provided for each section should provide further sources of information. Due to the many different aspects of microbiology, there are, inevitably, a few omissions from this book, notably the use of microbes in molecular biology. Other books in the *Instant Notes* series, including *Instant Notes in Molecular Biology* and *Instant Notes in Biochemistry*, will help to fill these gaps. However, for all topics related to general microbiology we hope that you will find this book a useful revision aid and a stimulus for further study.

Kate Graeme-Cook and Jane Nicklin

Acknowledgments

We would like to thank our families, colleagues (especially Martin Ingrouille), neighbors and friends for their support and help while we were writing this book. Thanks are also due to the staff at BIOS for their encouragement and assistance during this time. Finally, we would like to thank David Hames for asking us to write the book and whose constructive criticisms greatly improved its general style.

We would like to dedicate this book to our parents.

A1 THE MICROBIAL WORLD

Key Notes

What is a microbe?	The word microbe (microorganism) is used to describe an organism that is so small that, normally, it cannot be seen without the use of a microscope. Viruses, bacteria, fungi, protozoa and some algae are all included in this category.
Prokaryotes and eukaryotes	There are many differences between prokaryote and eukaryote cells. The main features are the presence of a nucleus, organelles, such as mitochondria and chloroplasts, and complex internal membranes in eukaryotes. Bacteria are prokaryotes; all other microbial cells are eukaryotes.
The importance of microbiology	Microbes are essential to life. Among their many roles, they are necessary for geochemical cycling and soil fertility. They are used to produce food as well as pharmaceutical and industrial compounds. On the negative side, they are the cause of many diseases of plants and animals and are responsible for the spoilage of food. Finally, microbes are used extensively in research laboratories to investigate cellular processes.

What is a microbe?

A **microbe** or **microorganism** is a member of a large, extremely diverse, group of organisms that are lumped together on the basis of one property – the fact that, normally, they are so small that they cannot be seen without the use of a microscope. The word is therefore used to describe **viruses**, **bacteria**, **fungi**, **protozoa** and some **algae**: the relative sizes and nature of these are shown in *Table 1*. However, there are a few exceptions, for example, the fruiting bodies of many fungi such as mushrooms are frequently visible to the naked eye; equally, some algae can grow to meters in length. Generally, microbes may be considered as fairly simple organisms. Most of the bacteria and protozoa and some of the algae and fungi are single-celled microorganisms, and even the multi-celled microbes do not have a great range of cell types. Viruses are not even cells, just genetic material surrounded by a protein coat, and are incapable of independent existence.

The science of microbiology did not start until the invention of the microscope in the mid 16th century and it was not until the late 17th century that Robert Hooke and Antoine van Leeuwenhoek made their first records of fungi, bacteria and protozoa. The late 19th century was the time when the first real breakthroughs on the role of microbes in the environment and medicine were made. Louis Pasteur disproved the theory of **spontaneous generation** (that living organisms spontaneously arose from inorganic material) and Robert Koch's development of **pure culture** techniques (see Topic D9) allowed him to show unequivocally that a bacterium was responsible for a particular disease. Since then the science has grown dramatically as microbiology impinges on all aspects of life and the environment.

Table 1.   Types of microbes, their sizes and cell type

Microbe	Approximate range of sizes	Nature of cell	Section of book
Viruses	0.01–0.25 μm	Acellular	K
Bacteria	0.1–10 μm	Prokaryote	D,E,F
Fungi	2 μm–>1 m	Eukaryote	G,H
Protozoa	2–1000 μm	Eukaryote	J
Algae	1 μm–several meters	Eukaryote	I

Prokaryotes and eukaryotes

Within the microbial world can be found two different categories of cell type, **prokaryote** and **eukaryote**. Bacteria are prokaryotes: they lack a distinct nuclear membrane, the organelles associated with energy generation, such as mitochondria and chloroplasts, and complex internal membranes, such as endoplasmic reticulum and Golgi apparatus, which are found in eukaryotes. A comparison of the main features of these two categories of cell is shown in *Table 2*, but other differences do occur. Although the basic mechanisms of DNA replication (Topic C2), RNA synthesis (Topic C4) and protein synthesis (Topic C7) are the same in both prokaryotes and eukaryotes, there are differences in the components and enzymes involved. These are discussed in the appropriate topics.

The importance of microbiology

Microbes impinge on all aspects of life; just a few of these are listed below.

- **The environment.** Microbes are responsible for the cycling of carbon, nitrogen and phosphorus (geochemical cycles), all essential components of living organisms (Topic F1). They are found in association with plants in symbiotic relationships, maintain soil fertility and may also be used to clean up the environment of toxic compounds (bio-remediation; Topics H6 and I4). Some microbes are devastating plant pathogens (Topic H7), which destroy important food crops, but others may act as biological control agents against these diseases.
- **Medicine.** The disease-causing ability of some microbes such as smallpox (Variola virus; Topic K8), cholera (*Vibrio cholera* bacteria; Section F3) and malaria (*Plasmodium* protozoa, Topic J7) is well known. However, microorganisms have also provided us with the means of their control in the form of antibiotics (Topic F7) and other medically important drugs.
- **Food.** Microbes have been used for thousands of years, in many processes, to produce food, from brewing and wine making, through cheese production and bread making, to the manufacture of soy sauce (Topic F2). At the other end of the scale, microbes are responsible for food spoilage, and disease-causing microbes are frequently carried on food (Topic F5).
- **Biotechnology.** Traditionally microbes have been used to synthesize many important chemicals such as acetone and acetic acid (Topic F2). More recently, the advent of genetic engineering techniques has led to the cloning of pharmaceutically important polypeptides into microbes, which may then be produced on a large scale.
- **Research.** Microbes have been used extensively as model organisms for the investigation of biochemical and genetical processes as they are much easier to work with than more complex animals and plants. Millions of copies of

the same single cell can be produced in large numbers very quickly and at low cost to give plenty of homogeneous experimental material. An additional advantage is that most people have no ethical objections to experiments with these microorganisms.

Table 2.   The major differences between prokaryote and eukaryote genetic and cellular organization

Prokaryotes	Eukaryotes
Organization of the genetic material and replication	
DNA free in the cytoplasm	DNA is contained with a membrane bound nucleus. A nucleolus is also present
Only one chromosome	>1 chromosome. Two copies of each chromosome may be present (diploid)
DNA associated with histone-like proteins	DNA complexed with histone proteins
May contain extrachromosomal elements called plasmids	Plasmids only found in yeast
Introns not found in mRNA	Introns found in all genes
Cell division by binary fission – asexual replication only	Cells divide by mitosis
Transfer of genetic information occurs by conjugation, transduction and transformation	Exchange of genetic information occurs during sexual reproduction. Meiosis leads to the production of haploid cells (gametes) which can fuse
Cellular organization	
Cytoplasmic membrane contains hopanoids. Lipopolysaccharides and teichoic acids found	Cytoplasmic membrane contains sterols
Energy metabolism associated with the cytoplasmic membrane	Mitochondria present in most cases
Photosynthesis associated with membrane systems and vesicles in cytoplasm	Chloroplasts present in algal and plant cells
	Internal membranes, endoplasmic reticulum and Golgi apparatus present associated with protein synthesis and targetting
	Membrane vesicles such as lysosomes and peroxisomes present
	Cytoskeleton of microtubules present
Flagella consist of one protein, flagellin	Flagella have a complex structure with 9+2 microtubular arrangement
Ribosomes – 70S	Ribosomes – 80S (mitochondrial and chloroplast ribosomes are 70S)
Peptidoglycan cell walls (eubacteria only: different polymers in archaebacteria)	Polysaccharide cell walls, where present, are generally either cellulose or chitin

B1 HETEROTROPHIC PATHWAYS

Key Notes

Nutritional types	Metabolism is divided into those pathways that are degradative (catabolic) and those that are involved in synthesis. Catabolic pathways often produce energy. Microbes that utilize organic molecules as a source of energy are called heterotrophs. Phototrophs obtain energy from light, and lithotrophs obtain energy from inorganic compounds.
Glycolysis	Most microbes utilize the glycolytic pathway for the catabolism of carbohydrates such as glucose and fructose. The products of this pathway are pyruvate, which can be further metabolized via the citric acid cycle, forming adenosine 5'-triphosphate (ATP) and the reduced form of nicotinamide adenine dinucleotide (NADH + H$^+$). This pathway is located in the cytoplasm of microbes and can function in the presence or absence of oxygen.
The Entner–Doudoroff pathway	The bacterial genera *Pseudomonas*, *Rhizobium* and *Agrobacter* substitute the Entner–Doudoroff pathway for the glycolytic pathway. This pathway is not as efficient in producing energy, with 1 mole of ATP being formed for each mole of glucose metabolized.
Pentose phosphate pathway	The pentose phosphate pathway produces NADPH + H$^+$ and sugars (4 C, 5 C). These are required for many synthetic reactions. When organisms are growing on a pentose (5 C) sugar, the pathway can be used to produce carbohydrates for cell-wall synthesis. Glyceraldehyde-3-phosphate formed by the pathway can be used to generate energy by glycolysis or by the Entner–Doudoroff pathway.
Citric acid cycle	The metabolism of pyruvate (formed by glycolysis) to CO$_2$ by the citric acid cycle is the major mechanism of ATP generation in the cell and is also an important source of carbon skeletons for biosynthesis. The fully functioning pathway requires oxygen; however, some organisms possess an incomplete cycle that can function in the presence or absence of oxygen but generates little or no energy.
Fermentations	NADH + H$^+$ produced by catabolic reactions such as the citric acid cycle can be oxidized by the electron-transport pathway in the presence of oxygen. However, in the absence of oxygen, many microbes utilize fermentation reactions to reoxidize NADH + H$^+$. Microbial fermentations are characterized by the end products formed. *Clostridia* are unusual in that they form ATP from the fermentation of amino acids by the Stickland reaction.
ATP yields	The citric acid cycle is the most efficient mechanism for generating ATP from glucose in the presence of oxygen. For microbes that live in environments where oxygen is absent or only present intermittently, ATP generation is less efficient.
Related topics	Bacterial taxonomy (D1) Bacterial growth and cell cycle (D8)

Nutritional types

Metabolism in all cells is divided into **catabolic** (those pathways involved in breakdown of organic molecules for energy and the production of small compounds that may be used for synthesis) and **anabolic** (pathways involved in synthesis) processes. In all organisms these pathways are balanced as the energy required for anabolic processes is produced by catabolic pathways. In mammalian cells, energy production has been maximized by the use of oxygen and thus the cell is usually well supplied with energy; however, in microbes this is not always the case. Microbes can be divided into metabolic classes which relate to the sources of energy they use. The three groups are **heterotrophs** which utilize organic molecules as a source of energy (these are also called **chemo-organotrophs**), **phototrophs** which obtain energy from light, and **lithotrophs** which obtain energy from inorganic compounds. Carbon for cell synthesis is obtained from organic molecules; however, some microbes, including the phototrophs, fix CO_2.

Glycolysis

The **majority** of microbes utilize the **glycolytic pathway** (also known as the **Embden–Meyerhof pathway)** for the catabolism of carbohydrates such as glucose and fructose (*Fig. 1*). This series of reactions occurs in the cytoplasm of microbes and can operate either anerobically (in the absence of oxygen) or aerobically (in the presence of oxygen).

The overall equation for this pathway is

$$\text{Glucose} + 2ADP + 2P_i + 2NAD^+ \rightarrow 2 \text{ pyruvate} + 2ATP + 2NADH + 2H^+$$

Pyruvate formed by glycolysis can be further metabolized in the presence of oxygen to generate energy via the **citric acid cycle** or can be used for synthesis of other compounds such as amino acids. Adenosine triphosphate (ATP) can be used directly to drive uptake of substrates or can be used to drive synthetic reactions. $NADH + H^+$ can be used to produce energy via **oxidative phosphorylation** (a method of ATP formation that requires electron transport; see Topic B2) or can be used as a source of H^+ for reduction reactions. Some organisms such as the bacteria *Clostridia* utilize **inorganic pyrophosphate** (PP_i) in place of ATP as a source of energy to drive the formation of pyruvate from phosphoenolpyruvate and for the conversion of fructose-6-phosphate into fructose-1,6-bisphosphate.

The Entner–Doudoroff pathway

A **minority** of bacteria including *Pseudomonas*, *Rhizobium* and *Agrobacter* substitute the **Entner–Doudoroff** pathway for the glycolytic pathway (*Fig. 2*). The pathway yields 1 mole each of ATP, $NADPH + H^+$ and $NADH + H^+$ for each mole of glucose **metabolized**. The products of this pathway, like those of glycolysis, can be used for a variety of functions; however, the $NADPH + H^+$ formed is used for synthetic reactions.

Pentose phosphate pathway

The **pentose phosphate pathway** or **hexose monophosphate pathway** may operate at the same time as glycolysis or the Entner–Doudoroff pathway. This pathway can also operate either in the presence or absence of oxygen. The pentose phosphate pathway is an important source of energy in many microorganisms; however, its major role would seem to be for biosynthesis. The basic outline of this pathway is shown in *Fig. 3*. The pathway produces $NADPH + H^+$

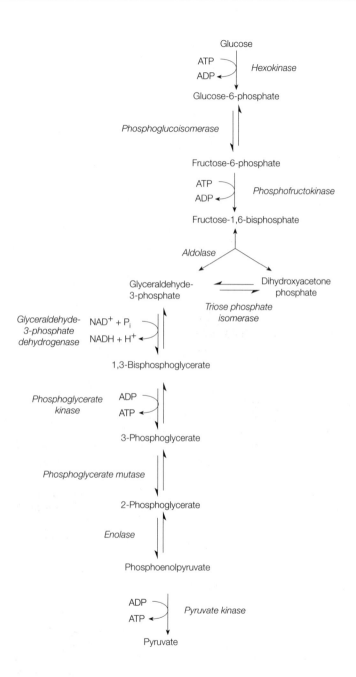

Fig. 1.   The glycolytic pathway.

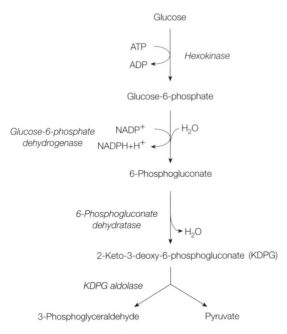

Fig. 2.   The Entner–Doudoroff pathway.

and sugars (4C, 5C), which are required for the synthesis of aromatic amino acids and nucleotides. When organisms are growing on a pentose (5 C) sugar, the pathway can also be used to produce carbohydrates for cell-wall synthesis. **Glyceraldehyde-3-phosphate** can be used to generate energy via the glycolytic/Entner–Doudoroff pathways.

Citric acid cycle   Although energy is obtained from the breakdown of pyruvate by one of the previous pathways, a significantly greater yield can be achieved in the presence of oxygen from the further oxidation of pyruvate to CO_2 via the **citric acid cycle** (*Fig. 4*) also known as the **tricarboxylic acid cycle**. Pyruvate does not enter this pathway directly, it must first undergo conversion into acetyl co-enzyme A (acetyl CoA):

$$\text{Pyruvate} + NAD^+ + CoA \rightarrow \text{Acetyl CoA} + NADH + H^+$$

This reaction is catalyzed by **pyruvate dehydrogenase**, a large complex containing three enzymes. Acetyl CoA can also be produced by the catabolism of lipids and amino acids as well as a wide range of carbohydrates. ATP can be formed from $NADH + H^+$ by **oxidative phosphorylation** (see Topic B2). This pathway is also an important source of carbon skeletons for use in biosynthesis. Citric acid cycle enzymes are widely distributed in most microbes and other microorganisms. Functional and complete cycles are found in most aerobic microbes, algae, fungi and protozoa; however, in **facultative organisms** (those that can grow in the presence or absence of oxygen) the complete citric acid cycle would only be functional in the presence of oxygen. Many anaerobic organisms have an incomplete cycle, which is used for the production of synthetic precursors.

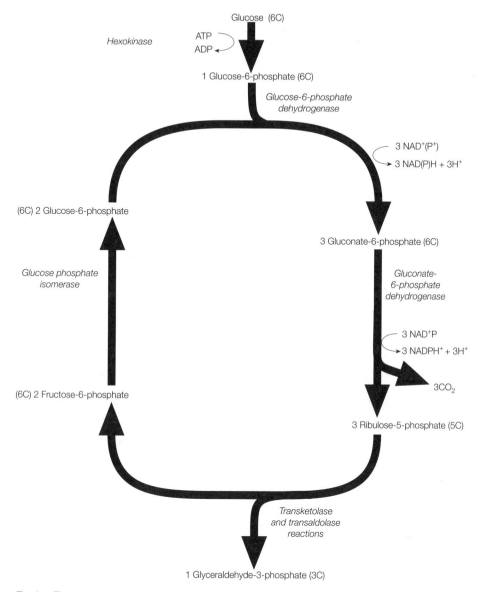

Fig. 3. The pentose phosphate cycle.

Fermentations

A major product in the catabolic pathways described previously is $NADH + H^+$. In the presence of oxygen this can be oxidized by the electron-transport pathway (see Topic B2). However, in the absence of oxygen, this must be oxidized back to NAD^+. Many microbes utilize derivatives of pyruvate as electron and H^+ acceptors and this allows $NADH + H^+$ to be re-oxidized to NAD^+. This process may lead to an increase in ATP synthesis, an important factor for organisms growing in the absence of oxygen (**anaerobes**). Such pathways are commonly termed fermentation reactions. **Lactic acid fermentation** is a common type of fermentation characteristic of lactic acid bacteria and some *Bacillus* species. Lactic acid fermentation can be divided into **homolactic fermentation**, where

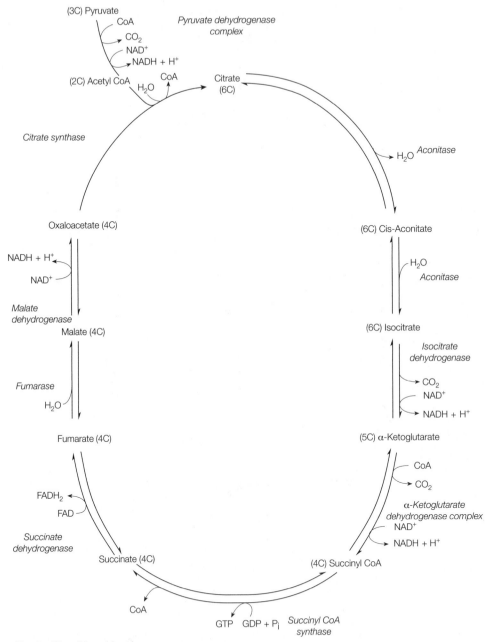

Fig. 4.   The citric acid cycle.

all the pyruvate is reduced to lactic acid, and **heterolactate fermentation** where other products may be formed, such as ethanol and CO_2. **Formic acid fermentation** is associated with the *Enterobacteriaceae*. Formic acid fermentation can also be divided into **mixed acid fermentation**, which results in the production of ethanol and a mixture of acids such as acetic, lactic and succinic. This is a characteristic of *E.coli* and *Salmonella*. The second type is **butanediol fermentation** where 2,3-butanediol, ethanol and smaller amounts of organic acids are formed. These are characteristic of *Serratia* and *Enterobacteriaceae*. These reactions are useful for the identification of members of the *Enterobacteriaceae*. A different type of bacterial fermentation is that of amino acids by *Clostridia*. These reactions can produce ATP by oxidizing one amino acid and using another as an electron acceptor in a process called the **Stickland reaction**.

ATP yields

ATP yields of the major catabolic pathways are shown in *Table 1*. The citric acid cycle is the most efficient mechanism for generating ATP from glucose; however, this requires oxygen as most of the ATP is generated by oxidative phosphorylation. For microbes that live in environments where oxygen is absent or only present intermittently, ATP generation is less efficient and often the limiting factor for growth.

Table 1.   ATP yields of the major catabolic pathways

Pathway	Yield of ATP (moles) produced per mole of glucose
Glycolysis	2
Entner–Doudoroff	1
Pentose phosphate	No direct production of ATP
Citric acid	34
	(most of the ATP is formed from the re-oxidation of NADH + H+ by oxidative phosphorylation)
Fermentations	2 or 3 if acetate formed
	(the use of alternative electron and H+ acceptors to reoxidize NADH + H+ will increase ATP production)

B2 ELECTRON TRANSPORT AND OXIDATIVE PHOSPHORYLATION

Key Notes

Electron-transport chain

NADH + H$^+$ and FADH$_2$ formed by catabolic reactions are used to produce ATP by the action of an electron-transport chain which is composed of a series of electron carriers. In bacteria, electron transport occurs in the inner cell membrane. In other microbes, electron transport occurs in the inner membrane of the mitochondria. Most bacterial electron-transport chains are branched unlike those in mammalian mitochondria. All electron-transport pathways function in a similar manner, requiring a series of oxidation and reduction reactions. The oxidation of a molecule involves the loss of electrons, and reduction involves the addition of electrons. Since electrons are conserved in chemical reaction, oxidations must be coupled with reduction reactions (redox reactions). The oxidation–reduction potential (redox potential) of a compound is a measure of its affinity for electrons. The difference in redox potential between NADH + H$^+$/NAD and ½O$_2$/H$_2$O drives the movement of electrons through a series of electron carriers from NADH + H$^+$ to O$_2$. Energy is released as electrons move between carriers. This can also be linked to ATP formation.

Formation of ATP

The chemiosmotic hypothesis is a widely accepted theory that explains how ATP is produced by electron transport. ATP production requires that H$^+$ move across the membrane where electron transport occurs, producing a transmembrane H$^+$ gradient. The H$^+$ moves back across the membrane via ATP synthase. The movement of H$^+$ is coupled with large releases of energy associated with electron transport, such that when NADH + H$^+$ is the electron donor and oxygen is the terminal electron acceptor, H$^+$ movement across the membrane occurs at three sites. In microbes with short pathways or where NADH + H$^+$ is not the electron donor, less ATP is produced.

Anaerobic respiration

Many bacteria, and some protozoa and fungi, live in environments devoid of oxygen. To utilize electron transport for ATP synthesis, nitrate, sulfate and CO$_2$ are used instead of oxygen (anaerobic respiration). These pathways, however, produce less ATP than aerobic respiration. Many microbes can use nitrate as an electron acceptor. The reduction of nitrite to N$_2$ is known as denitrification and is performed by *Pseudomonas* and some *Bacillus* species. Other groups that use alternative electron acceptors include methanogens which are obligate anaerobes and reduce CO$_2$ or carbonate to methane (methanogenesis) and those that can reduce sulfate to sulfide (e.g. *Desulfovibrio*).

Related topics

Bacterial cell structure (D2)
Transport across the membranes (D5)

Bacterial movement and
  chemotaxis (D6)

Electron-transport chain

NADH + H$^+$ and flavin adenine dinucleotide (reduced form) (FADH$_2$) formed by the catabolism of organic molecules are used to produce ATP by the action of an electron-transport chain that is composed of a series of electron carriers which transfer electrons to a **terminal electron acceptor** such as oxygen. This final reduction is performed by a **terminal oxidase**. Oxygen is not the only useful electron acceptor. In bacteria, electron transport occurs in the inner cell membrane, but in algae, fungi and many protozoa, electron transport and oxidative phosphorylation occur in the inner membrane of the mitochondria. Most bacterial electron-transport chains are different to that found in mammalian mitochondria (*Fig. 1a*) and many like *E.coli* are branched (*Fig. 1b*). Some chains are short and this reduces their capacity for ATP production (*Fig. 1b*). Electrons can enter bacterial electron-transport chains at various points (*Fig. 1b*) and this increases the number of substrates that can be used for ATP synthesis.

All electron-transport pathways function in a similar manner, requiring a series of oxidation and reduction reactions. The **oxidation** of a molecule involves the loss of electrons, and **reduction** involves the addition of electrons. Since electrons are conserved in chemical reactions, oxidation must be coupled with reduction (**redox reactions**). The **oxidation–reduction potential** (**redox potential**) of a compound is a measure of its affinity for electrons. Redox potentials are measured relative to hydrogen; thus, a positive redox potential indicates that the compound has a greater affinity for electrons than has hydrogen and would accept electrons from hydrogen. A negative redox potential indicates a lower affinity and thus the molecule would donate electrons to hydrogen. The difference in redox potential between NADH + H$^+$/NAD$^+$ (–0.32 V) and O$_2$/H$_2$O (+0.82 V) drives the movement of electrons through a series of electron carriers which are arranged to accept electrons from a carrier with a more negative redox potential and donate to the next carrier which has a more positive redox potential. Energy is released as the electrons move between carriers. If the energy release is large this can be coupled with the movement of H$^+$ across the membrane, which can generate ATP. The number of sites where this can occur depends on the difference between the redox potential of the substrate and the final electron acceptor.

Formation of ATP

The **chemiosmotic hypothesis** is a widely accepted theory that explains how the movement of protons is linked to the formation of ATP. According to this hypothesis the electron-transport pathway is organized such that when electrons move through the chain, protons are transferred from one side of the membrane to the other via a series of pumps. When NADH + H$^+$ is the electron donor and oxygen is the terminal electron acceptor, proton movement occurs at three sites, which are associated with large free energy changes. The sites are NADH dehydrogenase, cytochrome bc$_1$ complex and the terminal oxidase (*Fig. 2*). In microbes with short pathways or where electrons enter at other sites in the chain (and have a more positive redox potential than that of the NADH + H$^+$/NAD$^+$ couple) less ATP is produced. The pumping of H$^+$ across the membrane produces a **proton motive force** (**PMF**), which is produced as a result of the uneven distribution of H$^+$ across the membrane (the membrane acts as a barrier as it is impermeable to H$^+$) (*Fig. 2*). When H$^+$ are transported back into the cell energy is released and this drives the synthesis of ATP. This process is catalyzed by ATP synthase (ATPase) which allows H$^+$ to move through the membrane, and the energy released is coupled with ATP synthesis

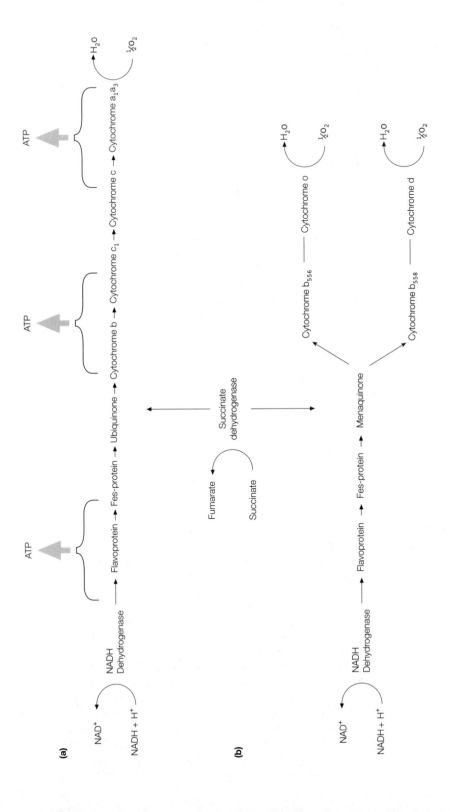

Fig. 1. Components of the electron-transport chains of: (a) mitochondria (b) E. coli.

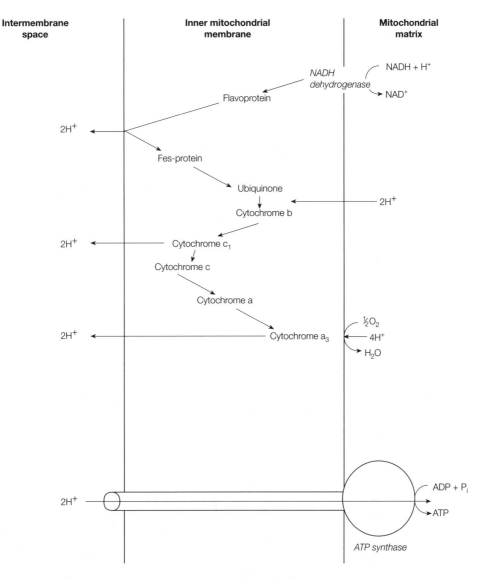

Fig. 2. The chemiosmotic hypothesis. In this scheme for eukaryotic mitochondria, sites of H⁺ movement and the proposed mechanism of ATP synthesis are shown.

(*Fig. 2*). PMF can be used directly to drive some processes such as active transport and rotation of bacterial flagella (see Topic D6).

Anaerobic respiration

Many microbes live in environments devoid of oxygen. To utilize electron-transport mechanisms for ATP synthesis, alternative inorganic electron acceptors such as nitrate, sulfate and CO_2 are used instead of oxygen (*Fig. 3*). Those organisms that are facultative will use oxygen for aerobic respiration when available. Microbes that use nitrate as an electron acceptor reduce nitrate to nitrite and ammonia by the action of **nitrate and nitrite reductase**:

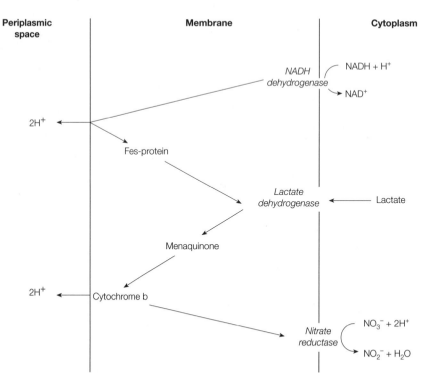

Fig. 3. Anaerobic respiration to nitrate in E. coli.

$$NO_3^- + NADPH + H^+ \rightarrow NO_2^- + NADP^+ + H_2O$$

$$NO_2^- \rightarrow [NH_2OH] \rightarrow NH_3$$

This is a largely inefficient method for the production of ATP; thus, a large amount of nitrate is required to produce sufficient ATP for the cell. Nitrite can also be reduced to N_2 by a more efficient mechanism. This process is known as denitrification and is performed by *Pseudomonas* and some *Bacillus* species:

$$2NO_3^- + 10e^- + 12H^+ \rightarrow N_2 + 6H_2O$$

Other groups that use alternative electron acceptors include **methanogens**, which are obligate anaerobes and reduce CO_2 or carbonate to methane (methanogenesis):

$$HCO_3^- + H^+ + 4H_2 \rightarrow CH_4 + 3H_2O$$

and those that can reduce sulfate to sulfide (*Desulfovibrio*):

$$SO_4^{2-} + 8e^- + 8H^+ \rightarrow S^{2-} + 4H_2O$$

Anaerobic respiration is much less efficient than aerobic respiration; however, the yield of ATP by this mechanism is much greater than that obtained by fermentation alone.

B3 AUTOTROPHIC METABOLISM

Key Notes

Chemolithotrophs

Microbes that obtain energy by the oxidation of inorganic compounds are termed chemolithotrophs or chemoautotrophs. These microbes are autotrophic (they fix CO_2); however, some are capable of heterotrophic metabolism if organic compounds are available. Carbon dioxide is fixed by the majority of these organisms using the Calvin cycle. The energy required for CO_2 fixation must be obtained from the oxidation of inorganic molecules. Many of the chemolithotrophic bacteria have considerable ecological significance, for example *Nitrosomonas* and *Nitrobacter* play major roles in the nitrogen cycle.

Photosynthesis

Photosynthesis in microbes can be either anoxygenic (does not form oxygen as a product of photosynthesis) or oxygenic (forms oxygen). Photosynthesis in algae is always oxygenic. There are four groups of microbes that carry out anoxygenic photosynthesis, and the *Cyanobacteria* perform oxygenic photosynthesis. In *Cyanobacteria*, the photosynthetic apparatus is localized in membranous systems called thylakoids. The pigments involved in the trapping of light energy by microbes are chlorophylls, carotenoids and phycobiliproteins (accessory pigments). These pigments are arranged in highly organized systems termed photosystems.

Oxygenic photophosphorylation

When photosystem I (PS I) in *Cyanobacteria* transfers light energy to the chlorophyll P700 the molecule undergoes a change in reduction potential (becomes more negative) and donates an electron to another chlorophyll molecule. This electron is then transferred to a ferridoxin and then on through either cyclical and non-cyclical routes. In the cyclical route the electron returns to P700 with the formation of one ATP. The non-cyclical route involves the light activation of chlorophyll P680 (PS II) and another electron transfer. This route generates 1 ATP, 1 NADPH + H$^+$ and $\frac{1}{2}O_2$.

Anoxygenic photosynthesis

Green and purple bacteria differ from *Cyanobacteria* in that most are strict anaerobes. These organisms use H_2, H_2S and elemental sulfur as electron donors and possess different light-harvesting pigments than those observed in the *Cyanobacteria*. These microbes lack PS II and exhibit cyclic electron transport, which can be used to generate ATP. Green sulfur bacteria, however, exhibit a form of non-cyclic photosynthetic electron flow in order to reduce NADP.

Dark reactions

When growing with CO_2 as carbon source, *Cyanobacteria* and the purple bacteria use the Calvin cycle for fixation. In the green bacteria, a reductive citric acid cycle is used to fix CO_2 into acetyl CoA.

Related topics

Algal nutrition and metabolism (I2)   Bacterial growth and cell cycle (D8)

Chemolithotrophs Microbes that obtain energy by the oxidation of inorganic compounds are termed **chemolithotrophs** or **chemoautotrophs**. These microbes are **autotrophic** (i.e. they fix CO_2); however, some are capable of heterotrophic metabolism if organic compounds are available. Most of these organisms use the **Calvin cycle** to fix CO_2. To fix 1 mol of CO_2 by this pathway, 3 mol of ATP and 2 mol of $NADPH + H^+$ are required. This is obtained from the oxidation of inorganic molecules. Less energy is available from these oxidation reactions than from the oxidation of glucose to CO_2. Typically, one ATP is formed by these oxidations; thus, these organisms must oxidize large quantities of material to generate sufficient ATP and NADPH for fixation. Many of these organisms have considerable ecological significance in the cycling of nitrogen or other elements, for example, the nitrifying bacteria *Nitrosomonas* oxidizes ammonia to nitrite:

$$NH_4^+ + 1\tfrac{1}{2}O_2 \rightarrow NO_2^- + H_2O + 2H^+$$

Nitrite can be further oxidized by *Nitrobacter* to yield nitrate:

$$NO_2^- + \tfrac{1}{2}O_2 \rightarrow NO_3^-$$

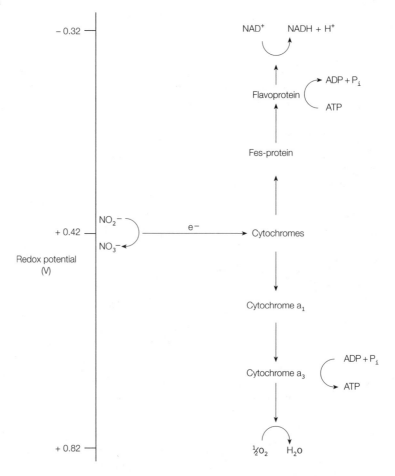

Fig. 1.   *Electron transport in* Nitrobacter. *ATP is required to drive reversed elctron transport for the production of NADH + H⁺.*

The complete oxidation of ammonia to nitrate by these two genera is called **nitrification**. Energy released by nitrification is used to produce ATP. These organisms also require reducing power (NADH + H$^+$/NADPH + H$^+$) for cellular functions. Because ammonia and nitrate have a more positive redox potential than NAD$^+$, electrons cannot be donated directly for the production of NADH + H$^+$. To overcome this problem ATP is used to drive **reversed electron flow** (*Fig. 1*). The net yield of ATP and NADH + H$^+$ from such reactions is very low. Chemolithotrophs can overcome this inefficiency by the oxidation of large quantities of inorganic material.

Photosynthesis

Photosynthesis in microbes can be divided into **anoxygenic** (does not form oxygen as a product of photosynthesis) and **oxygenic** (forms oxygen). The *Cyanobacteria* are microbes that perform oxygenic photosynthesis. This group, previously called blue–green algae, include unicellular forms (e.g. *Synechoccus*) and filamentous forms (e.g. *Anabaena*). The photosynthetic apparatus of the *Cyanobacteria* is localized in intracytoplasmic membrane systems called thylocoids, which are covered with phycobilisomes containing the light-harvesting components. These pigments trap light energy. Light-harvesting pigments can be divided into **chlorophylls**, which absorb light in the red (> 640 nm) and blue (< 440 nm) regions of the spectrum, and the **carotenoids**, **phycobiliproteins**, **phycoerythrin** and **phycocyanin**, which are termed **accessory pigments** because they absorb light at wavelengths where chlorophylls do not function efficiently (470–630 nm). These pigments are arranged in a highly organized way to maximize energy transfer from light termed **photosystem I** (PS I) which absorbs light of 700 nm and **photosystem II** (PS II) which absorbs light at 680 nm. Photosynthesis in algae is similar to that of *Cyanobacteria* (Topic I2).

Oxygenic photophosphorylation

Electron flow in *Cyanobacteria* is initiated when PS I transfers light energy to the chlorophyll P700. As a result, the reduction potential of the molecule becomes more negative. It can then donate an electron to another chlorophyll. The electron is then transferred to a ferridoxin and can travel through one of two pathways:

(i)   The cyclical route (*Fig. 2a*) involving electrons moving through a series of electron carriers and returning to P700 with ATP generated via the formation of a transmembrane proton gradient (see Topic B2). This process is termed **cyclic photophosphorylation** and only requires one photosystem;

(ii) In the **non-cyclical** route (*Fig. 2a*), during the movement of electrons from PS II to PS I, ATP is generated. This is called **non-cyclic photophosphorylation**. The non-cyclic route generates one ATP and one NADPH + H$^+$ for each pair of electrons that are transported. As three ATP and two NADPH + H$^+$ are required for each CO_2 fixed, the transport of two pairs of electrons would yield two ATP and two NADPH + H$^+$, with the extra ATP generated by cyclic phosphorylation, which operates independently.

Anoxygenic photosynthesis

Green and purple bacteria differ from *Cyanobacteria* in that most are strict anaerobes and do not use water as an electron source. These organisms use H$_2$, H$_2$S and elemental sulfur as electron donors and possess different light-harvesting pigments (**bacteriochlorophylls**) which absorb light at longer wavelengths: bacteriochlorophyll *a* (absorption maximum at 775 nm) and bacteriochlorophyll *b* (790 nm). These changes are thought to reflect the available wavelengths of light

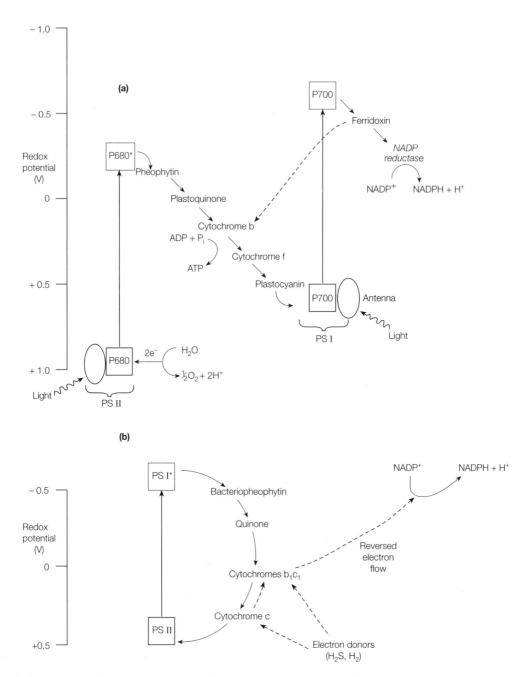

Fig. 2.   *Photosynthetic electron flow in: (a)* Cyanobacteria; *(b) anoxygenic photosynthesis. *Denotes a molecule in an excited state.*

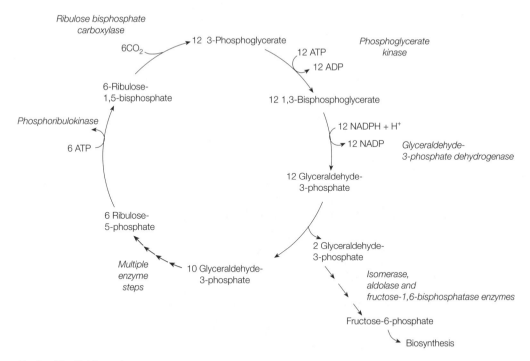

Fig. 3. The Calvin cycle.

within their ecological niches. Both groups lack PS II (*Fig. 2b*); thus, they do not produce oxygen and NADPH + H$^+$ by photosynthesis. Both exhibit cyclic electron transport which can be used to generate ATP. These organisms are unable to synthesize NADPH + H$^+$ directly by photosynthetic electron movement. Green sulfur bacteria exhibit a form of non-cyclic photosynthetic electron flow in order to reduce NAD$^+$. Purple bacteria do not possess such a mechanism; thus, they generate NADPH + H$^+$ either by reverse electron transport (see Topic B2) or in the presence of H$_2$, NADH$^+$ + H$^+$ can be produced directly as H$_2$ has a reduction potential more negative than NAD$^+$.

Dark reactions   When growing with CO$_2$ as carbon source, *Cyanobacteria* and the purple bacteria use the Calvin cycle for fixation (*Fig. 3*). The overall stoichiometry of the reaction is:

$$6CO_2 + 18ATP + 12NADPH + 12H^+ + 12\,H_2O \rightarrow$$

$$glucose + 18ADP + 18P_i + 12\,NADP^+$$

To synthesize one glucose the cycle must operate six times.

In the green bacteria a reductive citric acid cycle is used and CO$_2$ is converted into acetyl CoA. The result is that 2 CO$_2$ are reduced to acetyl CoA, which can be converted into a number of metabolic intermediates.

B4 BIOSYNTHETIC PATHWAYS

Key Notes

Carbohydrates

Synthesis of glucose from non-carbohydrate precursors is termed gluconeogenesis. The pathway involved is the reverse of glycolysis except for three points where the glycolytic reaction is irreversible. At these points the glycolytic enzyme is replaced.

Synthesis of amino acids

Bacteria and algae can assimilate nitrogen as ammonia, nitrate and nitrogen. Fungi and protozoa can assimilate nitrogen only as ammonia or nitrate. Only ammonia can be incorporated directly. The simplest mechanisms involve the formation of alanine and glutamate; these amino acids can be used to synthesize a number of new amino acids by transamination reactions (reactions involving the transfer of amino groups). Nitrate is more oxidized than ammonia and must first be reduced to ammonia before it can be incorporated by bacteria; this process is termed assimilatory nitrate reduction. Nitrogen can also be utilized if reduced to ammonia (nitrogen fixation).

Synthesis of lipids

Lipids are made up from long-chain fatty acids or their derivatives. These fatty acids contain on average 18 carbon atoms and may be saturated (no double bonds present in the chain) or unsaturated (contain one or more double bonds). Some may also be branched. The synthesis of fatty acids is catalyzed by the action of the fatty acid synthetase complex. Acetyl CoA and malonyl CoA are the building blocks for synthesis. Unsaturated fatty acids are formed in aerobic bacteria and most eukaryotes via the action of desaturase enzymes on the saturated fatty acid. In anaerobic and facultative organisms such as *E.coli* and *Clostridium* sp., double bonds are formed during fatty acid synthesis.

Synthesis of purines, pyrimidines and nucleotides

Purines and pyrimidines are cyclic nitrogenous compounds. Purines (adenine and guanine) consist of two joined nitrogen-containing rings, and pyrimidines (uracil, cytosine and thyamine) have only one. A pyrimidine or purine joined to a pentose sugar (either ribose or deoxyribose) is called a nucleoside. A nucleotide (building blocks of DNA or RNA) is a nucleoside with one or more phosphate groups attached to the sugar.

Related topics

Heterotrophic pathways (B1)
Electron transport and oxidative
 phosphorylation (B2)

Structure and organization of
 DNA (C1)
Bacterial growth and cell cycle (D8)

Carbohydrates Synthesis of glucose from non-carbohydrate precursors is termed **gluconeo-genesis** and is performed by organisms that do not synthesize glucose from CO_2. The pathway involved is the reverse of glycolysis except for three points where the glycolytic reaction is irreversible. At these points the glycolytic enzyme is replaced such that:

1. The conversion of pyruvate into phosphoenol pyruvate is catalyzed by two enzymes, **phosphoenol pyruvate carboxykinase** and **pyruvate kinase** (glycolytic enzyme pyruvate kinase).
2. The conversion of fructose-1,6-bisphosphate is catalyzed by **fructose bisphosphatase** (glycolytic enzyme phosphofructokinase).
3. The conversion of glucose-6-phosphate into glucose is catalyzed by **glucose-6-phosphatase** (glycolytic enzyme hexokinase).

Fructose can also be synthesized by this route.

The synthesis of other sugars can occur by simple rearrangement such as the production of mannose by **mannose-6-phosphate isomerase**:

$$\text{Fructose-6-phosphate} \leftrightarrow \text{Mannose-6-phosphate}$$

The synthesis of several sugars requires that precursors are bound to nucleoside diphosphates. For example, **uridine diphosphate glucose** (UDPG) is required for the synthesis of galactose. Nucleoside diphosphate (ADP linked) sugars are also required for the synthesis of polysaccharides:

$$\text{ATP} + \text{Glucose-1-phosphate} \rightarrow \text{ADP-Glucose} + PP_i$$

$$(\text{Glucose})_n + \text{ADP-glucose} \rightarrow (\text{Glucose})_{n+1} + \text{ADP}$$

Glucose-1-phosphate is formed from glucose-6-phosphate by the action of **phosphoglucomutase**.

Synthesis of amino acids Bacteria and algae can assimilate nitrogen in a variety of forms (e.g. ammonia, nitrate and nitrogen). Fungi and protozoa cannot fix nitrogen and assimilate only nitrate or ammonia Ammonia can be incorporated directly as it is more reduced than other forms. The simplest mechanisms found in all micro-organisms involve the formation of alanine and glutamate:

$$\text{Pyruvate} + NH_3 + \text{NAD(P)H} + H^+ \leftrightarrow \text{L-Alanine} + \text{NAD(P)}^+ + H_2O$$

$$\alpha\text{-Ketoglutarate} + NH_3 + \text{NAD(P)H} + H^+ \leftrightarrow \text{Glutamate} + \text{NAD(P)}^+ + H_2O$$

The formation of glutamate catalyzed by **glutamate dehydrogenase** occurs in many bacteria and fungi. These amino acids can then be used to synthesize a number of new amino acids by transfer of their α-amino group via **transamination reactions**. Nitrate must first be reduced to ammonia before it can be incorporated. This process is termed **assimilatory nitrate reduction**. and requires two enzymes, **nitrate** and **nitrite reductase** (see Topic B2). The ammonia formed by these reactions can be incorporated into amino acids. Nitrogen can also be utilized if reduced to ammonia (**nitrogen fixation**). Bacteria such as *Azotobacteria*, *Clostridium*, *Rhizobium* and some *Cyanobacteria* can fix nitrogen. This process requires the enzyme nitrogenase and a mechanism of electron transport (*Fig. 1*). The overall reaction is

$$N_2 + 6H^+ + 6e^- + 12\text{ATP} + 12H_2O \rightarrow 2NH_3 + 12\text{ADP} + 12P_i$$

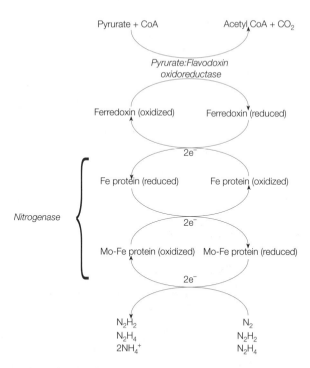

Fig. 1.   *Mechanism of nitrogen fixation by nitrogenase. Two electrons must be transported three times to produce ammonium (NH_4^+). Nitrogenase consists of iron (Fe) and molybdenum–iron (Mo–Fe) proteins.*

Synthesis of lipids   Lipids are the major components of membranes. They are made up from long-chain fatty (organic) acids. The fatty acid chains in lipids contain, on average, 18 carbon atoms and are saturated (no double bonds present in the chain) or contain only one double bond (unsaturated). Some may also be branched. The synthesis of fatty acids is catalyzed by the action of the **fatty acid synthetase complex**. Acetyl CoA and malonyl CoA (malonyl CoA is synthesized from acetyl CoA by the addition of CO_2 in a reaction driven by ATP) are the building blocks for synthesis. The pathway involved is shown in *Fig. 2*. Unsaturated fatty acids are formed in aerobic bacteria and most eukaryotes via the action of desaturase enzymes on the saturated fatty acid. In anaerobic and facultative bacteria (e.g. *E. coli* and *Clostridium* sp.) double bonds are formed during the initial stages of fatty acid synthesis.

Synthesis of purines, pyrimidines and nucleotides   Purines and pyrimidines are cyclic nitrogenous compounds. Purines (adenine and guanine) consist of two joined rings, and pyrimidines (uracil, cytosine and thyamine) have only one. A pyrimidine or purine joined to a pentose sugar (either ribose or deoxyribose) is a **nucleoside**. **Nucleotides** are the building blocks of DNA or RNA (see Topic C1) and consist of a nucleoside with one or more phosphate groups attached to the sugar. Purines are synthesized from seven different compounds. The sources of carbon and nitrogen in the purine ring are shown in *Fig. 3a*. Pyrimidine biosynthesis starts with the condensation of aspartic acid and carbamyl phosphate by **aspartate carbamyltransferase**. The product of this reaction, carbamoylaspartate is then converted into a common

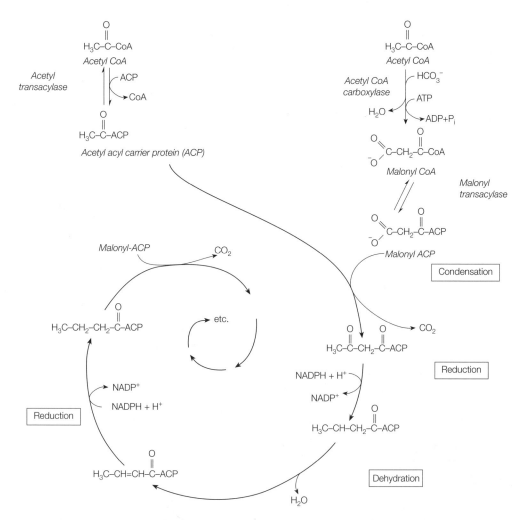

Fig. 2.   Synthesis of fatty acids. The cycle is repeated until the appropriate chain length is achieved.

intermediate of pyrimidine biosynthesis, orotic acid (*Fig. 3b*). After synthesis of the pyrimidine skeleton a nucleotide is produced by the addition of ribose-5-phosphate.

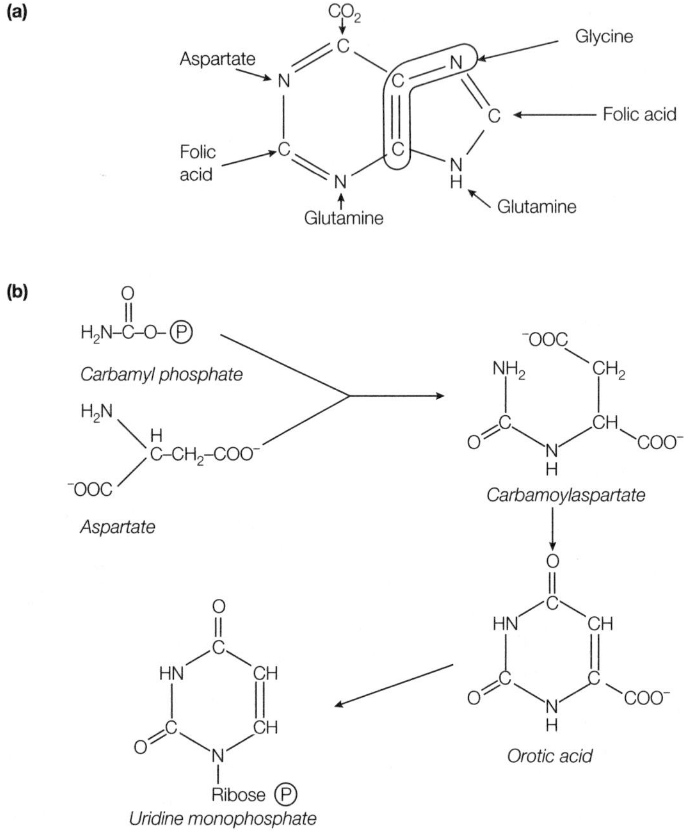

(a)

(b)

Fig. 3.   *Purine and pyrimidine synthesis: (a) sources of purine carbon and nitrogen; (b) synthesis of pyrimidines.*

C1 STRUCTURE AND ORGANIZATION OF DNA

Key Notes

DNA structure	DNA is made up of deoxyribonucleotides joined by 3′5′-phosphodiester bonds between the deoxyribose sugars. One of four nitrogenous bases, adenine (A), guanine (G), cytosine (C) or thymine (T), is attached to the sugar. Two strands of DNA are held together by specific hydrogen bonding between A and T and G and C to form a right-handed DNA helix. The specific pairing of A:T and G:C is called complementary base-pairing.
DNA conformation	Most DNA in the cell is in the B-form, which is a right-handed helix with 10 base pairs per turn.
DNA measurement and description	DNA can be measured by weight, in Daltons, or by length, in micro-meters or number of base pairs. The sequence of the DNA may be described as a sequence of the bases in a 5′→3′ direction; for example, 5′-AGCTTATTCCG-3′.
DNA packaging	DNA is negatively supercoiled, which places the molecule under torsion and reduces its volume. In prokaryotes, DNA is packaged with proteins and RNA to form a nucleoid. In eukaryotes, DNA is wound round histone proteins to form nucleosomes which are further twisted to form 30 nm chromatin fibers. At cell division the chromatin becomes condensely packed to form the metaphase chromosome.
Haploid/diploid	Bacteria have only one chromosome, and are termed haploid, whereas eukaryotes are generally diploid, having two copies of each chromosome, one from each parent.

Related topics	Biosynthetic pathways (B4)	Plasmids (E5)
	DNA replication (C2)	Structure of bacteriophage (E7)
	Transcription (C4)	Cell division and ploidy (G3)

DNA structure

Deoxyribonucleic acid (DNA) is the genetic material in all living cells; however, some viruses and bacteriophage contain RNA as their genetic material (Topics E7 and K1). The basic repeating subunit of the DNA molecule is a **deoxyribonucleotide**, which consists of a five-carbon deoxyribose sugar, phosphate and one of four nitrogenous bases. The bases are **cytosine (C)** and **thymine (T)** which are monocyclic **pyrimidines**, and **adenine (A)** and **guanine (G)** which are purines having two rings (*Fig. 1*). The nucleotides are held together by **phosphodiester bonds** between the 5′-carbon on one deoxyribose sugar and the 3′-carbon on the next (*Fig. 2*) to form the backbone of the molecule. The

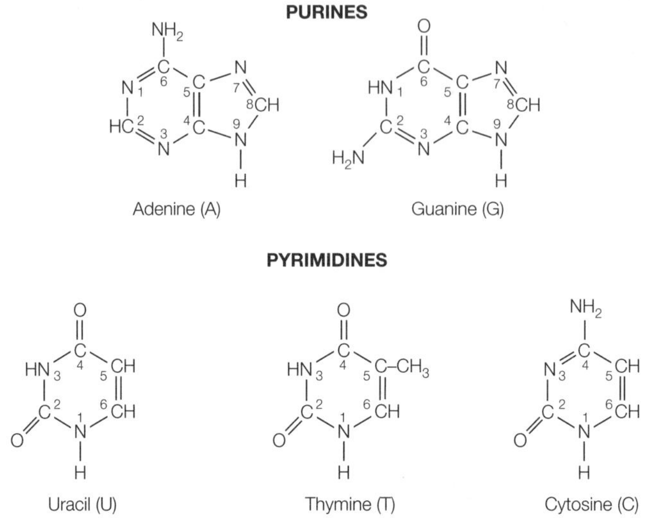

Fig. 1.   The structure of bases found in DNA and RNA.

nitrogenous bases are attached to the 1'-position of the sugar by a glycosidic bond. It is the order of the bases on the DNA strands that carries the genetic information. Two strands of DNA, running in opposite directions (**anti-parallel strands**), are held together in a **double helix** by hydrogen bonding between the nitrogenous bases in a highly specific manner (*Fig. 2*). C is always opposite G and A is always opposite T because these pairs of bases are capable of forming hydrogen bonds between them. C forms three hydrogen bonds with G and A forms two hydrogen bonds with T. These A:T and G:C base-pair interactions are called **complementary base-pairs**. DNA replication (Topic C2) and the transfer of genetic information in the DNA into the functioning components of the cell (Topic C4) are possible because of this specific complementary base-pairing. Another point to note is that the base-pairing is such that a purine is always opposite a pyrimidine, thus ensuring that the DNA helix is a constant width of approximately 2 nm. Finally, although hydrogen bonds are non-covalent and comparatively weak they can form a very strong interaction if there are large numbers of them, as is the case in DNA molecules; however, the base pairs can be broken apart when required.

DNA conformation   Most of the DNA in the cell is in what is called the B-form. This is a right-handed helix with 10 base pairs per turn (*Fig. 3*). Other forms of structure that may be found are the A-form, which has 11 base pairs per turn and is associated with DNA–RNA structures, and Z-DNA, which is a left-handed helix associated with particular sequences of alternating G:C and C:G pairs. The two distinctive features of the DNA molecule are the **major** and **minor grooves** which play an important role in the binding of proteins to the DNA.

DNA measurement and description   DNA is measured in different ways depending on the circumstances. Weight is measured in Daltons and length in micrometers or the number of base pairs (bp). The *E.coli* chromosome, for example, is circular, about 4.2×10^6 bp long which is equivalent to 1.2 mm. This holds the information for between 2000 and

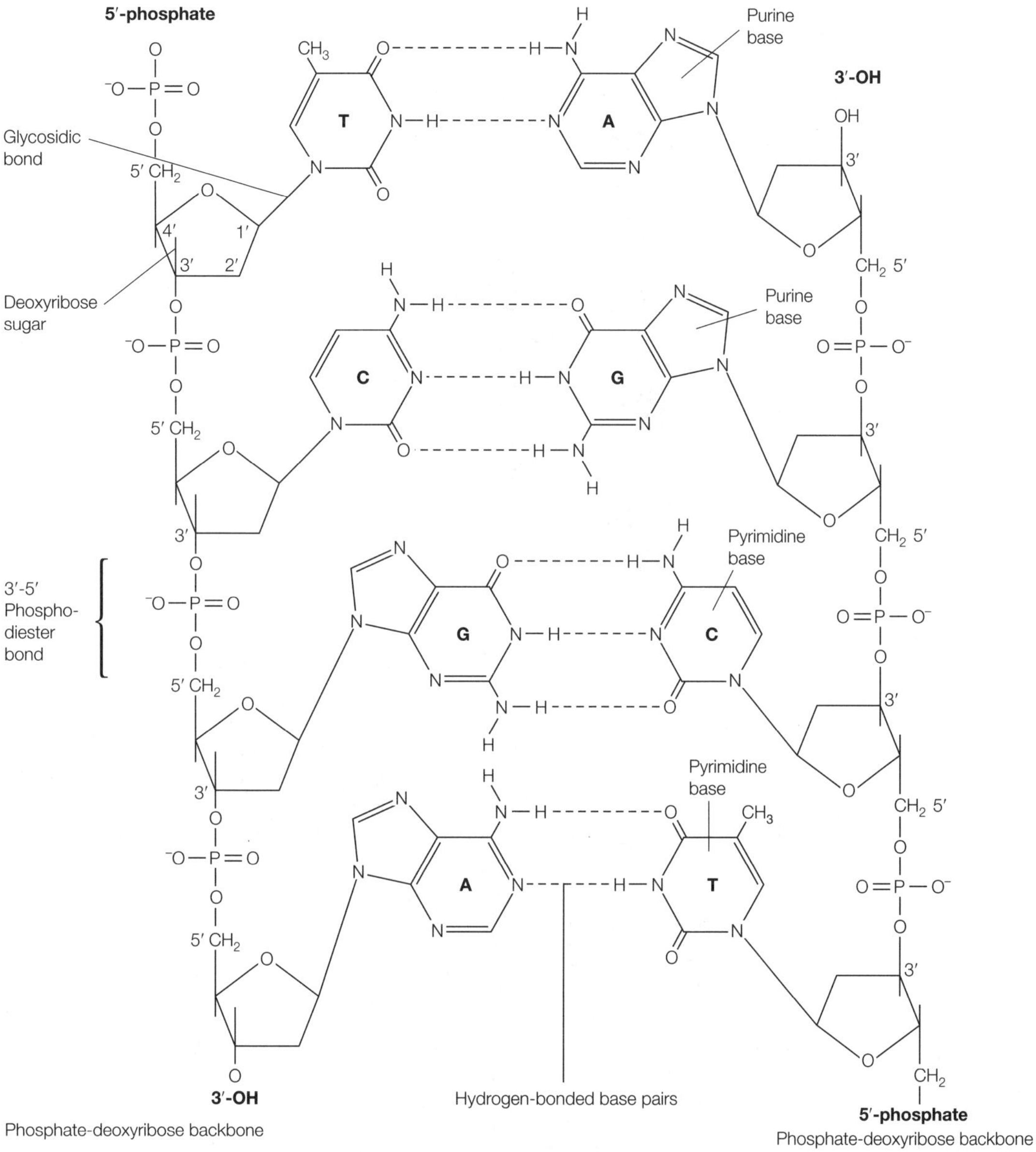

Fig. 2.   A diagram showing the structure of a DNA molecule. Note the anti-parallel nature of the DNA strands and hydrogen-bonded base pairs.

3000 genes. Another way of describing a DNA molecule is the sequence of the bases in the molecule in a 5′ → 3′ direction; for example 5′-ATTCGACGT-3′ may describe a short sequence. Only the sequence of one strand is required as the other can be worked out from complementary base-pairing. Although DNA sequencing of a whole organism is a costly and time-consuming task, we now know the complete sequence for a number of eubacteria, an archaebacterium and for a yeast, *Saccharomyces cerevisiae*.

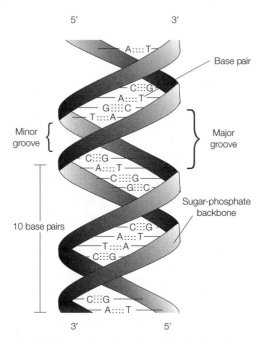

Fig. 3.   A schematic view of the DNA double helix.

DNA packaging   The length of DNA in an *E. coli* cell is over 1 mm when the cell itself is only a 1.7 μm × 0.65 μm cylinder. The density of the DNA is therefore approximately 10 mg ml^{-1}. In order to achieve this density the DNA is organized into a much tighter molecule by the introduction of **supercoiling** (twists) into the DNA molecule which puts it under torsion. The enzymes that introduce supercoils are called **topoisomerases**. Natural DNA is negatively supercoiled in that the coiling is in the opposite direction to the DNA right-handed helix, resulting in an underwound DNA helix. The further packaging of the DNA is dependent on the nature of the cell. In prokaryotes, the DNA is associated with proteins and RNA molecules, which hold it into about 40–50, independently super-coiled, loops to form a structure sometimes referred to as the **nucleoid**. In eukaryotes, the chromosomes, made up of DNA packaged with specialist proteins, are located in the nucleus: the DNA–protein complex is called **chromatin**. The first level of chromatin organization consists of DNA wrapped round a number of basic proteins called **histones** to form a chain of 11 nm beads called **nucleosomes** (*Fig. 4*). This is then further wound into a helical coil called a solenoid, which contains six nucleosomes per turn to give a **30 nm chromatin fiber** (*Fig. 4*). A third level of organization is seen in eukaryotic cells which are about to divide where the DNA is associated with non-histone scaffold proteins in a tightly packed condensed structure called the metaphase **chromosome** (*Fig. 4*).

Haploid/diploid   The differences that can seen between prokaryotic and eukaryotic DNA organi-zation is because of the increased complexity of the eukaryotic cells. Bacteria are **haploid** (monoploid), having only one copy of one chromosome which is found free in the cytoplasm in the cell. In eukaryotic cells, the bulk of the DNA

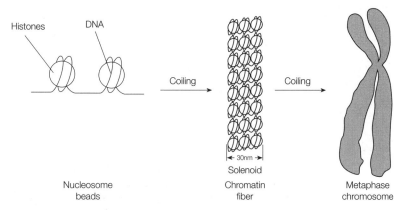

Fig. 4. Levels of chromatin structure in eukaryotic chromosomes.

is found, as a number of chromosomes, in the nucleus of the cell, therefore separated from the cytoplasm by the nuclear membrane. Each chromosome is present as two copies (diploid) because of the combination of genomes from two parents (sex). In eukaryotes this has led to the requirement of a very complex mechanism for cell division called mitosis to ensure that each offspring receives a copy of each chromosome (Topic G3) and a special mechanism of cell division (meiosis; Topic G3) for the production of haploid gametes for reproduction. The other main feature of eukaryotes is the presence of DNA in the mitochondria and chloroplasts of the cell, which tend to have the characteristics of prokaryotic cells rather than eukaryotic ones (Topic G2).

C2 DNA REPLICATION

Key Notes

Overview

DNA replication is the synthesis of new strands of DNA using the original DNA strands as templates. Complementary base-pairing between the bases on the precursor deoxyribonucleotide triphosphates and the exposed bases on the DNA molecule is responsible for the selection of the correct nucleotide to insert into the growing strand. DNA polymerase is the enzyme that forms the phosphodiester bonds.

The replication bubble

Initiation of DNA replication occurs at one site, *oriC*, on the bacterial chromosome. The helix is opened by a DNA helicase to expose the bases, and DNA synthesis starts on both strands. Replication is bidirectional forming a pair of replication forks that move around the chromosome until they meet at a termination site. Toposiomerases remove the positive supercoiling created by the unwinding of the helix for replication.

DNA polymerase

DNA polymerases synthesize DNA in a $5' \rightarrow 3'$ direction. They also have exonuclease activity which can remove mismatched bases (proof-reading). Because DNA polymerase can only add a nucleotide to an existing 3'-OH group two problems are created: (i) a primer is required to start DNA synthesis, and (ii) one strand must be synthesized in small fragments rather than continouously.

Primers

The primers to initiate DNA synthesis are short stretches of RNA synthesized by a RNA polymerase called primase.

Leading and lagging strands

The strand of DNA synthesized continuously is called the leading strand. The other strand (the lagging strand) is synthesized as a series of short fragments called Okazaki fragments. A complex of proteins called a primosome are responsible for the repeated initiation of DNA synthesis on this strand.

Eukaryote replication

The main differences between prokaryote and eukaryote replication are: (i) there are multiple origins of replication on one chromosome; (ii) different polymerases are used for the leading and lagging strand; (iii) a telomerase is required to replenish the telomers at the ends of the chromosome.

Rolling circle replication

Replication of DNA by the rolling circle mechanism produces concatemers of DNA that are essential for the life-cycle of some bacteriophage and viruses.

Related topics

Bacterial growth and cell cycle (D8)
DNA repair mechanisms (E4)
F plasmids and conjugation (E6)

Replication of bacteriophage (E8)
Cell division and ploidy (G3)

Overview

DNA replication is carried out be enzymes called DNA polymerases using deoxyribonucleoside triphosphates (dNTPs) as precursors (Topic B4). The process starts with the separation of the two strands of the DNA helix to expose the bases. New strands of DNA are synthesized using the old strands as templates. Complementary base-pairing between the bases on the incoming nucleotides and the template DNA strand (A:T; G:C) dictates which nucleotide will be inserted into the growing DNA chain (*Fig. 1*). This is called **semi-conservative replication** as, after the whole DNA molecule has been copied, each new DNA helix consists of one original DNA strand and one newly synthesized DNA strand. The role of the DNA polymerase is to check that the hydrogen bonding is correct and then to form a phosphodiester bond between two adjacent nucleotides. As the basic mechanism for DNA replication is the same for all organisms, this will be described first in relation to bacterial DNA replication. Differences found in eukaryotes will be discussed subsequently.

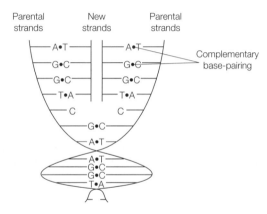

Fig. 1. Semi-conservative replication of DNA showing complementary base-pairing.

The replication bubble

The first stage of replication involves the identification of the origin of replication. There is only one on the chromosome of bacteria called *oriC*. A complex of proteins binds to this site, opens up the helix and initiates replication. DNA synthesis occurs in both directions (bidirectional) and on both strands, creating a **replication bubble** which appears as a θ (theta) structure in electron micrographs (*Fig. 2*). The two sites at which DNA synthesis occurs are called the **replication forks**. As replication proceeds the replication forks move round the molecule opening up the DNA strands, synthesizing two new complementary strands and rewinding up the DNA behind until the two replication forks meet at a termination site. The two completed circles of DNA are still linked and have to be separated by an enzyme **topoisomerase II** which makes a transient break in both strands of one of the molecules, allowing them to disconnect.

Other enzymes/proteins associated with DNA replication are:

● **DNA helicase** which unwinds the DNA helix to allow access to the bases.
● **Single-stranded binding proteins** that bind to the DNA strands, preventing them from base-pairing before they have been copied.

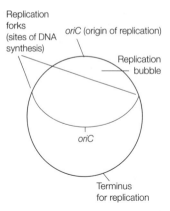

Fig. 2.   Diagram showing θ structure as a result of bidirectional replication of a circular bacterial chromosome.

- There is a problem in that as the helix is opened up the DNA ahead of the replication fork would become overwound because the DNA ends are fixed (the chromosome is circular) and therefore cannot rotate relative to each other. This is solved by **topoisomerase I** which makes a break in one of the DNA strands which allows the DNA to rotate round the intact strand thus releasing the torsion.

DNA polymerase   The enzyme responsible for normal DNA synthesis in bacteria is DNA polymerase III. DNA polymerase I also has a role in DNA replication and is the enzyme that is involved in DNA repair (Topic E4). The role of DNA polymerase II is unknown as yet. These enzymes function by adding a new nucleotide to the 3'-OH end of a growing polynucleotide chain (*Fig. 3*) with the release of pyrophosphate synthesizing a new strand of DNA in a 5' → 3' direction. DNA polymerases also have exonuclease activity which allows the sequential removal of nucleotides in both the 5' → 3' and 3' → 5' directions. The ability to remove incorrectly inserted nucleotides confers proof-reading properties on DNA polymerase III. Immediately after the enzyme has inserted a nucleotide into the growing chain the enzyme can detect if the base-pairing is correct or if a mismatch has occurred. If there is a mismatch, the 3' → 5' exonuclease activity removes the nucleotide and the polymerase can then replace it with the correct one. This proof-reading ability ensures that the error rate (mutation rate) in DNA synthesis is therefore less than 10^{-8} per base pair.

There are several features of DNA replication that are due to the nature of all DNA polymerase enzymes.

1. DNA polymerases cannot start from scratch – they must have a 3'-OH group to add a nucleotide on to. These enzymes therefore require some sort of **primer** to start DNA synthesis.
2. DNA can only be synthesized in one direction, 5' → 3'. As the two strands of DNA are antiparallel and are both copied at the same time at the replication fork, only one strand can be synthesized continuously. The other strand must be synthesized in short fragments.

Newly synthesized strand

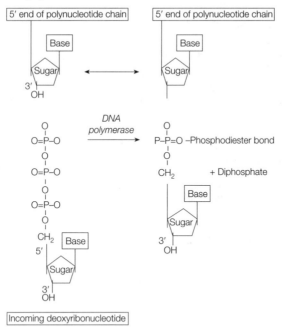

Fig. 3.   Diagram showing the joining of a deoxyribonucleotide to a growing DNA strand by DNA polymerase. Base-pairing to the complementary DNA strand is not shown.

Primers

The problem of how to start DNA synthesis is solved by using a specialist RNA polymerase called a **primase** to lay down a short stretch of ribonucleotides (primer) complementary to the DNA template at the start point for DNA synthesis. RNA polymerases do not require a 3′-OH group to start RNA synthesis so can start from scratch. The short piece of RNA provides the neces-sary 3′-OH group to which DNA polymerase III can join a deoxyribonucleotide and start to synthesize the DNA strand (*Fig. 4*). Synthesis continues until the DNA polymerase III reaches a primer. DNA polymerase III is released and DNA polymerase I binds in its place. DNA polymerase I has a 3′ → 5′ exonu-clease activity which confers the ability to remove the ribonucleotides one by

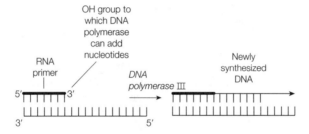

Fig. 4.   An RNA primer is required to initiate DNA synthesis.

one while its polymerase activity allows it to replace these with deoxyribonu-cleotides. The final gap in the deoxyribonucleotide chain is sealed by a specialist enzyme, **DNA ligase**, which can join adjacent 5'-phosphate and 3'-OH groups.

Leading and lagging strands

As DNA can only be synthesized in the 5' → 3' direction, only one strand can be synthesized continuously. This is called the **leading strand**. The other strand, called the **lagging strand**, is synthesized as short pieces, approximately 1000 bases long called **Okazaki fragments**. On this strand there is, therefore, constant reinitiation of DNA synthesis by a complex of proteins called the **primosome**, as described in the section above, with the consequent joining of the fragments by the removal of the RNA primers by DNA polymerase I and sealing of the gaps by DNA ligase (*Fig. 5*).

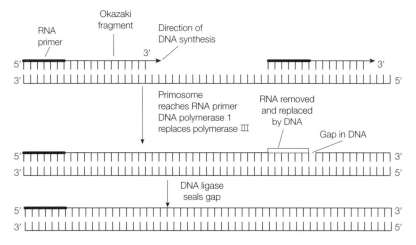

Fig. 5. The joining together of Okazaki fragments by DNA polymerase I and ligase on the lagging strand of DNA synthesis.

Eukaryote replication

The main differences between prokaryotic and eukaryotic replication are mainly as a result of the larger size of the genomes and the fact that they are linear. Some of the main ones are listed below:

- There are many origins of replication on one chromosome: approximately one every 3–300 kb of DNA.
- Leading and lagging strands are made by different DNA polymerases. The Okazaki fragments are shorter.
- Because of the need to reinitiate DNA synthesis constantly on the lagging strands there are short segments at the ends (**telomeres**) which do not get copied. The chromosomes would therefore gradually shorten with time. This problem is remedied by a specialist RNA-containing enzyme, a **telomerase**, which extends the ends of chromosomes by adding numerous hexanucleotide repeat sequences. These then act as templates for DNA replication.

Rolling circle replication

An alternative mechanism for DNA replication is used by some bacteriophage (Topic E8) and viruses (Topic K7) and in the transfer of plasmid DNA between cells by conjugation (Topic E6). Called rolling-circle replication, one strand of

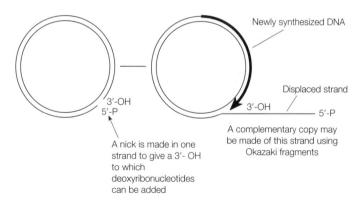

Fig. 6.   Diagram showing the mechanism of circle replication.

the DNA is nicked to provide a 3'-OH group. Helicase and single-stranded binding proteins create a replication fork, and the nicked strand is displaced. Deoxyribonucleotides are added to the 3'-OH group to create the leading strand and the displaced strand acts as a template for lagging strand synthesis (*Fig. 6*). In some cases of phage, viral and plasmid replication the strand may remain single. This process can continue for longer than one round of replication, if required, producing long linear molecules of repeated genome lengths called **concatamers**. These are essential for the life-cycle of some bacteriophage and viruses where the cutting on the concatamers into genome lengths is essential for the correct assembly of the virus particles.

C3 RNA MOLECULES IN THE CELL

Key Notes

Structure of RNA   RNA is a polymer of ribonucleotides with a similar structure to DNA except that it contains a ribose sugar instead of deoxyribose and uracil instead of thymine. RNA is normally single-stranded but can form secondary structures by base-pairing within the molecule

RNA molecules in the cell   RNA molecules in the cell can be split into two groups. Comparatively long-lived stable RNAs have a number of different structural and functional roles in the cell including ribosomal (rRNAs) and transfer RNAs (tRNAs). Messenger RNA (mRNA) molecules that act as templates for protein synthesis have a much shorter life.

Catalytic RNA molecules   Some RNA molecules have enzymic activity. The most common of these are self-splicing introns found in a range of microbes.

Related topics   Transcription (C4)   Translation (C7)

Structure of RNA   RNA is similar in structure to DNA (Topic C1) with the following differences:

- The sugar is ribose instead of deoxyribose, the difference being the presence of an OH group on the 2′-carbon. This gives the RNA molecule the capacity to have enzymic activity (*Fig. 1*).
- The pyrimidine thymine is replaced with uracil which also base-pairs with adenine (see Topic C1, *Fig. 1*).
- RNA is single-stranded; however, it is capable of forming double-stranded secondary structures, some of which may be considerably complex, by base-pairing within itself (see the structure of tRNA, Topic C7, *Fig. 3*)
- Unusual nucleosides such as inosine and pseudouridine are found in stable RNA molecules formed by the post-transcriptional modification of nucleosides.
- The precursors for the synthesis of RNA are ribonucleoside triphosphates (Topic B4).
- An RNA sequence can be written as the base sequence in the 5′ → 3′ direction: for example, ACGUCCAUG.

RNA molecules in the cell   There are a number of different RNA molecules in the cell, the most abundant of which are stable RNAs associated with the protein synthetic apparatus. **Ribosomal RNAs (rRNAs)** help to maintain the structure of the ribosomes and play a role in their function (see Shine–Dalgarno sequences, Topic C4). Prokaryote and eukaryote ribosomes contain rRNA molecules that differ in size

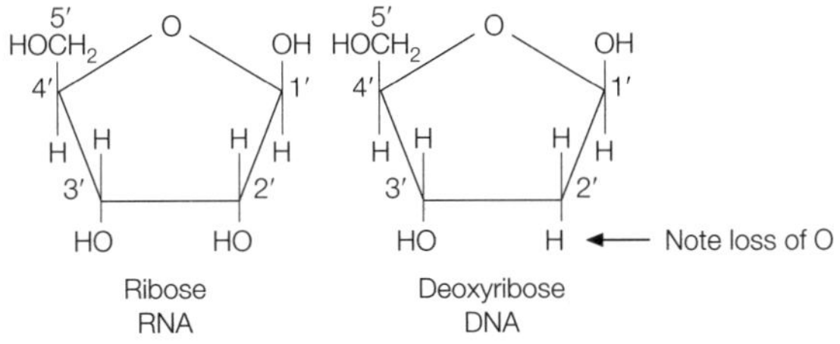

Fig. 1.   Structures of sugars in nucleic acids.

Table 1.   Classes of rRNA molecules in prokaryotic and eukaryotic cells

Organism	Ribosomal subunit	Svedberg coefficient	Size (approximate number of nucleotides)
Prokaryotes	Large	23S	2900
		5S	120
	Small	16S	1500
Eukaryotes	Large	28S	4200
		5.8S	160
		5S	120
	Small	18S	1900

and number. These various classes of rRNA are distinguished by their size and sedimentation behavior (Svedberg coefficient, S), which reflects the molecules density, mass and shape (see also Topic C7). The sizes of rRNA molecules in prokaryote and eukaryote cells are shown in *Table 1*. The sequences of rRNA molecules in an organism are highly conserved. Consequently, sequences of regions within the 16S RNA (prokaryotes) and the 18S RNA (eukaryotes) have been used to measure evolutionary relationships between different organisms.

Transfer RNAs (tRNAs) are the adapter molecules that convert a sequence of bases into an amino acid sequence (Topic C7). Other RNA molecules are found associated with a number of large molecules including spliceosomes and telomerases in eukaryotes and in the maintenance of DNA packaging in bacteria. Finally, the major group of RNA molecules are the **messenger RNAs** (mRNAs) produced by **transcription** from DNA, which act as the templates for protein synthesis. These RNAs have comparatively short half-lives compared with the stable RNA molecules.

Catalytic RNA molecules

In the early 1980s it was shown that some RNA molecules (sometimes called **ribozymes**) have enzymic activity. The most common of these are self-splicing introns which have the ability to remove themselves from a piece of transcribed RNA joining the two adjacent exons together (see Topic C4). Ribozymes have been found in a wide range of microorganisms including protozoa, mitochondria of fungi, chloroplasts of algae and some bacteriophage.

C4 TRANSCRIPTION

Key Notes

Overview

Transcription is the synthesis of messenger RNA (mRNA) and stable RNA molecules from a DNA template by RNA polymerase, using ribonucleoside triphosphates as precursors. Each mRNA carries the sequence of one gene or in the case of bacterial operons, a number of genes. Regulation of gene expression may occur at the level of transcription.

DNA-dependent RNA polymerases

Bacterial cells contain only one RNA polymerase whereas eukaryotes have at least three. In bacteria a holoenzyme consisting of five subunits (α_2, β, β', σ) is responsible for accurate initiation of transcription. The sigma (σ) subunit (factor) is released to give the core enzyme (α_2, β, β') which is responsible for elongation.

Stages of transcription

There are three stages of transcription: initiation, which involves the binding of RNA polymerase to promoter sequences in the DNA that signal the start site for transcription; elongation, which is the synthesis by RNA polymerase of a sequence of ribonucleotides that is complementary to the sequence of one of the strands of the DNA; and termination of RNA synthesis at the end of the gene or the operon.

Promoters

Promoters consist of conserved sequences necessary for the initiation of transcription. There are many different types of promoters. Many *E. coli* promoters contain two conserved regions at −35 and −10 bases before the start site for transcription. The first transcribed base is normally a purine.

Termination

There are two types of terminator sequence in *E. coli*. Some require a protein factor, rho (ρ), for accurate termination of transcription; the others rely on the formation of hairpin structures in the transcribed RNA causing termination.

Transcription in eukaryotes

The promoter sites for RNA polymerases I and II are located before the start site for transcription. RNA polymerase III recognizes sites within the gene itself. Eukaryotic RNA polymerases require the presence of additional transcriptional factors for DNA binding and the initiation of transcription.

RNA processing

Although in bacteria some structural RNA molecules may be processed post-transcription to give a functional molecule, mRNAs do not undergo any further processing. Eukaryotic RNAs undergo a number of processes to produce a mature molecule. Introns, non-coding transcribed sequences, are removed by splicing, a methylated guanine cap is added to the 5′ end of the RNA and a poly (A) tail to the 3′ end.

Related topics   DNA replication (C2)   Control of gene expression (C5)

Overview

The sequence of bases in the DNA that codes for a single polypeptide chain or stable RNA molecule is called a **gene**. Proteins cannot be synthesized directly from the DNA but require the production of an RNA intermediate called messenger RNA (mRNA). The synthesis of RNA is called **transcription** and the enzyme responsible is **RNA polymerase**. The mechanism of RNA synthesis is similar to DNA replication (Topic C2) in that a sequence of ribonucleotides is polymerized, which is complementary to the sequence of bases on the DNA molecule, but there are a number of important differences:

- Only one strand of the DNA is copied.
- The precursors for RNA synthesis are ribonucleoside triphosphates (NTPs).
- In eukaryotes the mRNA molecule carries the information for one gene product, but, in bacteria, more than one gene may be found on one mRNA molecule. These are **polycistronic mRNAs** and are due to the organization of the bacterial genome. Gene products that are required together are often clustered into one transcriptional unit, allowing the opportunity to coordinately regulate their synthesis. These units are called **operons** (see Topic C5).
- RNA polymerase binds to specific sites that signal the start of a gene or operon called a **promoter**.
- RNA polymerase does not require a primer.
- Transcription provides an opportunity to regulate the production of a gene product in that additional protein factors may activate or inhibit RNA synthesis of a particular gene (see Topic C5).

DNA-dependent RNA polymerases

Only one DNA-dependent RNA polymerase is found in eubacteria but at least three are present in eukaryotic cells, numbered I, II and III, which are responsible for different types of transcription in different regions of the nucleus (*Table 1*). Archaebacteria contain a number of RNA polymerases which show greater similarity to eukaryote RNA polymerase II. The complete *E. coli* RNA polymerase, called the **holoenzyme**, consists of five subunits, two α subunits (α_2), one each of β, β' and σ. The σ subunit (**sigma factor**) is required for the accurate initiation of transcription by binding to the promoter, but is released immediately after the start of transcription to give a **core** enzyme consisting of the α_2, β, β' subunits which is capable of polymerizing the RNA chain. A bacterium can have more than one σ factor that recognizes different promoter sequences. This provides a useful mechanism for the regulation of gene expression as found in the regulation of *Bacillus subtilis* sporulation. All the genes required for sporulation need a set of specific sigma factors for their transcription. These sigma factors are only synthesized when the conditions are right for sporulation.

Table 1.   Distribution and function of RNA polymerases

	Prokaryote	Eukaryote		
	RNA Polymerase	RNA Polymerase I	RNA Polymerase II	RNA Polymerase III
Location	Cytoplasm	Nucleolus	Nucleus	Nucleus
Products	All RNAs	28S, 18S, 5.8S rRNAs	mRNAs and most small nuclear RNAs (snRNAs)	tRNAs, 5S rRNA, one snRNA, other structural RNAs

Unlike prokaryotic RNA polymerases, eukaryotic RNA polymerases require additional transcription factors to facilitate the initiation of RNA synthesis.

Stages of transcription

Transcription consists of three stages:

● **Initiation.** RNA polymerase holoenzyme identifies and binds to the promoter. The RNA polymerase opens up the DNA helix, exposing the bases, to create a single-stranded template. This structure is called a **transcription bubble** (*Fig. 1*). The first two ribonucleotides can then enter the complex and base-pair to the template strand, and RNA polymerase forms the phospho-diester bond between them with the release of pyrophosphate.

● **Elongation**. The sigma factor is released from the enzyme and the core RNA polymerase moves down the DNA template, unwinding a small section of the DNA at a time and polymerizing a complementary copy of RNA. RNA is synthesized in the 5′ → 3′ direction like DNA. As the transcription bubble moves down the DNA molecule the newly synthesized RNA is dissociated from the DNA and the DNA helix is reformed behind the transcription bubble.

● **Termination**. Specific sequences in the DNA signal the end of the gene. Transcription ceases and RNA polymerase and the completed mRNA chain are released from the DNA.

The strand of DNA that acts as the template for RNA synthesis is called the **antisense** strand as the mRNA produced has the same sequence as the non-template strand called the **sense** strand.

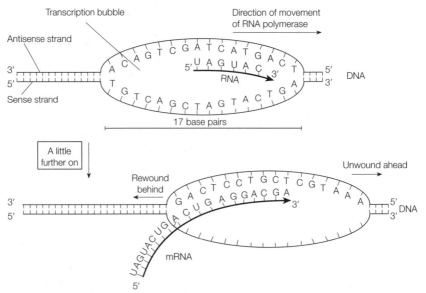

Fig. 1. Transcription bubble showing direction of RNA synthesis. RNA polymerase moves along the DNA molecule inserting ribonucleotides into the growing RNA chain. The enzyme unwinds the DNA ahead and rewinds it behind.

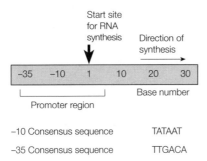

-10 Consensus sequence   TATAAT

-35 Consensus sequence   TTGACA

Fig. 2.   Consensus sequences of E. coli *promoters.*

Promoters

There are many different promoter sequences, depending on the organism and how the gene is regulated (see Topic C5). Promoters consist of conserved sequences which are required for the binding of RNA polymerase and the initiation of transcription. Some of the best studied promoters are those of *E. coli*. In many of these promoters, two sets of conserved DNA sequences have been identified before the start sites of transcription. They are called the –10 and –35 sequences as the convention is to number the first transcribed base as +1; therefore, regions before the start site which are not transcribed are given a negative number. The consensus sequences are shown in *Fig. 2*. Promoters which are frequently transcribed, called strong promoters, have sequences at the –10 and –35 sites which are very close to these consensus sequences. Weak promoters show greater differences to the consensus. The first base in *E. coli* mRNAs is usually a purine (>90%). Regulatory regions that control transcription are also found in this upstream region before the start site for transcription.

Termination

Elongation continues until the RNA polymerase reaches a signal that indicates the end of the RNA molecule. A number of termination signals have been found in bacteria, some require the presence of an additional protein factor, the rho factor (ρ). Other termination signals are due to the presence of certain sequences in the RNA molecule which once synthesized form hairpin structures by base-pairing. These secondary structures in the RNA cause the RNA polymerase to pause. This hairpin is usually followed by a run of Us (uracil) in the RNA which will be comparatively weakly base-paired with the DNA strand (A–U base pairs have only two hydrogen bonds) and consequently the mRNA–DNA–enzyme complex dissociates (*Fig. 3*).

Transcription in eukaryotes

In eukaryotes the promoter sites for RNA polymerases I and II are located at the 5' end of the gene but the polymerase III promoter lies within the gene. Most RNA polymerase II promoters have a region at –25 called the TATA box which has a consensus sequence of TATAAA. However, RNA polymerase II can also transcribe a number of genes that lack this sequence. These are generally house-keeping genes which encode proteins that are needed for routine function of the cell and are normally expressed constantly at low levels. In contrast to prokaryotes, RNA polymerase II requires the assistance of a number of additional general transcription factors to aid binding and initiation of transcription.

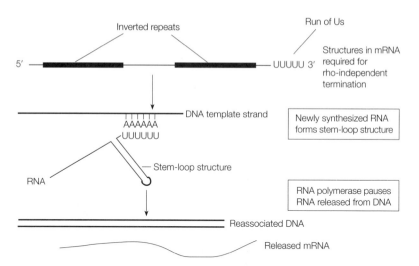

Fig. 3.   Model for rho-independent termination of transcription.

RNA processing   In bacteria, protein-encoding mRNAs do not require any further processing prior to translation. In fact, protein synthesis normally starts as soon as the ribosome-binding sites (the start signals for translation) have been transcribed and are free of RNA polymerase. It is not unusual to see multiple ribosomes attached to the mRNA while it is still being transcribed. However, some structural rRNA and tRNA molecules are synthesized as precursor molecules containing a number of these RNAs, which require cleavage by specific ribonucleases to produce the finished molecules.

In eukaryotes, the vast majority of the protein-encoding genes do contain sequences, called **introns**, which are not found in the translated proteins. After transcription, these sequences are spliced out of the transcript to leave just the protein-encoding regions, the **exons**, in the mRNA (*Fig. 4*). Eukaryote mRNAs are also further modified by the addition of a **methylated guanine cap** to the 5′ end and a string of adenylate residues to the 3′ end, the **poly (A) tail**. These

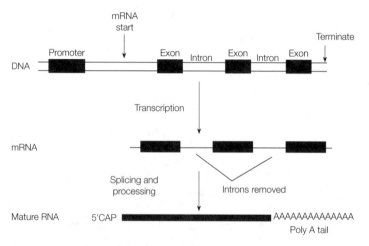

Fig. 4.   The processing of mRNA in eukaryotes.

modifications help to stabilize the mRNAs and prevent their degradation by ribonucleases. Unlike prokaryote, transcription and translation take place in two separate compartments of the cell, and the complete mRNA molecule must leave the nucleus before translation starts.

C5 CONTROL OF GENE EXPRESSION

Key Notes

Why?	Not all genes in the cell need be expressed at all times. Cellular regulatory systems allow the control of gene expression so that their products are only synthesized when required. Regulation can occur at all stages in the synthesis of proteins in the cell, but much of it is at the level of transcription which gives rapid and sensitive control.
General features of regulation	Constitutive genes are expressed at all times, whereas inducible genes are expressed when required. Regulation can be specific, controlling just one gene, or operon, or global, affecting a large number of genes. Regulation requires some way of detecting the current situation and a way of responding by controlling transcription. The control may be either negative, where the gene is normally switched on unless a repressor is present, or positive, in which an inducer is required to allow transcription to occur. Genes or operons may be controlled by more than one regulatory system.
Lactose operon	The *lacZ, lacY* and *lacI* gene products are required for the use of lactose as a carbon source. Gene expression is controlled by two regulatory systems. A negative control system, which switches off the genes in the absence of lactose, and a positive control system, called catabolite repression, which switches on the genes when glucose is not present in the surrounding medium.
Negative control	In the absence of lactose, the lactose repressor protein binds to an operator (O) site in front of the *lac* genes and inhibits transcription. If lactose is present, a product of lactose, allolactose, acts as the inducer molecule. It binds to the *lac* repressor and prevents it binding to the operator site, thereby allowing transcription to occur.
Positive control	A number of operons associated with carbon utilization cannot be transcribed unless a catabolite activator protein (CAP) is bound to the operator. The CAP requires the presence of cAMP in order to bind to the DNA and activate transcription. cAMP acts as the signal that glucose levels are low in the cell and therefore alternative carbon sources are required.
The tryptophan operon	The *trp* operon is negatively controlled by a repressor protein which is only active if it is complexed with tryptophan (a corepressor). Gene expression is also controlled by attenuation, (premature termination of *trp* RNA synthesis in the presence of tryptophan). The combined effect of the two regulatory systems is to reduce transcription by about 700-fold.

Two-factor control systems	These systems consist of a sensor component, which detects the environment around or within the cell, and an effector protein that regulates gene expression in response to that environment. The sensor protein modulates the activity of the effector protein by phosphorylation.
Regulation in eukaryotes	Regulation in eukaryotes involves *trans*-acting transcriptional factors that interact with control sequences near the promoter regions called *cis*-acting elements. These regulate gene expression in response to nutritional and environmental signals but are also important during development and differentiation.
Related topic	Transcription (C4)

Why?

Many of the genes in the cell encode functions that are not always required. For example, if bacteria are growing in a glucose-containing medium, with all the amino acids provided, they do not need to produce the enzymes that allow them to use alternative carbon sources such as lactose or cellulose: neither do they need to express the genes involved in the synthesis of their own amino acids. Regulatory systems therefore exist in the cell to control the expression of genes in response to their nutritional or physical environment, preventing the wastage of vital energy. This regulation of **gene expression** can occur at virtually any level in the flow of genetic information in the cell: transcription, mRNA processing, mRNA turnover, translation or enzyme function. However, in most cases, this regulation is at the level of transcription, where regulatory proteins modulate the binding of RNA polymerase to promoters and consequently the amount of RNA synthesis, giving rapid and sensitive control.

General features of regulation

- Some genes, such as those that encode essential RNAs and enzymes for key metabolic processes, are expressed all the time. Such genes are described as **constitutive**. Genes that are expressed only when required are called **inducible**.
- Regulation can be specific, controlling just one gene or operon, or **global**, in which a wide range of genes or operons are regulated together in response to some nutritional or physical signal.
- Regulatory systems consist of a means to detect the current situation of the cell and a way to cause a response by switching transcription on or off.
- Regulation may be either negative or positive (*Fig. 1*). In negative control, genes are normally switched on unless switched off by a **repressor**. This normally occurs in the case of biosynthetic genes where the presence of the final product of a metabolic pathway switches off the expression of the genes encoding that pathway. In the case of positive control, an activator protein (**apoinducer**) is required for transcription to occur.
- Genes may be controlled by more than one regulatory system which responds to different signals.

It should be pointed out that there are many different mechanisms of gene regulation acting on a wide range of genes; examples of some of the gene regulatory systems in *E. coli* are shown in *Table 1*. In this topic the regulation of

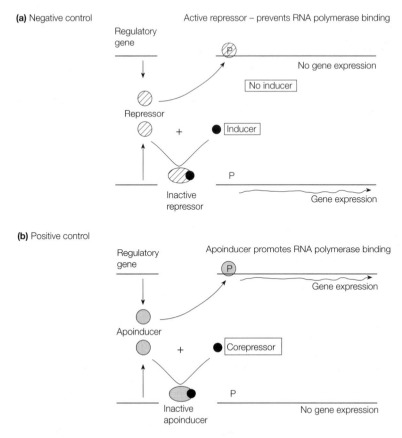

(a) Negative control   Active repressor – prevents RNA polymerase binding

Regulatory
gene

No inducer

No gene expression

Repressor

+ ● Inducer

P

Inactive
repressor

Gene expression

(b) Positive control

Regulatory
gene

Apoinducer promotes RNA polymerase binding

Gene expression

Apoinducer

+ ● Corepressor

P

Inactive
apoinducer

No gene expression

Fig. 1. (a) Positive and (b) negative control of gene expression. P, promoter region.

Table 1. *Some of the global regulatory systems found in* E. coli

Global control system	Signal
Anerobic respiration	Lack of oxygen
Catabolite repression	cAMP concentrations
Heat shock	Increase in temperature
Nitrogen utilization	NH_3 limitation
Osmotic stress	High osmolytes
Oxidative stress	Oxidizing agents
SOS response	DNA damage
Stringent response (starvation)	ppGpp; pppGpp

only three different sets of genes will be described: the lactose (*lac*) operon; the tryptophan operon and osmotic regulation of two outer membrane porins, OmpF and OmpC. These provide examples of the major types of regulation in the bacterial cell, showing how the regulatory proteins can respond to the environmental or nutritional signals and interact with the promoter regions of the controlled genes.

Lactose operon
The *lac* operon consists of three structural genes, *lacZ*, *lacY* and *lacA*, which are transcribed together from a single promoter, P_{lac}. The genes code for β-galactosidase, the enzyme that splits lactose into glucose and galactose; lactose permease, the lactose transport protein; and lactose transacetylase, whose role is unknown, respectively (*Fig. 2*). These proteins are only required if lactose is present in the medium and glucose is absent. Consequently, the *lac* operon is controlled at two different levels. A negative control system, requiring a repressor (LacI), switches off the genes in the absence of lactose, and a positive, global control system, called **catabolite repression**, is required to allow gene expression but only when the level of glucose drops below a certain concentration.

Negative control
The DNA sequence in front of the *lac* operon contains two regulatory regions, the promoter (P_{lac}) where RNA polymerase binds (see Topic C4) and an operator site (O_{lac}), the site where regulation occurs. There is also a gene, *lacI*, which encodes a regulatory protein, the *lac* repressor. In the absence of lactose, the repressor protein binds to the operator site, preventing transcription. In the presence of an inducer molecule (a signal that lactose is present), the repressor becomes inactivated so that it no longer can bind to the operator and transcription can now proceed (*Fig. 2*). The natural inducer of the *lac* operon is allolactose which is produced by β-galactosidase from lactose. This indicates that a small amount of transcription of the *lac* genes does occur even in repressing conditions, which allows the transport of lactose into the cell and its conversion into an inducing molecule.

Positive control
Catabolite repression controls a wide range of operons involved in using carbon sources other than glucose. Transcription of these operons cannot occur unless an activator protein, the **catabolite activator protein (CAP;** sometimes called the cAMP receptor protein, CRP**)** is bound to the operator. However, the CAP protein cannot bind to DNA unless it is associated with an **effector** molecule, cyclic AMP (**cAMP**). Glucose inhibits the synthesis of cAMP, so cAMP concentrations only rise when glucose levels drop; hence, cAMP is a signal that the glucose concentration is low. The presence of CAP–cAMP bound to the DNA in front of the *lac* promoter stimulates RNA polymerase binding and consequently transcription.

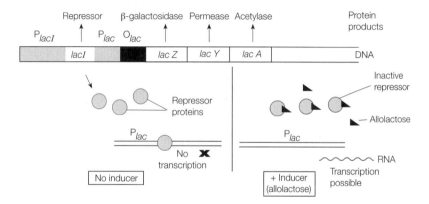

Fig. 2.   Regulation of the lac *operon. The presence of CAP-cAMP bound to promoter DNA is also necessary for transcription of* lac *genes.*

Thus, the *lac* operon can only be transcribed if the repressor is switched off by the presence of an inducer, allolactose, and if an activator protein, CAP–cAMP, is bound to the DNA.

The tryptophan operon

The *trp* operon encodes enzymes involved in tryptophan biosynthesis so these proteins are not required when tryptophan is present in the medium. This operon is regulated at a number of different levels. First, it is another example of negative control but in this case the repressor, sometimes called the **apo-repressor**, is unable to prevent transcription unless a **corepressor** is present. The corepressor is tryptophan, which is acting as a signal that there is no need to produce the amino acid as it is already present. In the absence of tryptophan the repressor is inactive and the genes are expressed. Tryptophan gene expression is also controlled by a mechanism called **attenuation** in which transcription actually starts but is stopped prematurely in the presence of tryptophan. This mechanism of gene control is common to a number of biosynthetic operons and is based on the fact that transcription and translation are coupled in bacteria (Topic C4) and that RNA can fold into secondary structures by base-pairing within the molecule if not prevented by the presence of ribosomes. The mRNA sequence before the translation start site for the first structural gene *trpE* contains a number of interesting features (*Fig. 3a*):

- A short sequence of translatable mRNA, *trpL*, which codes for a short leader peptide that contains two tryptophan residues.
- Four regions of mRNA (1–4) that are capable of forming two different sets of structures by base-pairing; region 1 with 2 and region 3 with 4, or region 2 with 3. The base-pairing between 2 and 3 is more stable.
- If regions 3 and 4 base-pair they from a hairpin loop followed by a run of Us which can act as a transcription termination signal (see Topic C4).

In the presence of tryptophan, transcription can start, and the leader peptide gene, *trpL*, is transcribed and translated. The ribosome therefore covers part of the region 2 (*Fig. 3b*) preventing it from base-pairing with region 3. Region 3 is therefore free to base-pair with region 4, forming a transcription termination signal. RNA polymerase stops synthesis and transcription is attenuated. In the absence of tryptophan, the ribosome is stalled at the tryptophan codons as there is no tRNAs loaded with the amino acid and translation cannot continue (see Topic C7). In this case the full region 2 is available to base-pair with region 3 (*Fig. 3c*). Region 3 is therefore not available to base-pair with region 4 and the termination signal is not formed. RNA synthesis therefore continues to transcribe the full set of genes.

The combined effects of the two regulatory systems for tryptophan is to reduce transcription by about 700-fold. Repression accounts for up to a 70-fold decrease in initiation of RNA synthesis, and attenuation reduces the level of full transcripts by another 10-fold.

Two-factor control systems

Many of the regulatory systems in which cells sense and respond to environmental signals consist of two components: a **sensor** protein in the membrane, which detects the situation, and an effector protein (**response regulator protein**) in the cytoplasm. The sensor protein modulates the activity of the effector protein by phosphorylation. In many cases the effector protein is a DNA-binding protein that regulates transcription, and the presence or absence of the phosphate group dictates whether it will have a positive or negative effect on RNA

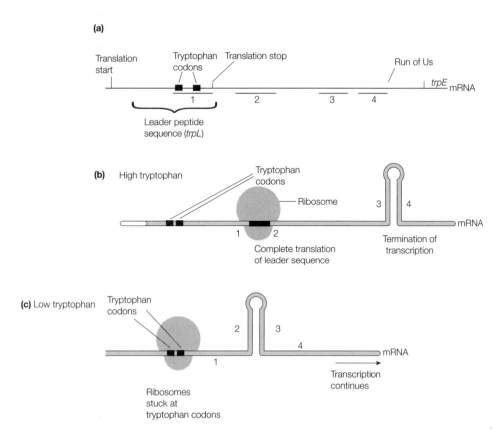

Fig. 3.   Regulation of the tryptophan operon. (a) Regions 1–4 in the RNA upstream from the start site trpE and the position of the tryptophan codons. Possible RNA secondary structures formed in the presence of (b) high tryptophan and (c) low tryptophan.

synthesis. One example of this type of control system is the osmotic regulation of the porins, OmpF and OmpC (see Topic D3), in the outer membrane of *E. coli*. OmpF is preferentially expressed in low osmolarity medium and OmpC at high osmolarities. These genes are regulated by EnvZ, a membrane sensor protein, and OmpR, a response regulator protein. At high osmolarity, EnvZ phophorylates OmpR which causes it to act as an activator of transcription of OmpC and a repressor of OmpF. At low osmolarity, OmpR is not phosphorylated, and the opposite occurs – OmpF is synthesized and OmpC is repressed.

Other mechanisms of regulation in prokaryotes include anti-termination of transcription used to control λ-lysogeny (Topic E7); antisense RNA, also involved in the osmoregulation of porins; control at the level of translation (regulation of ribosome proteins); and degradation of mRNA.

Regulation in eukaryotes

Genes in eukaryote cells, but especially in multicelled organisms, are controlled in response to the nutritional and environmental state of the cell and during development and differentiation. There are a number of other differences to prokaryotes that affect how genes may be controlled including the fact that transcription and translation happen in separate cellular compartments and that mRNA is more stable in eukaryotes. Control of gene expression can be at the level of transcription, mRNA processing and translation. Most transcriptional regulation is as a result of additional **transcriptional factors** (called *trans*-**acting factors**) binding to control sequences adjacent to the promoter regions (called *cis*-**acting elements**). The nature of the elements in the promoter region will dictate when that gene will be expressed and what factors can effect its expression. A second group of regulatory elements are **enhancers,** which in association with *trans*-acting factors, greatly increase the rate of transcription from a promoter. These sequences may be located a long distance (10–50 kb) from the gene.

C6 STRUCTURE OF PROTEINS

Key Notes

Proteins in the cell
Proteins in the cell consist of polypeptides of L-amino acids joined by peptide bonds in a sequence dictated by genetic information in the cell. There are 20 amino acids found in proteins, which differ in the nature of their side chains. The sequence of the amino acids in the protein dictates its structure, the nature of any changes made to the protein, its final destination in or outside the cell and its function.

Structure of proteins
There are four levels of structure in proteins: primary structure which is the sequence of amino acids in the polypeptide; secondary structure which is the result of hydrogen bonding between side chains of amino acids to form α-helices and β-sheets; tertiary structure which results from the spontaneous folding of the protein as a result of interactions between the amino acid side-chains, often controlled by chaperone proteins; and quaternary structure which occurs when more than one polypeptide makes up the functional protein.

Related topics
Translation (C7)

Proteins in the cell

Most of the enzymes and many of the structural components of the cell are proteins. These consist of one or more polypeptide chains arranged in a three-dimensional structure. The polypeptide chains are made up of sequences of L-amino acids, joined together by peptide bonds between the amino group of one amino acid and the carboxyl group of another (*Fig. 1*), in an order dictated by the genetic information in the cell. All proteins, in all organisms, are made up of the same 20 amino acids (*Fig. 2*). It is the nature and the order of the amino acids that gives each protein its unique structure and function and, equally, there is the capacity to create an enormously varied set of proteins that can carry out the multitude of functions in the cell. The structure of a protein has to be such that it can exist in the correct environment and also carry out its function. For example, proteins may be found either in aqueous environments such as the cytoplasm, or in hydrophobic regions such as the interiors of membranes. Proteins that have to exist in membrane interiors must have hydrophobic amino acids in positions where the protein is in contact with the lipid bilayer (Topic D3).

All proteins are synthesized initially as a sequence of amino acids, but subsequent to this the polypeptides are folded and may undergo many different types of modification to give the final proteins. However, it is important to note that all the information about the final fate of the protein is carried in the amino acid sequence. This includes:

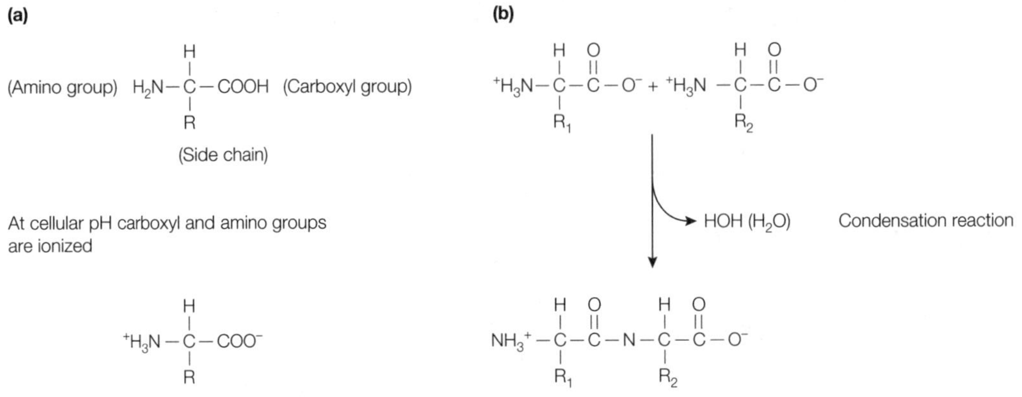

(a)

H
|
(Amino group) $H_2N-C-COOH$ (Carboxyl group)
|
R

(Side chain)

At cellular pH carboxyl and amino groups
are ionized

H
|
$^+H_3N-C-COO^-$
|
R

(b)

$$^+H_3N-\overset{\overset{\displaystyle H}{|}}{C}-\overset{\overset{\displaystyle O}{||}}{C}-O^- + {}^+H_3N-\overset{\overset{\displaystyle H}{|}}{C}-\overset{\overset{\displaystyle O}{||}}{C}-O^-$$
R₁   R₂

$\longrightarrow$ HOH (H_2O)   Condensation reaction

$$NH_3^+-\overset{\overset{\displaystyle H}{|}}{C}-\overset{\overset{\displaystyle O}{||}}{C}-N-\overset{\overset{\displaystyle H}{|}}{C}-\overset{\overset{\displaystyle O}{||}}{C}-O^-$$
R₁   R₂

Fig. 1.   (a) The basic structure of L-amino acids in the cell; (b) the formation of a peptide bond.

- its final structural conformation;
- whether additional groups are to be added – these include acetylation, hydroxylation, phophorylation, methylation, glycosylation or the addition of nucleotides;
- whether it is to be processed (cut) in any way;
- whether it is associated with cofactors such as NAD⁺ or other molecules;
- its final destination, for example, a membrane, secretion out of the cell or into an organelle (eukaryotes);
- its functional activity, for example, as an enzyme, transport protein, structural protein, signalling molecule, regulatory protein or nutritional protein.

Structure of proteins

There are four levels of protein structure; primary, secondary, tertiary and quaternary as shown in *Fig. 3*.

Primary
This is the linear sequence of amino acids joined together by the peptide bonds to form a polypeptide. Each polypeptide has an amino end (N-terminus) and a carboxyl end (C-terminus). The final structure of the protein is determined by the side-groups of these amino acids: whether they are polar or non-polar, charged or non-charged, dictates how they will interact with each other to fold the protein into its subsequent shape.

Secondary
Hydrogen bonds between amino acids in the polypeptide leads to the formation of regular structures in the majority of protein molecules. The two most common one are **α-helices**, rod-shaped structures and **β-sheets**. Fibrous proteins tend to have large amounts of these highly structured regions whereas globular proteins have smaller regions interspersed with random coils of amino acids.

Tertiary
Proteins spontaneously fold into tertiary structures dictated by the amino acid side chains which form a number of different types of weak bond that stabilize the structure:

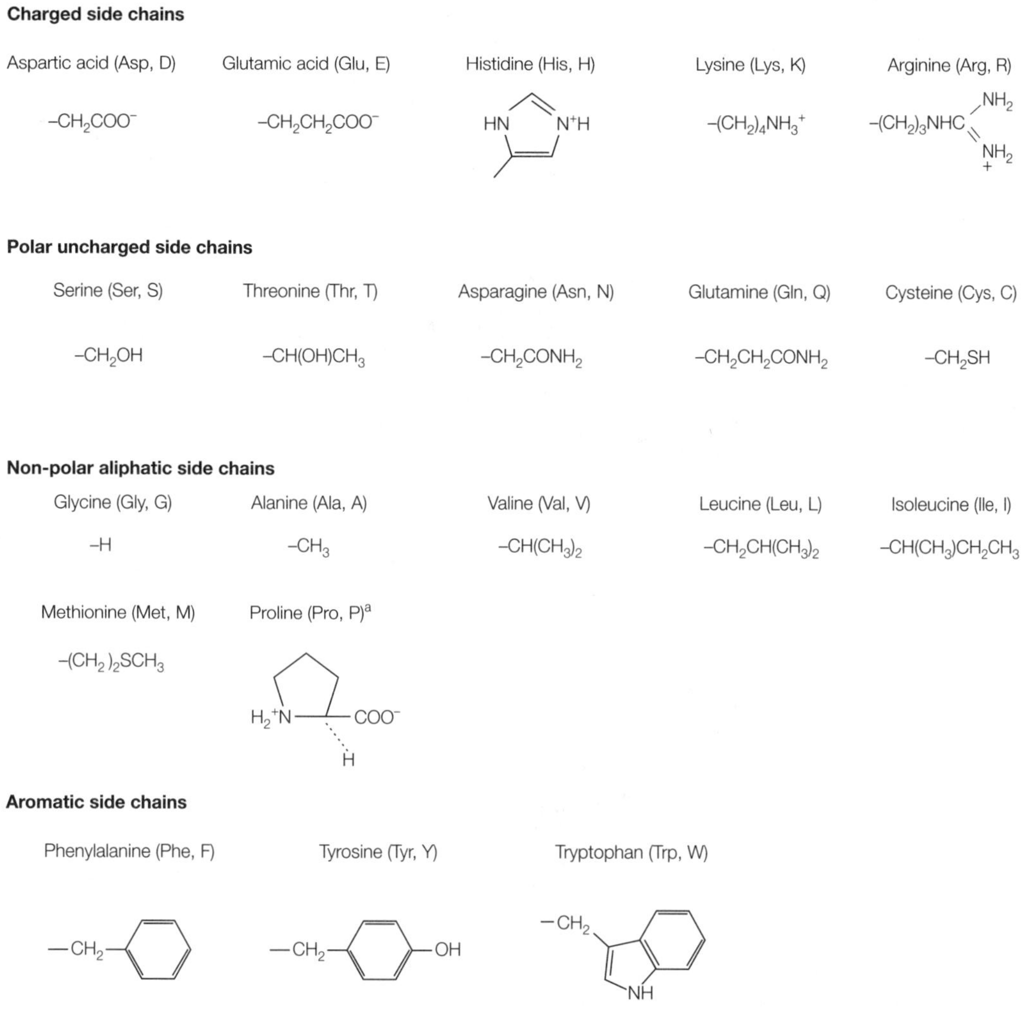
Charged side chains
Aspartic acid (Asp, D)
–CH2COO–
Glutamic acid (Glu, E)
–CH2CH2COO–
Histidine (His, H)
HN
N+H
Lysine (Lys, K)
–(CH2)4NH3+
Arginine (Arg, R)
–(CH2)3NHC
NH2
NH2
+
Polar uncharged side chains
Serine (Ser, S)
–CH2OH
Threonine (Thr, T)
–CH(OH)CH3
Asparagine (Asn, N)
–CH2CONH2
Glutamine (Gln, Q)
–CH2CH2CONH2
Cysteine (Cys, C)
–CH2SH
Non-polar aliphatic side chains
Glycine (Gly, G)
–H
Alanine (Ala, A)
–CH3
Valine (Val, V)
–CH(CH3)2
Leucine (Leu, L)
–CH2CH(CH3)2
Isoleucine (Ile, I)
–CH(CH3)CH2CH3
Methionine (Met, M)
–(CH2)2SCH3
Proline (Pro, P)a
H2+N
COO–
H
Aromatic side chains
Phenylalanine (Phe, F)
—CH2
Tyrosine (Tyr, Y)
—CH2
OH
Tryptophan (Trp, W)
—CH2
NH

Fig. 2. Side chains (R) of the 20 common amino acids found in the cell. The standard three-letter abbreviations and one-letter code are shown in parentheses. [a]*The full structure of proline is shown as it is a secondary amino acid. Reproduced from Turner* et al., Instant Notes in Molecular Biology, *1997, published by BIOS Scientific Publishers.*

- hydrophobic interactions between non-polar groups;
- ionic bonds between opposite charges;
- hydrogen bonds;
- Van der Waals forces due to interactions of electrons on clusters of amino acids.

There are also covalent bonds associated with this level of structure. The most common are **disulfide** bonds between cysteine residues, which can hold together two separated parts of the same chain or two different chains (quaternary structure). The protein may also have other atoms or molecules associated with it that contributes to its final function. The correct folding of proteins is often associated with **chaperone** proteins which prevent them folding into an incorrect shape before their synthesis is completed.

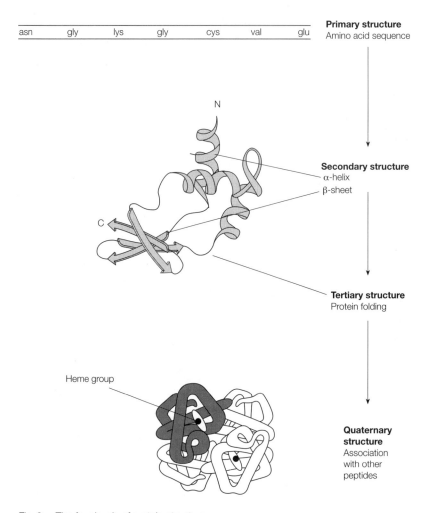

| asn | gly | lys | gly | cys | val | glu |

Primary structure
Amino acid sequence

N

Secondary structure
α-helix
β-sheet

C

Tertiary structure
Protein folding

Heme group

Quaternary structure
Association with other peptides

Fig. 3.   The four levels of protein structure.

Quaternary
Finally, if a protein is made up of more than one polypeptide chain, it is said to have quaternary structure. These chains may be different polypeptides such as the case of RNA polymerase (Topic C4) or multiple copies of the same one. The bonds that contribute to the quaternary structure are the same as those involved in the tertiary structure.

C7 TRANSLATION

Key Notes

The genetic code	A sequence of three bases, called a codon, in the mRNA codes for one amino acid. tRNA molecules act as the adapter molecules between the mRNA sequence and the amino acid sequence. They carry an anticodon which base-pairs with the codon sequence and a site at which an amino acid is attached. The process of matching the anticodon and codon and the synthesis of the amino acid chain occurs on the ribosome and is called translation.
Ribosomes	The role of the ribosomes is to hold the complex of mRNA and the amino acid loaded tRNAs, form the peptide bond between the amino acids and ensure accuracy of protein synthesis. They consist of a large and a small subunit made up of proteins and RNA. The tRNA molecules are held at the aminoacyl (A) site and the peptidyl (P) site within the ribosome.
Loading of tRNA molecules	Amino acids are attached to the correct tRNA molecule, called the cognate tRNA, by aminoacyl-tRNA synthetases to form aminoacyl tRNAs.
Protein synthesis	There are three stages of protein synthesis: initiation; elongation and termination. Initiation involves the identification of the first codon in the protein sequence, normally AUG, which codes for methionine. Various protein factors are also required and energy is provided by guanosine 5'-triphosphate (GTP) hydrolysis.
Prokaryotic initiation of translation	The small, 30S, ribosomal subunit binds to the mRNA so that the start codon (AUG) is located in the P site. The Shine–Dalgarno sequence on the mRNA is responsible for the correct positioning. The initiator tRNA carrying formyl methionine base-pairs with the AUG codon; the large, 50S, subunit then binds to form the complete ribosome, and the next aminoacyl-tRNA enters the A site.
Eukaryotic initiation of translation	The small, 40S, ribosomal subunit binds to the initiator tRNA which carries methionine. This complex then binds to 5' cap of the mRNA and moves to the first AUG in the sequence. The large, 60S, unit then binds to form the complete ribosome.
Elongation	A peptidyl transferase enzyme forms a peptide bond between the two adjacent amino acids, transferring the new amino acid on to the growing peptide chain. The ribosome then translocates (moves) down the mRNA in a 5'→3' direction so that the next codon is at the A site and the growing peptide chain is attached to the tRNA at the P site.

Termination	The codons UGA, UAA and UAG act as termination signals.
Wobble base-pairing	Some tRNA molecules can recognize more than one codon in the ribosome due to base-pairing which does not follow the normal rules (wobble base-pairing).
Related topic	Structure of proteins (C6)

The genetic code   A sequence of three bases in the mRNA, called a **codon,** codes for one amino acid. Each amino acid is coded for by between one and six codons and there are also codons that act as signals for the start and end of protein synthesis. This genetic code, as it is called, is shown in *Table 1* and is universal for all biological systems with a few exceptions in organelle genomes. Conversion of the base sequence in the mRNA into an amino acid sequence involves RNA adapter molecules called transfer RNAs (tRNAs). These have a sequence of three bases, called the **anticodon**, which can base-pair with the codon sequence in the mRNA and, attached to other end of the tRNA molecule, is an amino acid corresponding to that codon. The matching up of tRNA molecules to the mRNA and the subsequent joining together of the amino acids into a polypeptide chain is carried out on the **ribosomes**. The process of protein synthesis in the cell is called **translation**.

Table 1. The 'universal' genetic code; see Figure 2, Topic C6 for abbreviations

First position (5' end)	Second position				Third position (3' end)
	U	C	A	G	
U	Phe	Ser	Tyr	Cys	U
	Phe	Ser	Tyr	Cys	C
	Leu	Ser	Stop	Stop	A
	Leu	Ser	Stop	Trp	G
C	Leu	Pro	His	Arg	U
	Leu	Pro	His	Arg	C
	Leu	Pro	Gln	Arg	A
	Leu	Pro	Gln	Arg	G
A	lle	Thr	Asn	Ser	U
	lle	Thr	Asn	Ser	C
	lle	Thr	Lys	Arg	A
	Met	Thr	Lys	Arg	G
G	Val	Ala	Asp	Gly	U
	Val	Ala	Asp	Gly	C
	Val	Ala	Glu	Gly	A
	Val	Ala	Glu	Gly	G

Note: the boxed codons are used for initiation.

Ribosomes   The role of the ribosome is to hold the complex of mRNA and tRNAs together, form the peptide bonds between the amino acids and ensure the accuracy of protein synthesis. Ribosome particles consist of two subunits, one large and one

Table 2. The size of prokaryote and eukaryote ribosomes and their components

Whole ribosome	Prokaryotes (70S)		Eukaryotes (80S)	
Ribosomal subunits	30S (small)	50S (large)	40S (small)	60S (large)
Proteins	21	34	~ 33	~ 49
rRNA	16S	23S	18S	5S
		5S		5.8S
				28S

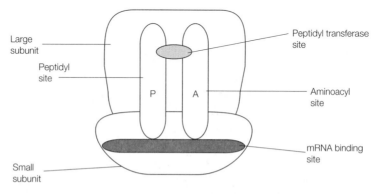

Fig. 1. The structure of a ribosome showing important functional sites.

small, made up of stable rRNA molecules and proteins. The size of ribosomes and the RNA molecules within them have traditionally been measured by their sedimentation coefficient in Svedberg units. This is the rate at which the particles sediment in a gradient, under centrifugal force, and takes into account the size, the shape and density of the particle. Although ribosomes from prokaryotes and eukaryotes look similar in structure there are differences in the size of the ribosomes (70S and 80S, respectively), their subunits and the composition of the proteins and rRNA within them (Table 2). This makes protein synthesis an ideal target for antibiotics as drugs that affect bacterial ribosomes will not affect eukaryotic cells (Topic F7). Within the ribosome there are two binding sites for tRNAs, the **amino-acyl (A) site** and the **peptidyl (P) site** as well as binding sites for mRNA (Fig. 1).

Loading of tRNA molecules

A tRNA molecule is a single strand of RNA (73–93 nucleotides long) folded into a cloverleaf secondary structure carrying the anticodon and amino acid binding site (Fig. 2). It contains some unique nucleotides, which are modified from the original A, U, G and C nucleotides post-transcriptionally, including pseudouridine and inosine. There is at least one tRNA molecule for each amino acid in the cell but not necessarily one for each codon (see wobble base-pairing).

The loading of the tRNA molecules with the correct amino acid is a key part of protein synthesis and it is the main point at which accuracy of translation is ensured. If the wrong amino acid is added to the tRNA molecule, it will subsequently be inserted incorrectly into the growing polypeptide chain, as the ribosome would not detect that the wrong amino acid was present. The enzymes

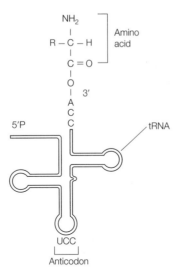

Fig. 2.   Structure of an aminoacyl-tRNA.

responsible for the addition of the amino acid to the tRNA are **aminoacyl-tRNA synthetases**. There is at least one enzyme for each amino acid. The enzyme binds the correct amino acid and the tRNA molecule and catalyzes the addition of the amino acid to the 3′ end of the tRNA molecule in two stages using ATP:

$$\text{Amino acid} + \text{ATP} \rightarrow \text{aminoacyl-AMP} + \text{P-P}$$

$$\text{Aminoacyl-AMP} + \text{tRNA} \rightarrow \text{aminoacyl-tRNA} + \text{AMP}$$

This process is sometimes called charging of the tRNA. The tRNA for an amino acid, called the **cognate** tRNA, is designated by a superscript (e.g. tRNA^{Ala}); once loaded with the amino acid it is written as Ala-tRNA^{Ala}.

Protein synthesis   Protein synthesis can be divided into three stages: initiation, elongation and termination. At initiation the start codon for the protein is recognized: normally it is AUG which codes for methionine, but very occasionally it may be GUG (valine). A complex is formed between the mRNA, the ribosome and the initiating tRNA. During elongation, amino acids are added sequentially to the growing peptide chain in accordance with the codon sequence in the mRNA. At termination, the end of the polypeptide chain is recognized and the complex of mRNA, polypeptide, tRNA and ribosomes breaks apart. Each stage requires a number of different factors to ensure the correct order of events, which are different between prokaryotic and eukaryotic translation. The energy for protein synthesis is provided by the hydrolysis of GTP.

Prokaryotic   The first stage of protein synthesis is the binding of the small (30S) ribosomal
initiation of   subunit to the mRNA so that the first AUG codon is positioned in the P site.
translation   The correct positioning is achieved by base-pairing between a short purine rich sequence (consensus sequence: 5′-GGAGG-3′), called the **Shine–Dalgarno** sequence, which is located in the mRNA, 8–13 nucleotides before the start site of translation, and a complementary sequence on the 16S RNA (consensus

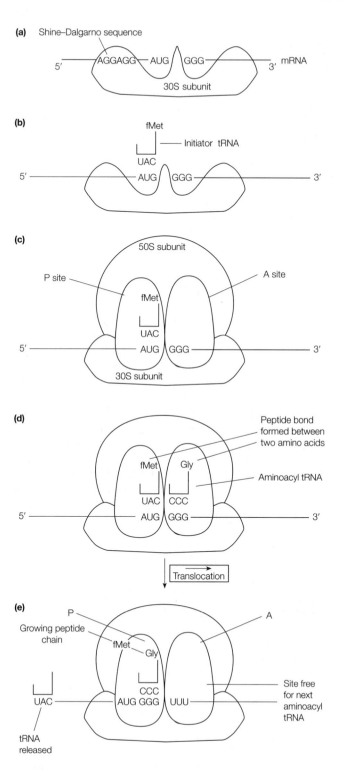

Fig. 3. Stages in the prokaryote protein synthesis.

sequence: 3'-CCUCC-5') in the small subunit of the ribosome (*Fig. 3a*). This mechanism, which is unique to prokaryotes, allows translation to start in the middle of a mRNA sequence as bacterial mRNAs frequently contain a number of genes which are translated independently (see Topic C4). Once the AUG is located at the P site the first tRNA, the initiator tRNA, can enter this site and its anticodon can base-pair with the AUG (*Fig. 3b*). The initiator tRNA is distinct from the tRNA which inserts methionine in response to AUG in internal sites in the polypeptide in that:

● it carries *N*-formylmethionine (fMet) instead of methionine (Met);
● it can enter the P site of the ribosome when the large subunit is absent.

Once the fMet–tRNA^{fMet} has bound to the AUG, the large (50S) ribosomal subunit joins to form the complete 70S ribosome (*Fig. 3c*) and the next tRNA can enter the A site to base-pair with the next codon (*Fig. 3d*). Three intitiation factors (IFs) and GTP hydrolysis are required for the formation of the initiation complex.

Eukaryotic initiation of translation

In eukaryotes the first AUG codon on the mRNA is the initiation codon, consequently the mechanism to identify the start site for protein synthesis is slightly different. The initiating tRNA carries Met rather than fMet and forms a complex with the small, 40S, ribosomal subunit before it binds to the ribosome. This complex binds to the 5'-methyl cap on the mRNA molecule then moves up the RNA until an AUG codon is located at the P site. The large, 60S, ribosomal subunit then binds and the next tRNA enters the completed A site.

Elongation

The ribosome detects that there is accurate base-pairing between the anti-codon and codon at the A site and then a **peptidyl transferase** enzyme, part of the ribosome, forms a peptide bond between the two adjacent amino acids. In doing this the amino acid on the tRNA at the P site is transferred on to the amino acid attached to the tRNA at the A site. The ribosome then moves in relation to the mRNA by a process called **translocation** so that the tRNA with the dipeptide attached is now at the P site. The empty tRNA is released and the A site is then free to accept a new tRNA molecule to base-pair with the new codon at the A site (*Fig. 3e*). Elongation factors, EF-tu, EF-ts, and EF-G, are required in prokaryotes to ensure that the process occurs in the correct order, and two molecules of GTP are hydrolyzed. The process of checking, synthesis of the peptide bond and translocation is repeated down the length of the mRNA molecule in a 5' → 3' direction. The polypeptide sequence grows from the amino end (N-terminus) to the carboxyl end (C-terminus).

Once the ribosome binding site and the AUG codon become available on the mRNA another ribosome can bind and start translation. Consequently, under the electron microscope, mRNA appears as a string of beads owing to the presence of multiple ribosomes. This structure is called a **polysome** or polyribosome.

Termination

There are three codons, UGA, UAA and UAG, which do not have a corresponding tRNA molecule. These act as termination signals indicating the end of an amino acid sequence. When the ribosome reaches such a codon it stalls as protein synthesis cannot continue. Release factors enter the empty A site and trigger the release of the synthesized polypeptide chain from the ribosome and the dissociation of the ribosome subunits.

Wobble base-pairing

Codons for an amino acid are often in families where only the first two bases in the sequence are important; for example, proline is coded for by CCA, CCU, CCC and CCG. Many tRNA molecules are therefore constructed to allow a mismatch, called a **wobble**, at the third position therefore allowing them to recognize more than one codon:

● Only in the ribosome can U and C base-pair with G at the third position of the codon.
● Some anticodons contain the unusual base inosine (I) in their anticodon which can base-pair with U, C or A at the third position in the codon.

D1 BACTERIAL TAXONOMY

Key Notes

Classification of bacteria	Classification of bacteria into groups based on common properties allows bacteria to be identified and named. The most important taxonomic rank is the species, and species are grouped further into genera. The name of a bacterium has two parts – the genus name followed by the species epithet, for example *Escherichia coli*.
Characters used in bacterial identification	Colony morphology, cell morphology, Gram stain reaction, growth characteristics, and biochemical, immunological and DNA tests are all used to classify and identify bacteria.
Current classification of bacteria	The prokaryote bacteria can be divided into two groups, the eubacteria and the archaebacteria. The eubacteria can then be further divided into at least 11 phyla based on 16S RNA sequence analysis.

Related topics	Bacterial cell structure (D2)	Taxonomy of protozoa (J1)
	Taxonomy (G1)	Virus taxonomy (K2)

Classification of bacteria

The main aim of bacterial taxonomy is to set up a **classification**: an ordering of bacteria into groups based on common properties. This classification can then be used to identify individual bacterial species or strains (**identification**) and name them (**nomenclature**). Classification is also used to study the relationships of different bacterial groups based on shared phenotypic properties and probably a common evolutionary history. A wide range of morphological, biochemical and molecular characters are measured and the information derived from this is used to create a hierarchical classification system, into which individual bacteria can be placed. The most important rank in the taxonomic hierarchy is called the **species**, which from the time of Linnæus (1707–1778), have been named in a similar way. Related species are grouped into a **genus** (pl. genera), which is usually based around some distinct phenotypic character, and the generic name is used as part of a binomial species name. So, for example *Escherichia coli* is in the genus *Escherichia* named after Escherich, who first isolated the bacteria in 1885. The specific epithet '*coli*' is in this case an indication as to where the organism was isolated (from i.e. the colon). However, note that the binomial name is a label not a description; it does not have to describe the species in any way. Note too that the genus name has a capital initial letter but the specific epithet (which cannot be used alone) does not, and that the whole binomial name is Latinized and therefore should be italicized or underlined. Normally, once the name of the organism has been written in full, the genus name can be reduced to a letter, or letters, as long as there can be no confusion with other genera. Thus, *Escherichia coli* will subsequently be written as *E. coli*. The higher taxonomic categories into which the organism can be ranked are, sequentially, family, order, class, division and kingdom, though the final three are not particularly useful in bacteriology. Similarly, species may be divided into subspecies and a culture of a particular organism is usually referred to as a **strain**. Strain is also used to describe a sub-group of a species that share some feature such as one type of O-antigen (Topic D3) or mutation (Topic E1). There are also informal names, not italicized, which are sometimes useful for describing groups of organisms such as the green-sulfur bacteria or the spirochetes.

The main impetus for bacterial classification has been the need to be able to identify bacterial disease-causing organisms and therefore these microbes are particularly well classified. However, it should be remembered that many other species have not been well characterized and many have not been isolated or described at all. The authority on bacterial classification is *Bergey's Manual of Systematic Bacteriology*, which contains as near as possible complete listings of all prokaryotic species and their distinguishing characteristics.

Characters used in bacterial identification

A list of the types of characters used to distinguish between bacteria is given below. This is not an exhaustive list and equally not all characters are used in either the classification or identification of every bacterial species. In many cases of well known microorganisms, a small set of tests may be all that is required to identify a microbe positively, especially if the site from which it has been isolated and the disease it caused are known. A comma-shaped, Gram-negative rod isolated from a patient excreting copious amounts of straw-colored, liquid stools may be enough to identify it putatively as *Vibrio cholera*, the causative agent for cholera. However, it may be necessary to do a wide range of tests to determine the name of an isolate if it is an undescribed microbe.

Colony morphology.   The shape, texture and color of bacterial colonies can be distinctive. For example, *Staphylococcus aureus* is so named as colonies are of a yellow color; '*aureus*' comes from the Latin for golden.

Cell shape, structures and reaction to stains.   Whether bacteria are Gram-positive or Gram-negative (Topic D3) is one of the most important clinical diagnostic tests. Gram staining and microscopic examination of bacteria give information as to the shape, size and arrangement of the bacterial cells (Topic D3) and in some cases this is enough to identify certain genera such as Streptococci which are Gram-positive cocci that grow in chains. Stains may also be used to show other morphological features (see Topic D3), which may be used for classification, such as the presence of:

- spores (malachite green stain);
- unusual cell walls (Ziehl Neelsen acid-fast stain);
- capsules (Indian ink stain);
- intracellular lipids (Sudan black);
- flagella (flagellar stain);
- metachromatic granules (Albert–Leybourne stain).

Growth characteristics.   The temperature, pH and O_2 requirements of an isolate are useful tests for the identification of bacteria.

Biochemical tests.   There is a wide range of tests available that measure various aspects of bacterial metabolism:

● what carbon and nitrogen sources the bacteria can use;
● the end products of their metabolic processes, such as acetoin, tested for in the Vogues–Proskaeur test;
● what enzymes the bacteria produce, such as decarboxylases, proteases and DNases;
● the presence of other molecules such as toxins (Topic F6), long-chain fatty acids or antibiotics.

Immunological tests.   Antibodies to cellular components such as the O-side chains of lipopolysaccharide (LPS) (Topic D3) or capsules are frequently used to distinguish between strains of one species.

DNA tests.   Comparisons of DNA content and sequence between strains are the most definitive methods for separating organisms into different groups and for measuring the evolutionary relatedness between them (**phylogenetic analysis**). Traditionally, DNA was analyzed by measuring the G-C content or the amount of DNA homology between strains. More recently, the comparison of the actual DNA sequences of conserved cellular molecules such as 16S ribosomal RNA (see Topic C3) or proteins such as cytochrome C, ATPase and elongation factor has been used. This information has been analyzed by a technique called **cladistics** to trace true phylogenetic lineages that have not been obscured by patterns of convergent evolution (Topic I1).

Current classification of bacteria

On the basis of 16S RNA sequences, it is now clear that there are (at least) three domains of organisms, the two prokaryotic groups, archaebacteria and eubacteria and the eukarya (*Fig. 1*). Within the eubacteria there are at least 11 groups, referred to sometimes as phyla, which can be represented on a phylogenetic

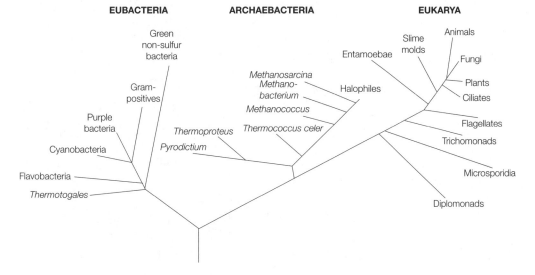

Fig. 1.   An overview of the phylogeny of the living world, showing the three major domains of organisms: the Eubacteria, the Archaebacteria and the Eukarya. Reproduced from Madigan et al., Brock Biology of Microorganisms, *1997, with permission from Prentice-Hall, Inc., Upper Saddle River, NJ.*

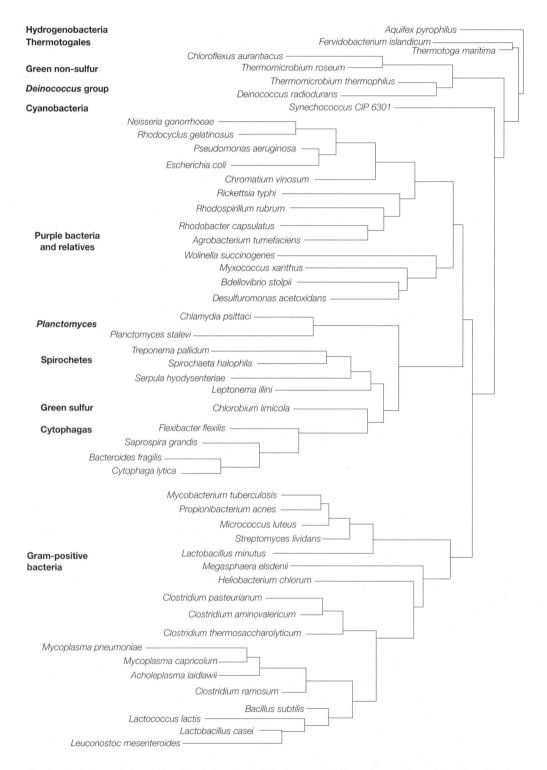

Fig. 2. A phylogenetic tree of bacteria derived from 16S ribosomal RNA sequences. Reproduced from Madigan et al., Brock Biology of Microorganisms, 1997, with permission from Prentice-Hall, Inc., Upper Saddle River, NJ.

tree *(Fig. 2)*. What can be seen is the wide diversity within the bacterial world and that many of the bacteria that may be familiar to us such as *E. coli, Bacillus* species and *Neisseria gonorrhoeae* are located within just a few of these phyla. It should be noted that the number of phyla will increase as more bacteria from diverse habitats are sequenced.

D2 BACTERIAL CELL STRUCTURE

Key Notes

Prokaryotic cells	Bacteria are prokaryotic cells whose single chromosome is not contained within a nuclear membrane. They lack many of the organelles and internal structures found in eukaryotic cells. The prokaryotes can be split into two groups, the archaebacteria and the eubacteria.
The cytoplasm and ribosomes	The main structures found in the aqueous cytoplasm of all bacterial cells are the ribosomes, the sites for protein synthesis in the cell. The archaebacterial ribosomes are the same size as those of eubacteria (70S), but differ in sensitivity to diphtheria toxin and some antibiotics.
Bacterial DNA	DNA is found, as a single circular chromosome, in the cytoplasm and is sometimes referred to as the nucleoid. Small extra-chromosomal circular pieces of DNA called plasmids are often found in bacteria.
Other internal bacterial features	Some bacteria contain structures associated with specialist functions. Inclusion bodies and lipid droplets are storage sites in the cell. Membrane systems are associated with the photosynthetic capabilities of some cells.
Endospores	Bacterial endospores are produced by a few genera including *Bacillus* and *Clostridium*. They consist of bacterial DNA surrounded by a number of layers of protein and peptidoglycan that are highly resistant to drying and heat.
Eubacterial cell wall and surface	The bacterial cytoplasm is surrounded by a plasma membrane. Surrounding this, eubacterial cells have, with a few exceptions, a rigid cell wall made of peptidoglycan, a substance not found in archaebacterial or eukaryotic cells. Gram-negative bacteria have an additional outer membrane. Other features which may be found associated with the cell surface, in some bacteria, are flagella, associated with cell movement, pili (or fimbriae), which have a role in adhesion, and extracellular polysaccharides (sometimes called glycocalyx) or proteins which help to protect the bacteria from their environment.
Archaebacterial cell wall and surface	The lipids in the plasma membrane of archaebacteria are branched and long and are ether-linked to glycerol. Cell-wall materials are very diverse and range from a peptidoglycan-like material, pseudomurein, to polysaccharides, protein and glycoproteins.
Related topics	Bacterial taxonomy (D1) Entry and colonization of human hosts (F5) Bacterial cell envelope (D3) Eukaryotic cell structure (G2)

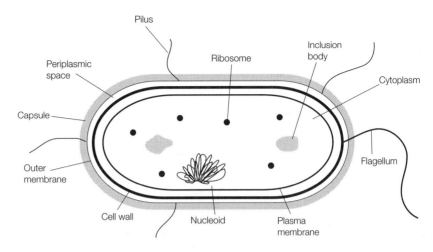

Fig. 1.   Diagram of a prokaryote cell (Gram-negative).

Prokaryotic cells   Bacteria are small (< 1 μm–50 μm width/diameter), single-celled, micro-organisms that belong to a group called prokaryotes so classified because their DNA is not enclosed within a nuclear membrane (Gr. *pro*, before; *karyote*, nucleus). Their internal cell structure is simple (*Fig. 1*), with most of the cellular complexity associated with the cell surface structures. Bacteria lack mitochondria and chloroplasts, the organelles associated with energy production in eukaryote cells and internal membrane structures such as the endoplasmic reticulum and Golgi apparatus (see Topics A1 and G2) although some photosynthetic bacteria do have internal membranes. In spite of this simplicity in structure there is a large amount of variation in the appearance of the cells when observed under the microscope. Two shapes of cell predominate, the coccus (spherical) and the bacillus, or rod, but there are also a wide range of other shapes (morphologies) as shown in *Table 1*.

Table 1.   Shapes of bacteria and characteristic associations of cells seen using light microscopy

Shape	Organization of cells	Examples
Cocci – spherical bacteria	In chains In grape-like clusters In pairs (diplococci) In cubes of cells (packets)	*Streptococcus* *Staphylococcus* *Neisseria* *Sarcina*
Rods/bacilli – rods may be very short (cocci-bacilli) or long	Singly In filaments of attached cells	*Pseudomonas* *Bacillus* sp.
Tightly coiled flexible bacteria – spirochaetes		*Treponema* *Borrelia*
Comma shaped – called a vibrio		*Vibrio*
Spirillum – long, curved, rigid rods		*Rhodospirillum*
Appendaged – cells have stalk or tube-like extensions		*Rhodomicrobium*
Pleomorphic – bacteria vary in shape		*Corynebacteria*
	Branching mycelium-like filaments	*Streptomyces*

Table 2.   *Some differences between eubacteria and archaebacteria*

Feature	Eubacteria	Archaebacteria
Cell wall	Muramic acid present	Muramic acid absent
Lipids	Ester-linked	Ether-linked
Methanogenesis	Absent	Possible
RNA polymerases	One	Several
Initiator tRNA	fMet	Met
Ribosomes	Sensitive to streptomycin and chloramphenicol	Resistant to streptomycin and chloramphenicol
	Resistant to diphtheria toxin	Sensitive to diphtheria toxin

Bacteria are ubiquitous. They are a highly successful and diverse group of organisms that can obtain energy and carbon from a wide range of sources and therefore can colonize every niche on our planet from deep ocean trenches to volcanic craters. In the 1970s, using DNA sequencing information, it was found that the group we know as the bacteria could be split into two, the **eubacteria** (Gr. *eu*, true and *bakterion*, a small rod) and the **archaebacteria** (Gr. *Archaios*, ancient) and it appears that these two groups evolved away from each other very early in the history of living things at about the same time that the first group of eukaryotic cells evolved (see Topic D1). Members of the eubacteria include some of the more familiar bacteria such as *Escherichia coli* and *Staphylococcus aureus* and are the prokaryotes that are best studied and understood. The archaebacteria are a very diverse group of organisms, which differ from the eubacteria in a number of features having, in particular, very different cell walls and membranes (see *Table 2* for a list of the important differences). This group includes bacteria that are capable of existing in extreme environments, such as hot springs (examples: *Sulfolobus* and *Pyrococcus*) and high salinity (*Halobacterium*), and the methanogens such as *Methobacterium*, which produce methane as a result of metabolism.

The cytoplasm and ribosomes

The cytoplasm of bacteria is aqueous, containing a cocktail of molecules, ribonucleic acid (RNA) and proteins necessary for the cell functions. The main structures, common to all bacteria, found in the cytoplasm are the **ribosomes**. The ribosomes which consist of a small and a large subunit, made up of a complex of proteins and RNAs, are the sites of protein synthesis in the cell. The ribosomes in prokaryotic cells, although similar in shape and function to those of eukaryotic cells, are different in the nature of the proteins and RNAs that make up their structure (Topic C7). This has proved to be very useful to the human population as antibiotics that act by inhibiting protein synthesis in bacteria are not effective against eukaryotic protein synthesis, thus allowing selective toxicity. This will be discussed in more detail in the topics on translation (Topic C7) and antibiotic action (Topic F7). Archaebacterial ribosomes are the same size as those of the eubacteria but, in some features, they are similar to eukaryotic ribosomes in that they are resistant to the antibiotics, streptomycin and chloramphenicol, and sensitive to the action of diphtheria toxin (*Table 2*).

Bacterial DNA

Bacterial DNA is located within the cytoplasm. It consists of a single chromosome which varies in size between different species of bacteria (the *E. coli* chromosome is 4×10^6 base pairs long). The DNA is circular, tightly supercoiled and associated with proteins which are similar to the histone proteins found in eukaryotic cells (Topic C1). Although the chromosome is not contained within a nuclear membrane, it is often seen as a discrete area within the cell in electron micrographs which may be referred to as the **nucleoid**. Some bacteria also contain small molecules of extra-chromosomal DNA called **plasmids**. These often carry genes which are not essential to the normal life of the cell but confer an advantage to the cell in certain situations such as antibiotic-resistant plasmids (Topic E5). The chromosome in archaebacteria, like that of the eubacteria, is a single, circular DNA molecule not contained within a nucleus, but the size of the DNA molecule is often smaller than that of *E. coli*.

Other internal bacterial features

Some bacteria contain structures associated with specialist functions. Granular structures, called **inclusion bodies,** can often be seen under the light microscope. These granules are usually for storage and may be membrane bound, such as poly-β-hydroxybutyrate (PHB) granules, or found scattered in the cytoplasm, such as polyphosphate granules (also called metachromatic granules). **Lipid droplets** can also be seen in some bacteria. An interesting inclusion body, the **gas vacuole**, is found in *Cyanobacteria* (blue–green algae) and other photosynthetic bacteria that live in an aqueous environment (Topic B3). This protein-surrounded vacuole provides buoyancy, allowing the bacteria to float near the surface of the water.

Although bacteria do not have organized intracellular membranes, invaginations of the plasma membrane called **mesosomes** are often seen in electron micrographs. There is some debate as to whether these mesosomes really exist and are not just an artefact of the fixation process for electron microscopy. Their function is thought to be in cell division, possibly in laying down new cell-wall material or in the replication of the chromosome and its subsequent distribution to daughter cells. Other, more specialist and complex, intracellular membrane systems are found in photosynthetic bacteria, such as *Chloroflexus* (green non-sulfur bacteria) and *Rhodopseudomonas* (purple bacterium), associated with the trapping of light energy (Topic B3).

Endospores

Finally, a number of bacteria including mainly the *Bacillus* and *Clostridium* genera produce a special reproductive structure called the **endospore**. This can be seen in the light microscope using either specialist spore stains such as malachite green or by phase-contrast microscopy. The spore has a number of layers surrounding the bacterial genetic material making it incredibly resistant to all kinds of environmental stress such as heat, UV irradiation, chemical disinfectants and drying (*Fig. 2*). As a number of important pathogenic bacteria are spore producers, sterilization measures must therefore be designed to remove these hardy structures as some can withstand boiling for several hours (Topic D9).

Eubacterial cell wall and surface

The cytoplasm of all bacteria is enclosed within a **plasma (cytoplasmic) membrane** external to which, in most cases of eubacteria, is a rigid **cell wall** made up of sugars and amino acids called **peptidoglycan**. The role of the cell wall is to protect the cell from lysis resulting from osmotic pressure and it also gives shape to the cell. Some bacteria, the mycoplasma, do not have cell walls and therefore are unable to survive outside an animal host which provides it with the right osmotic environment. Gram-negative bacteria have an additional

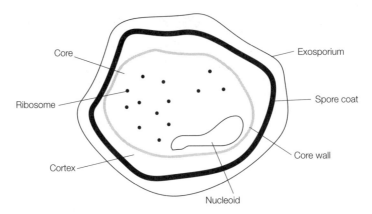

Fig. 2. Diagram of an endospore.

outer membrane containing LPS. External to this may be other layers of poly-saccharide or protein making up a capsule or slime layer. Layers external to the cell wall may be referred to as the cell **envelope.** The structure of the eubac-terial cell envelope is discussed in more detail in Topic D3.

A number of features are found on the surface of the bacteria which aid the survival of the cell in the environment. Thin hair-like proteinaceous appendages, several micrometers long, called **pili** (s. pilus) or **fimbriae** (s. fimbria) are often seen projecting from the surface of the cell in electron micrographs. These struc-tures are normally associated with the adhesion of bacteria to surfaces (Topic F5) although there is one particular type of pilus, the sex pilus, which is involved in the transfer of genetic material from one bacterium to another (Topic E6). Motile bacteria have one or more **flagella** (s. flagellum) which are slender, rigid structures up to 20 μm long. Rotation of the flagellum in relation to the cell allows the movement of the bacterium. Normally, bacteria move in response to external stimuli; an example of this is **chemotaxis** when bacteria move towards nutrients or away from harmful substances as a result of chemical gradients (see Topic D6).

Archaebacterial cell wall and surface

One of the distinctive features of archaebacteria is the nature of the lipids in the plasma membrane which, unlike the ester-linked lipids of eubacteria and eukaryotes are ether-linked to glycerol. They are also long chained and branched. Archaebacterial cell walls and envelopes show great diversity and complexity in structure. They do not contain peptidoglycan although some do have a similar compound called **pseudomurein** which contains *N*-acetyl-alosaminuronic acid in place of muramic acid (see Topic D3). Another common type of cell wall is the S-layer, a two-dimensional, paracrystalline, array of protein or glycoprotein on the cell surface. Others have thick polysaccharide walls outside their plasma membrane.

D3 BACTERIAL CELL ENVELOPE

Key Notes

The Gram stain	The Gram stain is one of the most commonly used stains for bacteria as it distinguishes between two groups of bacteria on the basis of their cell-wall structure. The envelope of Gram-positive cells consists of a plasma membrane surrounded by a thick peptidoglycan layer. Gram-negative bacteria have only a thin layer of exterior peptidoglycan but on the outside of this is a second, outer, membrane. Gram-positive bacteria retain a crystal violet and iodine complex on washing with alcohol and therefore appear purple under light microscopy. Gram-negative bacteria stain pink as they lose this complex and take up the paler carbol fuschin counter-stain.
Lipid bilayer	The membranes of the cell are made of phospholipids which are amphipathic molecules containing hydrophilic (water-loving) and hydrophobic (water-hating) regions. In the membrane, the phospholipids are arranged in a bilayer so that their polar headgroups, the hydrophilic regions, are on the surface of the membrane and their hydrophobic, fatty acid tails, are pointed into the interior of the membrane. Consequently, the membrane acts as a selective permeability barrier to bulky or highly charged molecules, which cannot easily pass through the hydrophobic interior of the lipid bilayer.
The plasma (cytoplasmic) membrane	The plasma membrane is a phospholipid bilayer in which are embedded integral proteins associated with transport, energy metabolism and signal reception. Other, peripheral proteins, are loosely associated with the membrane by charge interactions. The lipids and proteins in the membrane are mobile in relation to each other.
The bacterial cell wall	The cell wall is made up of a peptidoglycan, a polymer of N-acetyl glucosamine (NAG) and N-acetyl muramic acid (NAM) which has side chains of alternating D- and L-amino acids. This is a highly cross-linked molecule which gives the cell rigidity, strength and protection against osmotic lysis. Peptidoglycan contains many unique features such as D-amino acids, which makes it a useful target for antibiotics. Gram-positive cell walls also contain teichoic acids.
The periplasmic space	The periplasmic space is an aqueous region between the two membranes of Gram-negative bacteria. It contains proteins associated with nutrient transport and nutrient acquisition.
The outer membrane	Peculiar to Gram-negative bacteria, the outer membrane contains a number of unique structures. Braun's lipoprotein attaches the outer membrane to the peptidoglycan; porins allow the passive movement of nutrients through the membrane, and LPS gives protection to the cell. LPS, also known as endotoxin, is highly toxic to mammals.

| Extracellular layers | The surface of cells are often covered by layers of polysaccharide or proteinaceous material known as capsules or slime layers, depending on their density. The polysaccharide layers are sometimes called the glyco-calyx. These layers play a role in protecting the cell from desiccation and toxic compounds and also in the adhesion of bacteria to surfaces. |

Related topics	Transport across the membrane (D5)	Bacterial toxins and human disease
	Bacterial movement and chemotaxis	(F6)
	(D6)	Control of bacterial infection (F7)

The Gram stain   The eubacteria are frequently divided into two groups on the basis of their reaction to a stain devised by Christian Gram in 1884. The differential reaction to the staining procedure is because of the structure of the cell envelope in these two groups of bacteria. **Gram-positive** bacteria have a single membrane called the **cytoplasmic** (or **plasma)** membrane, surrounded by a thick layer of peptidoglycan (20–80 nm). The **Gram-negative** bacteria have only a thin layer of peptidoglycan (1–3 nm) but on the outside of this there is a further **outer membrane** which acts as an additional barrier (*Fig. 1*).

The procedure for the Gram stain is as follows. Fixed cells are stained with a dark stain such as crystal violet, followed by iodine which complexes with the stain in the cell wall of the bacteria. Alcohol is added, which washes the dark stain of crystal violet–iodine out of cells that have thin cell walls but not from those that have thick cell walls. Finally, a paler stain such as carbol fuschin, called a counter-stain, is added which stains the decolorized cells pink but is not seen on the dark staining cells that retained the first stain. The cells that retain the stain (with thick cell walls) are called Gram-positive and appear dark purple under light microscopy. The ones that lose the stain (with thin cell walls and an outer membrane) are called Gram-negative and stain pink or pale purple.

(a)   (b)

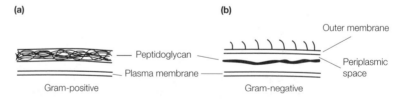

Fig. 1.   Structure of the cell surface of a Gram-positive (a) and a Gram-negative (b) bacterium.

Lipid bilayer   The plasma membrane is a semi-permeable lipid bilayer which acts as a barrier between the cytoplasm and the surrounding environment. The membrane is made of phospholipids (*Fig. 2*), such as phosphatidyl choline, which consist of a polar headgroup attached via glycerol to two long chains of non-polar fatty acids. Such molecules which consist of both polar and non-polar groups are called **amphipathic**. The polar headgroups are **hydrophilic** (Gr. *hydro*, water; *philic*, loving) and the fatty acids are **hydrophobic** (Gr. *phobis*, hating) which means that in an aqueous environment these molecules line up to ensure that

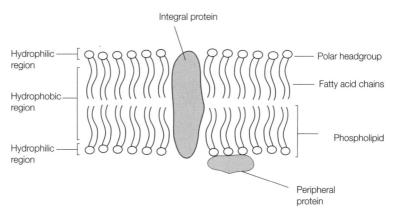

Fig. 2.   Diagram of the structure of the plasma membrane.

the polar headgroups are associated with the water molecules and the hydrophobic chains are tucked away in the interior of the bilayer away from the water. Hence, a bilayer can be formed, called a membrane (*Fig. 2*), about 5–10 nm thick. The interior of the membrane is highly hydrophobic, therefore acting as a barrier to anything that is bulky, such as glucose, or highly polar, such as ions. Small molecules like water can diffuse through and hydrophobic compounds such as benzene can also move through by dissolving in the hydrophobic interior. Bacterial membranes do not contain sterols such as cholesterol which provide stiffness in higher organisms. However, recently it has been shown that sterol-like molecules called **hopanoids** are found in bacterial membranes.

The plasma (cytoplasmic) membrane

Embedded in the lipid bilayer are proteins of various functions (*Fig. 2*) including transport proteins (Topic D5), proteins involved in energy metabolism (Topics B1 and B2), and receptor proteins that can detect and respond to chemical stimuli (Topic D6). **Integral** proteins are those that are fully associated with the membrane and may penetrate all the way through. These proteins therefore contain hydrophobic amino acids in the regions which are buried in the lipid (Topic C6). **Peripheral** proteins are ones that are only loosely associated by charge interactions with the positively charged polar headgroups of the phospholipids or on integral proteins and can be removed from purified membranes by washing with salt solutions. The lipids and proteins are fully mobile and move in relation to each other. This widely accepted model of membrane structure is called the fluid mosaic model.

The bacterial cell wall

Outside the plasma membrane is the cell wall made of sugars and amino acids called the peptidoglycan (sometimes called the murein layer). The role of the cell wall is to provide rigidity and strength, preventing the cell from osmotic lysis when placed in dilute environments. The structure of peptidoglycan from *E. coli* is shown in *Fig. 3*. It consists of long polymers of two sugar derivitives, NAG and NAM with side chains of four alternating D- and L-amino acids attached to the NAM. Rigidity is achieved by crosslinks between the amino acid chains; normally from the third amino acid in one chain to the fourth amino acid in another chain. The nature of the amino acid side chains, and the links that join them, vary between bacterial species; however, the third amino

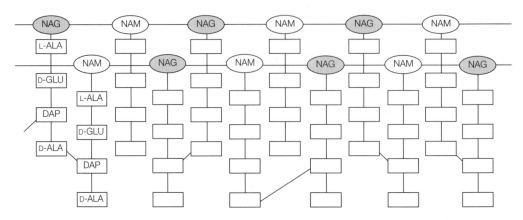

Fig. 3. Structure of the E. coli *peptidoglycan showing crosslinks between two strands. NAG: N-acetyl-glucosamine; NAM: N-acetylmuramic acid; D-Ala: D-alanine; DAP: meso-diaminopimelic acid; D-Glu; D-glutamic acid.*

acid is always a diamino acid (has two amino groups, so can form links) and the fourth amino acid is normally D-alanine (D-Ala). In *E. coli* and most other Gram-negative bacteria there is a direct link between the diamino acid *meso*-diaminopimelic acid and D-Ala. In Gram-positive bacteria the linkage often contains a bridge of amino acids: for example, in *Staphylococcus aureus* there is a five amino acid (pentapeptide) bridge of glycines between D-Ala and L-lysine. Multiple crosslinks within and between chains makes peptidoglycan a very strong and rigid structure.

There are several unique features of peptidoglycan including:

● the presence of NAM; this sugar is not found in eukaryotic cells;
● the presence of D-amino acids; L-amino acids are normally found in proteins.

These features make the peptidoglycan a target for antimicrobial agents that destroy prokaryotic cells specifically, but do not harm eukaryotic cells; an example of this is the antibiotic **penicillin** (Topic F7). **Lysozyme**, a natural antibacterial agent found in tears and natural secretions (Topic F4), breaks down the β(1→4) linkage between NAM and NAG. Removal of the cell wall under conditions where the osmolarity of the medium is the same as the inside of the cell (isotonic solution) results in the formation of round **protoplasts** (Gram-positives) or **spheroplasts** (Gram-negatives) which survive as long as the isotonicity is maintained. These structures lyse, however, if placed in a dilute medium, illustrating the importance of peptidoglycan to the cells' survival.

Gram-positive cell walls also contain large amounts of another polymer, called **teichoic acid**, made up of glycerol or ribitol joined by phosphate groups. D-Ala, glucose or sugars may be attached to the glycerol or ribitol and the polymers are attached either directly to the NAM in the peptidoglycan or to lipids in the membrane (in this case they are called **lipoteichoic** acids). The function of these molecules is unclear but they may have a role in maintaining the structure of the cell wall and in the control of **autolysis**.

Cell walls of a number of genera including *Mycobacterium*, *Corynebacterium* and *Nocardia* contain waxy esters of mycolic acids, which are complex fatty acids.

The periplasmic space

The outer membrane of Gram-negative bacteria acts as an additional barrier protecting the peptidoglycan from toxic compounds such as lysozyme which act on the cell wall. It creates an aqueous space between the two membranes called the **periplasmic space** which is thought to have a gel-like structure with a loose network of peptidoglycan running through it. Estimates as to the width of the peptidoglycan vary from 1–71 nm but it has proved to be difficult to obtain a real definitive value. The periplasmic space contains a range of proteins associated with:

- transport of nutrients (Topic D5) into the cell;
- enzymes that are involved in nutrient acquisition such as proteases;
- enzymes that defend the cell against toxic chemicals such as β-lactamases that destroy penicillin (Topic F7).

In Gram-positive cells these enzymes (called exoenzymes) are normally secreted into the surrounding medium. The presence of the outer membrane in Gram-negative bacteria allows the cell to keep the enzymes close to itself rather than losing them into the medium.

The outer membrane

The outer membrane of Gram-negative bacteria is made up of phospholipids but it also contains some unique features (*Fig. 4*).

- Pores formed by proteins called **porins** such as OmpF and OmpC that allow the passive diffusion of small molecules into the periplasmic space.
- An abundant small lipoprotein called Braun's lipoprotein that is covalently bound to the peptidoglycan and is embedded in the outer membrane by its hydrophobic lipid, therefore holding the peptidoglycan and outer membrane close together.
- LPS molecules are found in the outer leaflet of the outer membrane projecting into the surrounding medium.

LPS consists of three parts:

- Lipid A which is embedded in the membrane.
- A core polysaccharide which contains a number of unusual sugars including 2-keto-3-deoxyoctonate (KDO).
- A side chain of repeating sugars called the O-antigen or O-side chain. The

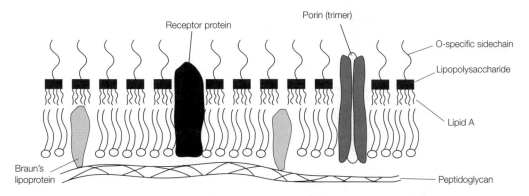

Fig. 4. Diagram showing the structure of the outer membrane of a Gram-negative bacterium.

composition of the O-side chain varies between strains and antibodies to the O-antigen are often used for typing bacteria in the laboratory.

LPS is responsible for a number of features of the Gram-negative bacterium:

● they cause a net negative charge on the surface of the cell;
● they may hinder the access of toxic molecules to the surface of the cell therefore play a protective role (Topic F5);
● the long side chains are capable of variation in structure therefore may play a role in allowing Gram-negative bacteria to evade the immune response (Topic F4).

Most importantly, the lipid A portion of the LPS molecule is highly toxic to mammals. Called endotoxin, its presence in the blood stream, even at very low concentrations, leads to toxic shock and death (Topic F6).

Extracellular layers

Different words are used to describe the layers of material often seen on the surface of cells. A well organized dense structure is called a **capsule** whereas if it is diffuse and easily lost it is called **a slime layer**. **Glycocalyx** is often used to describe both capsules and slime layers that are made up of polysaccharides; however, some bacteria have proteinaceous capsules. These additional surfaces layers have a number of functions:

● they act as permeability barriers to the cell surface;
● they protect the bacteria from phagocytosis;
● they protect the cell from desiccation;
● they aid in the attachment (adhesion) of bacteria to surfaces.

These roles will be discussed in more detail in Topic F5 on bacterial colonization of a host.

D4 BACTERIAL CELL-WALL SYNTHESIS

Key Notes

Peptidoglycan synthesis

Peptidoglycan found in Gram-positive and Gram-negative cell walls consists of long polysaccharide chains made of alternating NAM and NAG units. Pentapeptide chains are attached to NAM groups and these link the polysaccharide chains. The synthesis of peptidoglycan is a multi-step process requiring a number of enzymes and two carriers, bactoprenol and uridine diphosphate (UDP). Peptidoglycan synthesis is sensitive to antimicrobial agents, such as penicillins.

Cell-wall synthesis

The peptidoglycan layer in bacteria is a stress-bearing structure. Synthesis of new peptidoglycan during growth is complex. Controlled lysis by autolysins allows synthesis to occur at specific sites on the cell.

Related topics

Biosynthetic pathways (B4)
Bacterial cell structure (D2)

Bacterial cell envelope (D3)

Peptidoglycan synthesis

Peptidoglycan in bacterial cell walls consists of long polysaccharide chains made of alternating **NAM** (*N*-acetylmuramic acid) and **NAG** (*N*-acetylglucosamine) units. Pentapeptide chains are attached to NAM groups and these link the polysaccharide chains either by direct binding between pentapeptide chains or by the use of **pentaglycine bridges**. The synthesis is a multi-step process requiring several enzymes and two carriers; **bactoprenol** is a 55-carbon alcohol that attaches to NAM via a pyrophosphate group and carries short peptidoglycan units through the membrane. The other carrier is uridine diphosphate (UDP). The synthesis occurs in several stages (*Fig. 1*). UDP derivatives of NAM and NAG are synthesized in the cytoplasm (see Topic B4). Amino acids are sequentially added to UDP–NAM with the two terminal D-alanines added as a dipeptide. ATP is required to form the peptide bonds. The NAM–pentapeptide is transferred from UDP to bactoprenol located at the membrane surface. NAG (in the form of UDP–NAG) is added to the NAM–pentapeptide to form a **peptidoglycan repeat unit**. If a pentaglycine bridge is required this is added using a special glycyl–tRNA molecule. This is transported across the membrane by the bactoprenol carrier. The peptidoglycan unit is then added to the growing end of the peptidoglycan chain and the bactoprenol is recycled. Peptide crosslinks between the chains are formed by **transpeptidation**. The precise mechanism of transpeptidation varies between species. This process is sensitive to antimicrobial agents: **penicillin** inhibits transpeptidation and **bacitracin** blocks the recycling of bactoprenol. The enzymes inhibited by penicillins are also known as **penicillin-binding proteins**.

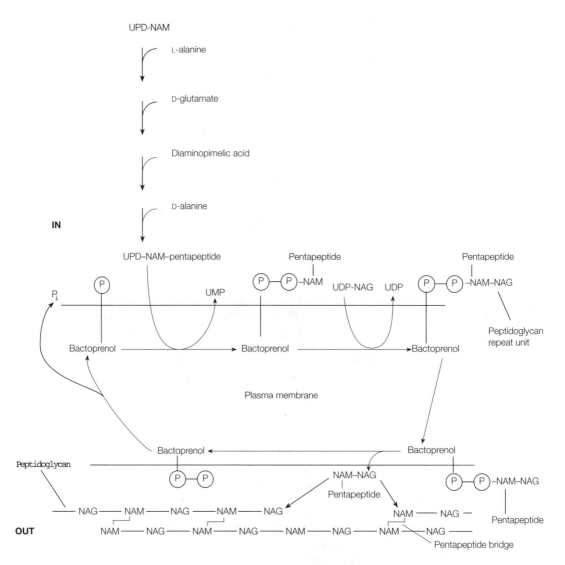

Fig. 1.   Peptidoglycan synthesis and incorporation of peptidoglycan repeat units in E. coli.

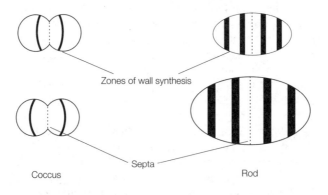

Fig. 2.   Pattern of cell-wall synthesis.

Cell-wall synthesis

The peptidoglycan layer in bacterial cell walls (see Topic D2) is a stress-bearing single sheet; thus, the growing bacterium must degrade this structure, just enough to allow new peptidoglycan to be incorporated. This digestion is accomplished by enzymes known as **autolysins**. Some degrade the polysaccharide chains, others cleave peptide bridges that link adjacent polysaccharide chains. If peptidoglycan is degraded in an uncontrolled way the growing cell will lyse; thus, the activity of autolysins is tightly controlled by the activity of **autolysin inhibitors**. The location of new cell-wall synthesis varies with species; however, there are some general patterns. In Gram-positive cocci (*Fig. 2*) synthesis occurs principally at the site of **septum formation** (point of formation of a cross wall prior to cell division). The second pattern of synthesis is seen in rod-shaped bacteria. Here, peptidoglycan synthesis occurs at the site of septum formation and at other sites scattered along the cylindrical portion of the rod (*Fig. 2*).

D5 TRANSPORT ACROSS THE MEMBRANE

Key Notes

Passive diffusion

The movement of gases and water into or out of the cell occurs by passive diffusion. This process is driven by concentration gradients. Net movement occurs down the gradient from high to low concentration.

Transport proteins

The intact cell membrane is an effective barrier to most polar molecules such as sugars and amino acids; this poses a problem for microorganisms. Movement of these molecules requires the action of membrane transport proteins which span the membrane. Three classes of these proteins are known. Uniporters move a single type of compound across the membrane. Symporters move two types of compounds simultaneously in the same direction. Antiporters move two compounds in opposite directions.

Facilitated diffusion

Facilitated diffusion is a mechanism for moving large, charged molecules through the cell membrane and is driven by a concentration gradient. However, this process uses membrane proteins to facilitate transport. Transport proteins show specificity in that they will only bind one type of molecule; this binding shows saturable kinetics (all substrate binding sites on the transport protein are filled at high substrate concentrations). This transport mechanism is influenced by temperature and pH.

Active transport

Active transport allows molecules to be taken up by cells against a concentration gradient. This is the most important method for the uptake of nutrients especially in environments where substrates may be present at low concentrations. Energy is required to drive this process. Active transport is affected by temperature and pH, and the process is saturable.

Group translocation

This differs from active transport in that substrates are chemically modified (usually by phosphorylation) during transport. The phosphoenolpyruvate glucose phosphotransferase system in *E. coli* is a good example of this type of system. Not all bacteria utilize this system.

Other transport mechanisms

Gram-negative bacteria can transport a range of large molecules via a modified active transport mechanism that utilizes substrate-specific binding proteins and an ATP requiring transporter. Iron is taken up by many Gram-negative bacteria by a chelating (iron-binding) transport system.

Related topics

Heterotrophic pathways (B1)
Electron transport and oxidative
  phosphorylation (B2)

Bacterial cell envelope (D3)

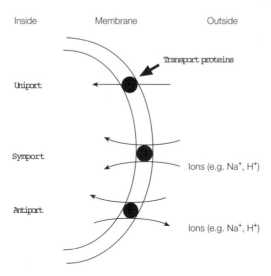

Fig. 1.   Classes of transport proteins.

Passive diffusion When substances move across a membrane without specifically interacting with proteins in the membrane it is termed passive diffusion. This type of movement is driven by a concentration gradient. Substrates (including gases) move from a region of high to one of low concentration. When equilibrium is reached net movement in one direction stops; however, where there is a mechanism within the cell for utilization of the substrate, such as respiration, equilibrium is never reached and the process continues. Molecules that move via this mechanism must be small and/or soluble in a hydrophobic (non-polar) environment. Such molecules include O_2 and CO_2.

Transport proteins The intact cell membrane is an effective barrier to most polar molecules such as sugars and amino acids; this poses a problem for microorganisms. Movement of these molecules requires the action of **membrane transport proteins** which span the membrane (**transmembrane**). Three classes of transport proteins have been identified (*Fig. 1*). **Uniporters** move a single type of compound across the membrane. **Symporters** move two types of compounds simultaneously in the same direction. **Antiporters** move two compounds in opposite directions. These transporters are important for both **facilitated diffusion** and **active transport**.

Facilitated diffusion This process is similar to that of passive diffusion except that membrane carriers (proteins) are required for the transport of molecules across the membrane. Molecules bind to a specific transport protein on the outside of the membrane and are released on the inside. The transport proteins are specific for a substrate or substrate type. This type of mechanism is only used by organisms for the uptake of compounds that are found at a concentration in the environment higher than that in the cell; however, rapid cellular utilization of the molecules stops equilibrium being reached. Because this mechanism is mediated by proteins, activity is affected by pH and temperature and can be **saturated** by substrate (all substrate binding sites on the transport protein are filled at high substrate concentrations).

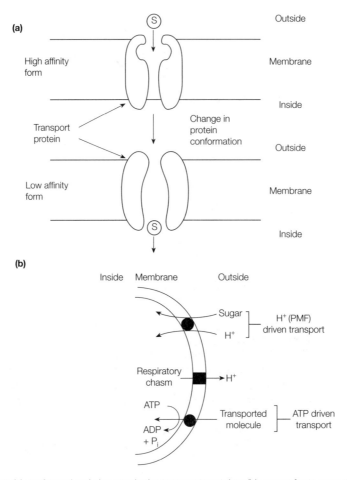

Fig. 2.   Active transport: (a) conformational changes in the transport proteins; (b) energy for transport. S, substrate.

Active transport   Active transport is an energy-requiring mechanism of uptake. The advantage of this type of transport process is that uptake occurs even when the concentration of the substrate in the environment is significantly lower than that in the cell. This enables cells to live in sites where substrates are at low concentration, the normal situation in nature. During transport, the affinity (strength of the interaction between the protein and the substrate) of the transport protein must change so that substrate can be released into the cell; on the outer surface of the membrane the protein exhibits a high affinity. As the substrate is transported across the membrane, conformational (structural) changes in the protein produce a low affinity form which allows release of the substrate (*Fig. 2a*). The transporters are also substrate specific. The energy for the process is obtained from proton motive force (PMF) (see Topic B2) or ATP hydrolysis (*Fig. 2b*). To utilize PMF the transporter must bind both H^+ and the substrate such that both are moved into the cell (symport). Lactose in *E. coli* is taken up by such a symport. However, some substrates are transported by an ion-dependent symport mechanism that functions only with a Na^+/H^+ antiport (molecules are moved in opposite directions). Glutamate uptake by *E. coli* is an example of such a mechanism.

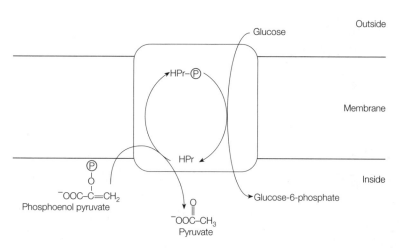

Fig. 3. Mechanism of the phosphoenolpyruvate:glucose phosphotransferase system. A small protein (HPr) acts as a carrier of high-energy phosphate.

Group translocation

This differs from active transport in that the substrate moved across the membrane is chemically modified during the transport. The modification usually involves phosphorylation. The **phosphoenolpyruvate: glucose phosphotranferase** system in *E. coli* is a well studied example. This mechanism proceeds in several steps utilizing one protein and three enzymes (*Fig. 3*). A phosphotransferase system is not present in all bacteria, *Pseudomonas aeruginosa* relies on active transport for uptake of sugars.

Other transport mechanisms

Transport of amino acids, small peptides, some sugars and organic acids in many Gram-negative bacteria is via a modified active transport mechanism that requires **substrate specific binding proteins** and an ATP-requiring transporter. In these organisms the outer membrane contains pores that allow movement of molecules into the periplasmic space (see Topic D2). The binding proteins are located in the periplasmic space, and substrate molecules bind to these. The complex formed then interacts with proteins embedded in the inner membrane. The transport of iron presents a unique problem for microorganisms as it is essential for growth but is not freely available in many environments. To overcome this, bacteria produce **chelating transport systems**. These require a chelating agent or **siderophore** (organic compound that binds iron) and a specific receptor protein (located in the outer membrane) that binds the siderophore–iron complex.

D6 BACTERIAL MOVEMENT AND CHEMOTAXIS

Key Notes

Motility	Many bacteria use a special structure, the flagellum, for motility. The bacterial flagellum is a long structure that is free at one end and attached to the cell at the other end.
Flagella structure	Flagella are composed of helically arranged protein subunits; the protein is called flagellin. The portion of the flagellum embedded in the membrane is surrounded by two pairs of rings (basal body). In Gram-negative bacteria, the outer pair of rings associated are attached to the LPS and peptidoglycan layers of the cell wall, and the inner pair of rings is located within or just above the plasma membrane. In Gram-positive bacteria, only one pair of rings is present.
Flagellar movement	Flagella are rigid structures which do not flex but rotate. The rotary motion of the flagellum is driven by the basal body, which acts like a motor. The direction of flagellar rotation determines the type of movement. Bacteria with a single flagellum move forward during counterclockwise rotation and tumble when the flagellum rotates clockwise. Where there are more than one flagellum, they behave as a single bundle during counterclockwise rotation and thus move forward; however, during clockwise rotation the flagella act independently and the organism tumbles.
Chemotaxis	Chemotaxis is the movement of an organism towards or away from a chemical. Positive chemotaxis is movement towards a chemical (attractant); negative chemotaxis is movement away from a chemical (repellent). Bacterial movement is controlled by the presence of these compounds such that where there is no gradient of attractant or repellent in the environment, the organism moves in a random way. However, in the presence of a concentration gradient, the net movement of the bacterium is in one direction. Bacteria detect the presence of a gradient through the action of membrane-bound chemoreceptors.
Related topics	Electron transport and oxidative phosphorylation (B2) Bacterial cell structure (D2) Bacterial cell envelope (D3)

Motility

Many bacteria use a special structure, the flagellum, for motility. The bacterial flagellum is a long thin (20 nm) structure that is free at one end and attached to the cell at the other end. The position and number of flagella are often used as characteristics for classification. **Polar flagella** are positioned at one or both ends of the cell. Where flagella are found at various sites around the cell these are termed **peritrichous**, and where several are located at one end of the cell they are termed **lophotrichous**.

Flagella structure

Flagella are composed of protein subunits; the protein is called **flagellin**. The structure of a typical flagellum is shown in *Fig. 1*. The portion of the flagellum embedded in the membrane is surrounded by two pairs of rings (basal body). In Gram-negative bacteria, the outer pair of rings is associated with the lipo-polysaccharide (LPS) and peptidoglycan layers of the cell wall, and the inner pair of rings is located within or just above the plasma membrane. In Gram-positive bacteria, which lack the outer LPS layer, only the inner pair of rings is present.

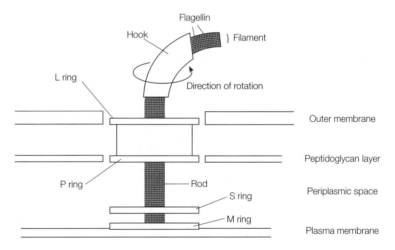

Fig. 1. Structure of the flagellum basal body and hook in a Gram-negative bacterium.

Flagellar movement

Flagella are rigid structures which do not flex but rotate. The rotary motion of the flagellum is driven by the basal body, which acts like a motor. The rings within the structure are thought to act together in generating rotational movement and the driving force for this comes from PMF (see Topic B2). Dissipation of the proton gradient releases energy which causes rotation of the flagellum. The direction of flagellar rotation determines the type of movement. Bacteria with a single flagellum move forward during counterclockwise rotation and tumble when the flagellum rotates clockwise. Flagella of peritrichous organisms behave as a single bundle of flagella during counterclockwise rotation and thus move forwards; however, during clockwise rotation the flagella act independently and the organism tumbles (*Fig. 2*).

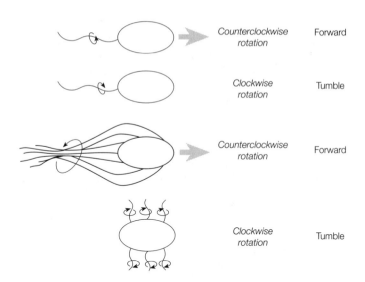

Fig. 2.   Effects of flagellar rotation on bacterial movement.

Chemotaxis

Chemotaxis is the movement of an organism towards or away from a chemical. **Positive chemotaxis** is movement towards a chemical (attractant, i.e. a substrate). **Negative chemotaxis** is movement away from a chemical (repellent). Chemotaxis is thus a response to chemicals in the environment and requires that the cell has some form of sensory system. Bacterial movement can be divided into runs where the organism moves in one direction (caused by counterclockwise flagellar rotation) and 'twiddles' where the organism randomly tumbles (caused by clockwise rotation of the flagella). Where there is no gradient of attractant or repellent the organism moves in a random way with a large number of twiddles; however, in the presence of a gradient, the runs become longer and twiddles less frequent as the organism experiences higher concentrations of attractant or repellent. The link between flagella movement and concentration of attractant or repellent involves the action of chemoreceptors which are proteins located in the periplasm. Although a chemoreceptor is fairly specific for the compound which it combines with, this specificity is not absolute. For example, the galactose chemoreceptor also recognizes glucose and fructose, and the mannose chemoreceptor also recognizes glucose. **Methyl-accepting chemotaxis proteins** (MCPs, also called **transducers**) are involved in translating the chemotactic signals from chemoreceptors to the flagellar motor.

In *E. coli* four types of MCP proteins have been identified – MCPs I, II, III and IV. Each MCP responds to different attractants and repellents. MCPs are so named because they become methylated or demethylated during chemotactic events. MCPs are transmembrane proteins that interact with attractants or repellents either directly or by way of chemoreceptors. A maximum of four methyl groups can be added to each MCP. Following sensory recognition of an attractant, an MCP becomes increasingly methylated and this causes the transmission of an excitatory signal to the flagellar motor, causing it to rotate in a specific direction. If rotation of the flagellum is counterclockwise, the cell continues to move. The longer the counterclockwise movement, the longer the run. When

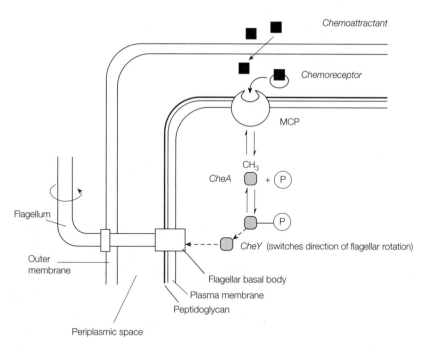

Fig. 3.   Interaction between a chemoattractant and flagellar rotation. CheA and CheY are small proteins involved in flagellar rotation; methylation of MCP affects phosphorylation of CheA, which in turn affects the activity of CheY.

the chemoreceptor can no longer sense an increasing concentration gradient, a process called **adaptation** occurs. This is a result of the complete methylation of the MCPs and leads to an approximately 100-fold decrease in sensitivity to the attractant/repellent. This action results in clockwise flagellar movement and the cell twiddles. Demethylation of the MCP returns it to a state in which it can excite the flagellar motor. In the presence of a repellent, increasing concentration causes increasing demethylation. The link between the MCPs and flagella motion shown in *Fig. 3* is not completely understood.

D7 GROWTH IN THE LABORATORY

Key Notes

Growth media	Microbes are grown (cultured) in the laboratory in either liquid or solid medium. Bacterial growth media should contain all the essential requirements for growth: an energy, carbon and nitrogen source as well as the essential ions, phosphate, sulfate, sodium, calcium, magnesium, potassium and iron, and various trace elements. Prototrophs require no additional ingredients in the medium, while auxotrophs require additional growth factors such as amino acids, vitamins and nitrogenous bases. The media used may be defined (synthetic) media where all the ingredients are known or complex media containing a mixture of undefined nutrients. Agar is used to solidify a liquid medium. Selective or differential media are used to detect the presence of particular groups of microbes.
Environmental conditions for growth	Each bacterial species has an optimum temperature, oxygen concentration, pH and water activity for growth and will survive in a range of conditions around these optima.
Related topics	Heterotrophic pathways (B1) Techniques used to study Autotrophic metabolism (B3) bacteria (D9)

Growth media

In the laboratory bacteria are normally grown (**cultured**) in either **liquid medium**, in flasks, bottles or large culture vessels called fermenters, or **solid medium**, in **Petri dishes,** which are round, normally plastic, sterile dishes. Introduction of microbes into or on to these media is called **inoculation**. The nature of the medium depends on the microorganism's natural environment and on the reason why it is being grown. Normally, to isolate large numbers of microbes or their products, liquid culture is used but solid medium is used for the isolation of individual bacteria (Topic D9) and for storage. As bacteria are a very disparate group of microbes that can exist in a wide range of environments, and can be split into a number of different nutritional groups with very different prerequisites for growth (see Topics B1 and B3), the ingredients in the medium will depend on the individual species. The common requirements of all bacteria are water, a source of energy, carbon, nitrogen, essential inorganic ions such as phosphate, sulfate, sodium, calcium, magnesium, potassium and iron and a number of trace elements (Zn, Mn, Mo, Se, Co, Cu, Ni, W). For heterotrophs, energy and carbon can be derived from the same molecule. Bacteria that can synthesize all they require from these basic ingredients are called **prototrophs** and most microorganisms that survive in the outside environment can do this. Microbes that have become adapted to life in a situation

Table 1. Examples of growth media for bacteria

Defined media		Complex medium Tryptone soya broth (TSB) for many different types of bacteria
For *Escherichia coli*[a]	For *Cyanobacteria*	
Glucose	NH_4Cl	Enzymic digest of casein (tryptone)
Na_2HPO_4	$NaNO_3$	Enzymic digest of soybean meal (peptone)
KH_2PO_4	Na_2HPO_4	Glucose
NH_4Cl	NaH_2PO_4	NaCl
NaCl	$MgSO_4$	K_2HPO_4
$MgSO_4$	$FeSO_4$	
$FeSO_4$	Trace elements	
$CaCl_2$ (optional)	$ZnSO_4$	
	H_3BO_3	
	$MnSO_4$	
	MoO_3	
	$CoSO_4$	
	$CuSO_4$	

[a]Normally enough trace elements are present in the water.

rich with nutrients such as the human body may require other growth factors to be provided, such as vitamins, amino acids or nitrogenous bases. These organisms are called **auxotrophs**. *Leuconostoc mesenteroides,* for example, requires over 40 additional growth factors. Bacterial growth media are therefore designed to suit the organism of interest and may be either **defined** or **complex media**.

● A **defined** or synthetic **medium** contains known ingredients which are usually the basic requirements for growth of a particular microbe. These are normally simple (minimal) media that contain the minimal requirements for growth. Examples of defined media for the photosynthetic *Cyanobacteria* (note the lack of a carbon source) and heterotrophic *E. coli* are shown in *Table 1*.

● A **complex medium** contains undefined ingredients such as proteolytic digests of meat (peptones) and meat and yeast extracts which provide enough ingredients (amino acids, vitamins, sugars and bases) to sustain the growth of a wide range of microbes. Complex media such as nutrient broth (NB) and tryptic soya broth (TSB; *Table 1*) are routinely used for cultivation of bacteria in the laboratory and are particularly useful for growing bacteria whose growth requirements have not been defined. Blood is frequently used as an additive to media, for the isolation of human pathogens, as it provides many of the essential nutrients for the growth of fastidious human pathogens such as *Streptococcus*.

● Specialist **selective** and **differential** media are frequently used to isolate particular groups of microbes especially in hospital laboratories. These are media designed to select for a particular group of microbes or to differentiate between two species. They therefore contain ingredients which suppress the growth of unwanted bacteria, encourage the growth of those of interest and, in the case of differential media, some means of distinguishing between two different types of microbe. A medium containing only acetate as a carbon source would be selective for organisms that grow on acetate. Blood in agar will allow the detection of those bacteria capable of lysing red blood cells such as β-hemolytic streptococci, seen as zones of clearing (hemolysis). MacConkey agar is one of many selective media for *E. coli* from environmental or medical samples. It contains bile salts and dyes that suppress the growth of Gram-positive bacteria but allows Gram-negative organisms to grow. It is also a differential medium in that it allows the distinction between lactose-fermenting organisms (e.g. *E. coli*) and other non-fermenters (e.g. *Shigella* sp). Lactose in the medium is fermented by *E. coli* producing acid which causes an indicator dye to change color to red. Colonies that ferment lactose are therefore red whereas non-fermenters are white.

Liquid medium (or **broth**) can be converted into solid medium by the addition of agar (1–2%) which is isolated from seaweed. It is a particularly useful polysaccharide polymer because, once it is melted by boiling, it does not harden until it is cooled to 40–42°C which allows the addition of temperature-sensitive ingredients, such as proteins, before it sets. Having set, it will not melt again until it has been heated to 80–90°C. Agar is also not normally degraded by bacteria. Petri dishes containing a nutrient agar are often referred to as agar plates.

Environmental conditions for growth

Temperature

Most bacteria have characteristic temperature ranges of growth with a maximum, minimum and optimum growth temperature. **Psychrophiles** such as *Bacillus psychrophilus* are bacteria that have become adapted to living at temperatures as low as –10°C and have an optimum growth temperature around 20°C. **Mesophiles** are bacteria that grow at temperatures between 15°C and 45°C with an optimum around 37°C. Eubacterial **thermophiles** typified by *B. stereothermophilus* grow between 30°C and 75°C with an optimum of 55°C, but there are also groups of archaebacteria that have been isolated from hot-air vents in the seabed, which are capable of growth at temperatures greater than 100°C, termed **hyperthermophiles**.

Oxygen concentration

Bacteria vary in their requirements for oxygen, depending on the nature of their metabolism (Topic B2). **Aerobes** are bacteria that are capable of growing in the presence of oxygen; **anaerobes** are bacteria that do not require oxygen for growth. Within this range are **obligate anaerobes** that die in the presence of O_2; **facultative anaerobes** such as *E. coli* which grow much better in the presence of O_2 but can grow anaerobically; **aerotolerant** anaerobes that ignore the presence of O_2 and grow equally well in its presence or absence; and **microaerophilic** bacteria that are damaged by normal atmospheric concentrations of 20% and only survive at much lower concentrations of O_2.

pH

Bacteria that can grow at high pH (8.5–11.5) are called **alkaliphiles. Acidophiles** are those that grow at low pH (0–5.5).

Water activity

Like all organisms microbes are sensitive to the osmolarity of the surrounding medium: at low osmolarity water will be accumulated in the cell and at high osmolarity water will be lost. Lysis of the cell at low osmolarity is prevented by the presence of the cell wall. Growth at high osmolarity is dependent on the ability of the microbe to maintain a high osmolarity within the cell without damaging cellular metabolism. Compounds such as betaine, choline and potassium ions (called compatible solutes) are used by bacteria to maintain this osmotic balance. Bacteria, such as *Staphylococcus aureus*, that are capable of growing in 3 M NaCl, are called **osmotolerant** whereas bacteria that have become adapted to growth at very high concentrations of salt (2.8–6.2 M NaCl), such as *Halobacterium*, are called **halophiles**.

In the laboratory, water activity and pH are controlled by the composition of the medium. Incubators are used to provide the correct temperature. Aeration to provide adequate O_2 levels is normally achieved by shaking liquid cultures or by providing large surface areas of solid medium. Anaerobic conditions can be obtained by simply filling a bottle to the top with medium so there is not much room for air, but for more stringent anaerobes a reducing agent such as thioglycolate can be added to the medium. Specialist anaerobic jars or growth cabinets where O_2 can be eliminated or replaced are also used.

D8 BACTERIAL GROWTH AND CELL CYCLE

Key Notes

How bacteria grow

Bacterial growth is exponential as cells divide by binary fission (1→2→4). The time it takes for the cells in the population to double, the doubling time (t_d), depends on the growth rate of the organism, media composition and environmental conditions.

Bacterial cell cycle

The cell cycle is the sequence of events between the formation of one cell and the next. The process can be divided into distinct phases: the C (chromosome replication), G (variable) and D (division) phases. The start of DNA replication is controlled by the mass of the cell. DNA segregation and division are controlled by the length of the cell.

Rapid growth

The cell cycle in *E. coli* must take a minimum of 60 min; however, in favorable conditions, *E. coli* can grow with a doubling time of approximately 20 min. A new round of DNA replication must therefore be initiated every 20 min although previous cell cycles have not been completed. Several cycles may therefore happen in the cell at the same time.

The bacterial growth curve

Bacterial growth in liquid medium follows a typical pattern often called the bacterial growth curve. Immediately after inoculation there is a lag phase where the bacteria adapt to the medium. Exponential (or log) growth follows which is seen as a straight line on a plot of $\log_{10}$ viable count of cells against time. Once nutrients become exhausted or toxic metabolites accumulate, the net increase in cell numbers stops. This is called the stationary phase, which may be followed after a period of time by the death phase in which the number of bacteria decreases due to cell lysis.

Continuous culture

Continuous bacterial growth can be maintained in systems, such as the chemostat, where fresh medium is constantly fed into a culture vessel. Removal of excess liquid by an overflow mechanism ensures that the culture volume is kept constant.

Related topic

DNA replication (C2)

How bacteria grow

When a bacterial cell is inoculated into (placed on or in) medium, containing all the essential ingredients for growth, the cell will accumulate nutrients, synthesize new cell constituents, grow in size, replicate its genetic material and, eventually, lay down new cell wall and divide in two. So one cell becomes two which, after another period of time, becomes four. This type of doubling growth is called **exponential growth** and the time between cell divisions is called the doubling or the generation time.

$$\text{Doubling time } (t_d) = \text{the time taken for the number of cells}$$
$$\text{in a population to double}$$

The speed at which a bacterium grows is called its specific growth rate and this is measured using the following equation:

$$\text{Specific growth rate } (\mu) = 0.69 / t_d$$

From this formula it can be seen that as the specific growth rate increases, the doubling time will decrease.

The rate at which bacteria grow and divide depends on the nature of the microbe, the ingredients of the medium in which it is grown and the environmental conditions. For example, *E. coli*, when grown in a rich medium with plenty of aeration at 37°C is capable of dividing every 20 min but the rate of cell division decreases if the bacterium is required to synthesize essential macromolecular precursors such as amino acids and bases (see Topic B4). In contrast, *Mycobacterium tuberculosis* has a maximum doubling time of about 18 h and will take much longer than *E. coli*, for example, to form colonies on an agar plate.

Bacterial cell cycle

The sequence of events extending from the formation of a new cell to the next division is called the **cell cycle**. In this cycle, a growing *E. coli* will double in length then divide into two cells of equal size, with each new cell containing at least one copy of the bacterial DNA. Consequently, during this time, a copy of the chromosome must be synthesized and the two chromosomes segregated into the two daughter cells. DNA replication occurs during the **C (chromosome replication) phase** and chromosome segregation occurs in the **G (gap) phase**, which may be of variable length (*Fig. 1*). Segregation of the chromosomes is achieved by attachment of the replicated DNA onto two adjacent sites on the membrane. Membrane growth between the sites pushes the chromosomes towards the poles of the cells. Finally, a cross wall (septum) is laid down between the two chromosomes and the cell divides into two (**D phase**). Cell division and DNA replication have to be coordinated. Initiation of DNA replication (see Topic C2) at the origin (*oriC*), a short adenine and thymine rich sequence, is dependent on the cell reaching a critical mass (initiation mass) and requires a number of protein initiation factors. DNA segregation and

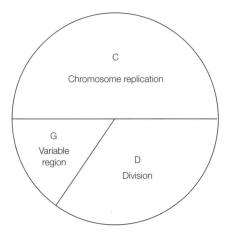

Fig. 1. Stages in the cell cycle of prokayotes.

division, however, are controlled by the length of the cell which must reach a particular threshold length before the chromosomes are partitioned and cell division initiated.

Rapid growth When conditions for growth are favorable, *E. coli* can grow with a doubling time (see Topic C3) of approximately 20 min. However, the time it takes to synthesize a complete copy of the *E. coli* chromosome is 40 min, under optimum conditions and segregation of the DNA and division takes another 20 min. Thus, the shortest cell cycle and, therefore, doubling time for *E. coli* should be 60 min. This is obviously not the case. For cells to divide faster than every 60 min, DNA replication must begin in one cycle and finish in another. When cells are growing fast (doubling time <60 min), initiation of replication occurs, as normal, producing two replication forks which move bidirectionally round the chromosome to the termination point (Topic C2). However, the origins on these new strands then initiate further rounds of replication before the previous round of DNA replication has finished (*Fig. 2*). Thus, when cell division occurs the DNA in the daughter cells is already replicating. The faster the cell growth rate, the more replication forks are formed such that the DNA in new cells may have multiple replication forks. That repeated DNA replication can occur without cell division indicates that the control of the two processes is not linked. This is in contrast to eukaryotic cells where the two processes are directly connected (Topic G3). The faster the rate of growth the more replication forks are formed such that the DNA in new cells may have multiple replication forks. That repeated DNA replication can occur without cell division indicates that the control of the two processes is not linked. This is in contrast to eukaryotic cells where the two are directly connected (Topic G3).

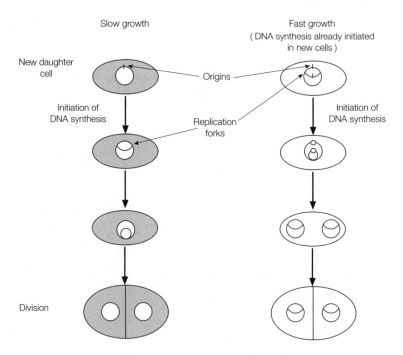

Fig. 2. Coordination of DNA synthesis and cell division in slow and fast growing bacteria.

The bacterial growth curve

The best way of producing large numbers of microbes or their natural products is to grow them in a liquid medium. Normally, the technique we use is called **batch culture** in which the bacteria are inoculated into flasks of a suitable medium and grown at an appropriate temperature and degree of aeration. Bacteria grown in this way show a particular pattern of growth which is referred to as the bacterial growth curve (*Fig. 3*). The number of viable bacterial cells is measured over time and is plotted as a graph of the $\log_{10}$ viable cell numbers against time, which is called a **semi-logarithmic** plot. A logarithmic scale is used to plot bacterial growth owing to the large numbers of cells produced and to reveal the exponential nature of bacterial growth. If an arithmetic scale is used to plot the increase in the number of cells, a curve of increasing gradient would be seen which is converted into a straight line when a logarithmic scale is used. The doubling time for the bacteria can be read directly from the graph. The bacterial growth curve reveals four phases of growth.

- **Lag phase.** When bacteria are first inoculated into a medium there is a period in which no growth occurs. During this phase the bacteria are adapting to the new environment, synthesizing new enzymes as required and increasing cell size ready for cell division. The length of this time depends on the nature of the inoculum. If this comes from a fresh culture in the same medium, the lag phase will be short, but if the inoculum is old or the medium has been changed (especially moving bacteria from a rich medium to a poor one) the lag phase will be longer.
- **Exponential (logarithmic) phase.** Once the bacteria start to divide, the numbers increase at a constant rate which reflects the doubling time of the bacteria. This is seen as a straight line part of the graph.
- **Stationary phase.** As the bacteria increase in numbers they use up all the available nutrients and produce toxic waste products. Eventually a point is reached where there is no net increase in cell numbers, seen as a flattening-off of the growth curve. During this state of equilibrium cells are still functioning. There is some cell death which is balanced by some small amounts of controlled cell division.
- **Death phase.** After a while the rate of cell death becomes greater than cell division and the number of viable cells drops. Cells lyse and the culture becomes less turbid.

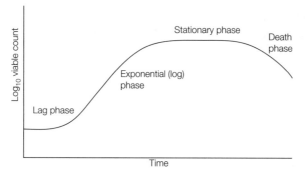

Fig. 3. A typical bacterial growth curve.

Continuous culture

The problem with batch culture is that eventually cell growth stops due to the exhaustion of nutrients or the accumulation of toxic products. If these problems can be prevented by replacing the spent medium with fresh, then bacterial growth can be maintained indefinitely. Continuous culture systems such as the **chemostat** do just that. They are flow systems in which fresh medium is constantly fed into a culture vessel which is kept at a constant volume by an overflow system that removes the excess of liquid (*Fig. 4*) so the bacteria are kept in the exponential phase of growth. Consequently, once the system has reached equilibrium, cell numbers are kept constant and a **steady-state** growth rate is reached. This growth rate can be manipulated by altering conditions such as medium composition and flow rate. A major advantage of these systems is in the provision of constant conditions for physiological studies. This is impossible in batch culture as the growth of the bacteria changes the environmental conditions such as pH and nutrient concentration of the medium.

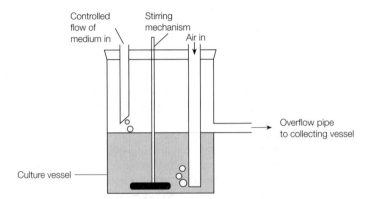

Fig. 4.   A simplified diagram of a continuous culture system.

D9 TECHNIQUES USED TO STUDY BACTERIA

Key Notes

Aseptic (sterile) technique

The methods used to protect the worker from the microbes with which they work, and microbial cultures from contamination with other microbes from the environment, are called aseptic or sterile techniques. Sterilization is the process that eradicates all microbes whereas disinfection implies only the removal of potential disease-causing organisms. Heat (wet and dry), chemicals, radiation and filtration are all used to sterilize materials used in microbiology.

Obtaining bacterial colonies

Pure cultures of bacteria are normally required in the laboratory where all the cells have arisen from the same original bacterium. Streaking, spread plates and pour plates are all methods used to inoculate bacteria on to agar plates so that the individual bacterial cells are separated from each other. After incubation at an appropriate temperature, each bacterium forms a colony, visible to the naked eye, which contains up to 10^9 bacterial cells.

Counting bacteria

The number of bacteria in a culture can be estimated by: counting the individual cells using some form of counting chamber (total counts); diluting and inoculating the culture on to solid medium so that each original cell develops into a colony which can be counted (viable count); measuring the light scattered by a culture using a nephelometer or spectrophotometer; weighing wet or dried cells; using biochemical methods to measure cell components; and using specialist electronic methods.

Related topics   Growth in the laboratory (D7)   Bacterial growth and cell cycle (D8)

Aseptic (sterile) technique

When handling any microorganisms in the laboratory it is very important to behave as if that microbe is potentially harmful. It is therefore essential that the methods used ensure that there is no or minimal contact between the worker, and others in the laboratory, with the microbe, either directly or as aerosols. Equally, it is important that the microbes being worked with are kept pure and are not contaminated with those from the environment. The skills and methods used to prevent these two occurrences are called **aseptic** or **sterile techniques**. This essentially means that all tools and media used are free from the presence of any living material (**sterile**) and experiments are carried out in such a way as to prevent the accidental introduction of unknown microbes (**contamination**). **Sterilization** is a process which ensures the complete eradication of all living cells and viruses whereas **disinfection** is a vaguer term normally used to indicate the removal of potentially harmful microorganisms but not necessarily all other microbes. Physical (heat and radiation) and chemical methods are used to

Table 1.   Methods of sterilization

Method	Effectiveness	Use
Moist heat		
Boiling/steam treatment	Kills most vegetative bacteria, fungi and viruses – not effective against endospores	Dishes, basin, heat resistant equipment
Autoclaving – 15 pounds per square inch (121°C)	Kills all vegetative cells and endospores. The length of time depends on the number of microbes present	Media, solutions, utensils, any item that can withstand heat and pressure
Pasteurization		
71.7°C for 15 sec	Heat treatment for liquids that kill pathogens but not necessarily all bacteria. Minimizes changes to flavor	Milk, juices, etc.
Tyndalization		
Three treatments of 90–100°C for 10 min with 24 h gaps in between	Vegetative cells are killed on day 1. Any spores germinate during the gap periods and are killed by the next heat treatment	Media that cannot be autoclaved but are relatively heat resistant
Dry heat		
Flaming	Red-hot direct heat	Inoculating loops
Hot air	200°C for 2 h will destroy cells and endospores	Glassware
Filtration		
Filters – pores size of 0.22 – 0.45 μm	Does not remove viruses	Liquids that cannot be heat treated
Radiation		
UV light – non-ionizing	Does not penetrate materials well	Work surfaces and air
γ-rays – ionizing	Penetrates well	Pharmaceuticals Plastic ware

sterilize or disinfect materials and the method used is dependent on the nature of the materials to be sterilized (see *Table 1*). Heat, either moist or dry, is one of the most commonly used techniques but this cannot be used for any heat labile materials such as proteins and some plastics. Due to the heat stability of bacterial spores (Topic D2), which cannot be removed by boiling, sterilization using moist heat must be done at higher temperature and pressure in an **autoclave**. These machines normally operate at 121°C and 15 pounds per square inch (psi) which is sufficient to kill most microbes with the exception of spongiform encephalopathic agents, which are resistant to these conditions. The length of time used for sterilization depends on the volume of the materials. Dry heat is routinely used in the laboratory to sterilize wire loops and remove microbes from the rims of culture vessels by **flaming** (heating) in a Bunsen flame. Heat-labile materials can be sterilized by filtration (liquids) or by radiation (solids). Chemicals such as sodium hypochloride (chlorine bleach), 70% alcohols and other commercial disinfectants are normally used to decontaminate surfaces.

Obtaining bacterial colonies
Bacteria normally grow as complex mixtures of microorganisms in their natural environments. However, in the laboratory we normally wish to work with **pure cultures** of one particular strain of bacteria. The routine way to isolate a single strain is to inoculate the culture on to to plates using a method called **streaking**. A wire (or sterile plastic) loop is used to streak bacteria on to plates of solid nutrient medium, as shown in *Fig.1*, in order to dilute the bacteria and disperse them over the surface of the medium. The plates are then incubated at an appropriate temperature and each bacterium replicates to form a **colony** that is visible to the naked eye. A colony contains up to 10^9 copies of the original bacterium. The appearance of the colony is often distinctive and can be used to distinguish one species from another by its morphology (see Topic D1). An alternative method of isolating single colonies is to use a **spread plate** method. In this < 200 bacteria in a liquid medium are spread over the surface of a plate using a glass spreader sterilized by flaming in alcohol. After incubation, individual colonies may be seen. This method is frequently used for counting the number of viable bacteria in a culture as described below. An alternative method is to use spread plates or pour plates where the microbes are mixed with molten agar which has been cooled to 45°C and then poured into Petri dishes.

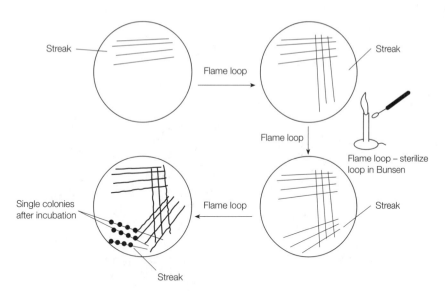

Fig. 1. Diagram showing how to streak a bacterial culture for single colonies.

Counting bacteria *Total counts*
The total number of cells in a culture can be determined directly by counting the number of cells under the microscope using specialist counting chambers. These chambers consist of a grid of known area etched on to a depression in a glass slide. A rigid coverslip is placed over the depression creating a precise volume between 0.02 and 0.1 μl. The number of microbes in that volume can be counted under the microscope. This method is difficult to use for small cells and does not distinguish between live and dead cells.

Viable counts

In this method the culture is inoculated on to solid medium using the spread-plate technique or pour-plate method. The number of live cells in the original culture can be estimated from the number of colonies growing on the plate. As the number of microbes in the original culture may be as high as 10^{10} and the ideal number of countable colonies on a plate is between 30 and 200, it is normal to dilute the culture in a series of 10-fold dilutions (1 ml of culture plus 9 ml of diluent) in order to achieve a countable number of cells in 0.1 ml. As the precise number of colonies in the original culture is normally unknown it is standard practice to spread 0.1 ml of a number of different dilutions on to plates so that one of the dilutions will have the correct number of colonies on the plate (*Fig. 2*).

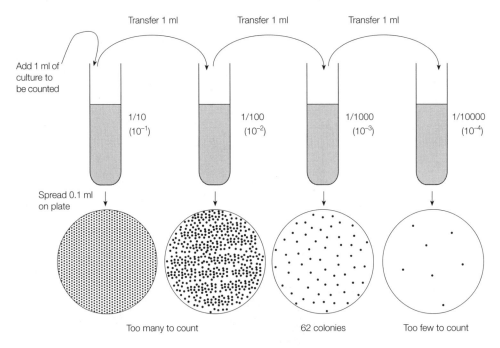

Number of colonies × dilution factor = cells (colony-forming units, CFUs)/ml of original cultures
62 × 10^3 × 10 (0.1 ml plated out) = 6.2 × 10^5 CFUs/ml

Fig. 2.   Diagram showing serial dilutions for the estimation of a viable count of a microbial culture.

Turbimetric measurements

A very useful method for estimating cell numbers in liquid culture is to measure the turbidity of a suspension of the microbe using a nephelometer (measures scattered light directly) or a spectrophotometer, which measures light lost from the culture due to scattering; this is called absorbance or optical density (OD). The amount of light scattered by the suspension is proportional to the number of cells present, but is also influenced by the size, shape and nature of the cells. A standard curve relating turbidity to cell numbers is therefore required for each individual species under specific growth conditions if an accurate estimation of cell numbers is required. However, for monitoring microbial growth it is often sufficient to measure the changes in turbidity or absorbance of the culture with time of incubation.

Other methods for monitoring growth

Wet and dry weight of cells; biochemical analysis of cellular components such as protein, nucleic acids or ATP; change of conductivity (used in the Coulter counter) and changes in electrical impedance are just some of the alternative methods for measuring microorganisms.

E1 MUTATIONS

Key Notes

Genotype/ phenotype	The genotype of an organism is the information stored in its genetic material: the sequence of the bases in its DNA. The phenotype of an organism is the expression of that genotype as it can be observed or measured; for example, whether or not it produces a capsule or is resistant to an antibiotic. The most common form of an organism is often referred to as the wild-type.
Mutation	Any change to the sequence of bases in the DNA of an organism is a mutation. The change may be a single-base change called a point mutation or involve a much larger sequence of DNA in which case it is called a multi-site mutation. DNA may be deleted, inserted, inverted, duplicated or substituted.
Point mutations	Replacement of a purine with another purine (or a pyrimidine with a pyrimidine) is called a transition. The change of a purine to a pyrimidine or vice versa is a transversion. The consequence of a point mutation on the phenotype of the cell depends on how it alters the protein sequence. The possibilities are same-sense, mis-sense, nonsense and frameshift mutations.
Isolation of mutants	Mutants resistant to toxic compounds can be selected for by growth in the presence of that agent. Mutations leading to altered growth requirements may be screened for by replica plating.
Replica plating	Colonies of bacteria are transferred, using a sterile velvet pad, to two sets of media; one on which the mutant will not grow and the other on which it will. The mutant colony is identified by its lack of growth on the former medium but purified from the latter.
Conditional mutants	Mutations which are expressed only under certain conditions are called conditional mutants. The most common type are temperature-sensitive mutants which only express their phenotype at high temperature.
Reversion of mutations	The effect of a mutation may be reversed by either a true back mutation to give the original phenotype or by a suppressor mutation at a different site, which overcomes the effect of the first mutation.

Related topics	Structure and organization of DNA (C1)	Translation (C7)
	DNA replication (C2)	Recombination and transposition (E3)
	Structure of proteins (C6)	DNA repair mechanisms (E4)

Genotype/ phenotype

The genetic information of an organism is the exact sequence of nucleotides in its DNA, often referred to as its **genotype**. This sequence dictates what proteins or RNA molecules can be produced by the organism, how the genes may be regulated and the functioning of the protein once it is synthesized. Not all proteins are made at the same time and the full genotype of an organism cannot be discovered unless the genome is sequenced. What is observable and measurable is the **phenotype** of the organism; the manifestation of the genotype in the form of characteristics such as the ability to produce a capsule or resistance to an antibiotic. Alterations to the genotype can therefore be seen as changes in the phenotype. Different nomenclatures/abbreviations are used for describing genes and mutations in different organisms, but in bacteria the following convention is used:

- Phenotype is described using a three-letter abbreviation of roman letters with the first letter capitalized. A superscript + or – is used to designate the presence or absence of this phenotype; for example, Phe⁻ indicates the inability to synthesize phenylalanine or Lac⁺ indicates the ability to utilize lactose as a carbon source. The superscripts 's' and 'r' may also be used to designate sensitive or resistant.
- Genotype is indicated by three lower case letters (usually reflecting the phenotype) with a fourth upper case letter to indicate specific genes involved, all of which are italicized, for example, *pheA* or *lacZ*.

Although, naturally, there may be a large number of individuals of a species with slightly different phenotypes, microbiologists and in particular microbial geneticists try to work with organisms that are genetically identical. Consequently, in the laboratory, we work with **pure cultures** where microbes

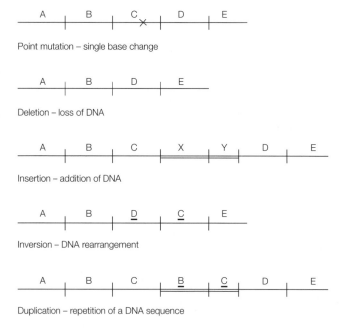

Fig. 1. Possible changes to DNA sequence that may result in mutation.

are descendent from the same original parent to minimize the genetic variability. That parent may often be referred to as the **wild-type** (Topic D9).

Mutation

A **mutation** is any change in the nucleotide sequence of the DNA in a cell. This change may be seen as an alteration in the phenotype of the cell but it may be silent if the change occurs in a non-essential part of a protein sequence or in a non-coding part of the genome. Mutations in essential genes such as those necessary for DNA replication may be lethal and therefore never isolated. The changes in the DNA sequence range from single base changes, normally called **point mutations**, to large rearrangements of the genome, sometimes called multi-site mutations, which involve short or long stretches of DNA and may affect a number of genes. These multi-site mutations may be deletions, insertions, inversions, substitutions and duplications of sequence as shown in *Fig. 1* and occur as a result of recombination or transposition in the genome as described in Topic E3.

Point mutations

Point mutations may be the change of a purine for another purine (A ↔ G) or a pyrimidine for another pyrimidine (C ↔ T), called **transitions,** whereas the changes of a purine to a pyrimidine and vice versa are called **transversions**. The effect of a base change in the genotype on the phenotype depends on the nature of the mutation and where it occurs on the genome. There are four types of point mutation illustrated in *Fig. 2* and described below.

1. **Same-sense mutations** are due to the redundancy in the genetic code; the same amino acid is inserted into the protein so no change in the phenotype is seen (see the genetic code in Topic C7).

Original sequence
5′-AUG CCU UCA AGA UGU GGG CAA-3′
Met Pro Ser Arg Cys Gly Gln

Same sense mutation – same amino acid inserted
5′-AUG CCU UCA AGA UGU GGA CAA-3′
Met Pro Ser Arg Cys Gly Gln

Mis-sense mutation – different amino acid inserted
5′-AUG CCU UCA GGA UGU GGG CAA-3′
Met Pro Ser Gly Cys Gly Gln

Nonsense, the creation of a termination codon – protein synthesis stops
5′-AUG CCU UCA AGA UGA GGG CAA-3′
Met Pro Ser Arg STOP

Frame shift mutation – from the point that a base is inserted or deleted the amino acids are altered

Deletion or insertion of one or two bases

5′-AUG CCU UCA AG U GUG GGC AA-3′
Met Pro Ser Ser Val Gly etc.

Fig. 2. The nature of point mutations as seen in the mRNA sequence and their consequence on protein sequence.

2. **Mis-sense mutations** occur when a different amino acid is inserted into the chain. The consequence of this change is dependent on whether the alteration is in an essential or non-essential part of the protein. In the first case, protein function may be altered or lost but in the latter no change to the phenotype may be observed.

3. **Nonsense mutations** result when the change in the base sequence alters an amino acid-encoding codon to a STOP codon (UAA, UAG, UGA). Protein synthesis terminates prematurely, leading to the production of a truncated protein.

4. **Frameshift mutations** occur when one or two nucleotides are deleted or inserted into the DNA sequence, therefore altering the translational reading frame and producing a completely altered amino acid sequence from the point of change.

Isolation of mutants

Although mutations occur spontaneously in bacteria this is at a very low frequency so it is usually desirable to use a mutagen such as UV light to increase the chance of isolating the mutation of interest (Topic E2). Identification of the presence of that mutation depends on the nature of the gene of interest, but generally these fall into three categories:

1. Mutants that are resistant to some toxic compound such as antibiotics or to the infection by a bacteriophage (Topic E7). These mutations can be **selected** by growing in the presence of the toxic agent. Only the resistant mutant strains will grow.

2. Auxotrophic mutants that are unable to synthesize a compound essential for growth such as an amino acid or vitamin (Topic D7). These mutants cannot be isolated directly but can be **screened** for by replica plating as described below.

3. Mutants that are unable to use particular substrates such as lactose or maltose for growth. These may also be identified using replica plating.

Replica plating

This is a method used to screen a large number of colonies for the presence of a particular mutation. Bacteria are spread on to plates, containing medium on which both the parent and the mutants will grow, at a dilution such that individual colonies can be seen on the plate (Topic D9). After incubation, the colonies are transferred to plates containing media that will allow the detection of the presence of the mutation, using a sterile velvet pad (*Fig. 3*). The nature of that media depends on the mutation required. In the case of auxotrophic mutations the colonies may be replica plated on to plates with and without a particular growth factor. Mutant strains that are unable to synthesize that factor will not grow on the minimal medium but will grow on the minimal medium that does contain the growth factor. Having identified the mutant colony by its lack of growth, it may then be picked off and purified from the plate on which it has grown.

Conditional mutants

There are many genes in the cell whose gene product is essential for cell growth such as proteins necessary for DNA replication. It is therefore impossible to isolate mutations in these genes, which totally inactivate the gene function, as the cell would die. **Conditional** mutants can be used in these circumstances. These are mutations which are expressed only under certain circumstances, the most common being high growth temperature. These **temperature-sensitive**

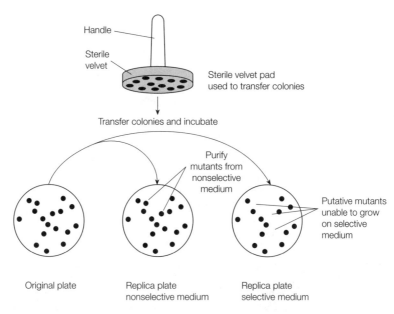

Fig. 3.   Replica plating for the isolation of mutant colonies.

mutants have a wild-type phenotype at low temperatures but when switched to a higher temperature the mutation is manifested which allows the effect of the mutation to be studied.

Reversion of mutations

The effect of a mutation can be reversed by a number of means. The simplest is by reversion where a **back** mutation occurs to give back the original base sequence (the wild-type). Selection for reversions is a good test to indicate if the original mutation was a point mutation as deletions do not revert easily.

Another way of achieving a wild-type phenotype is by a mechanism called suppression in which a second mutation occurs at a different point in the genome, which compensates for the effect of the first mutation. The second mutation may be at a different site in the same gene in which case it is called an **intragenic suppressor** or in a completely different gene when it is termed an **intergenic suppressor.**

E2 MUTAGENESIS

Key Notes

Spontaneous mutations	Mutations constantly occur in the genome nearly at random, though 'hot spots' exist where mutations can occur at much higher frequencies. Such spontaneous mutations are a result of tautomerization of bases, oxidative deamination of cytosine or strand slippage during DNA synthesis at sites where there are short repeated nucleotide sequences. The movement of, and recombination between, transposable elements may also be responsible for mutations in the genome.
Mutagens	Mutagens are agents that increase the frequency of mutation by interacting with the DNA molecule. Chemical agents include base analogs which are incorporated into the DNA molecule during replication, intercalating agents which insert between the base pairs of the helix, or DNA-modifying agents which alter the structure of DNA. The most important physical DNA-damaging agent is short-wave radiation especially UV light. Transposable elements are mutagenic agents and are responsible for a range of genomic rearrangements in the cell. Finally, in the laboratory, specific mutations can be introduced into DNA by *in vitro* mutagenesis when the DNA sequence is known.
Related topics	Structure and organization of DNA (C1) Mutations (E1)
	DNA replication (C2) DNA repair mechanisms (E4)

Spontaneous mutations

Mutations occur throughout the genome all the time, but some areas of the genome are more prone to change than others. These are called **hot spots.** The rates for mutations in a gene can vary from one mutation per gene every 10^4 rounds of replication to one every 10^{11} with an average of about one mutation per gene every 10^6 rounds of replication. Mutations can arise for a number of reasons, including errors made by DNA polymerase during DNA replication (Topic C2), physical damage to the DNA, recombination and transposition (Topic E3). However, it is important to be aware that DNA is protected by a large number of DNA-repair systems geared to minimize the mutation rate. One main cause of spontaneous mutations is due to the fact that bases can exist as different forms called tautomers, which are capable of forming different base pairs. So during replication, adenine, which normally base-pairs with thymine in its normal amino form, switches to its rare imino form (**tautomerism**) and base-pairs with cytosine meaning that a cytosine is inserted into the DNA instead of a thymine. If this is not repaired before the next round of replication the A–T (adenine–thymine) base pair will be changed to a G–C base pair as shown in *Fig. 1*.

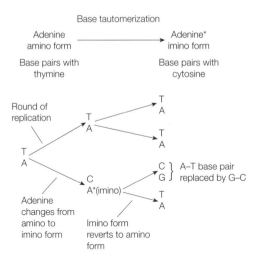

Fig. 1. Spontaneous mutation of an A–T base-pair to G–C as a result of adenine base tautomerization.

Spontaneous mutations can also arise as a result of slippage of DNA strands at short repeated nucleotide sequences during DNA replication, leading to the insertion or deletion of a short length of DNA, and as a result of actual damage to the DNA molecule itself. Very rarely, a base may be removed from a nucleotide leaving a gap called an **apurinic** or **apyrimidinic** site which will not be able to base-pair normally at the next round of replication. A common reason for this is the natural deamination of cytosine to form uracil. This is recognized by DNA repair systems as an error and the uracil is removed to leave an apyrimidinic site. Finally, a major cause of spontaneous mutation is transposable elements which can insert at random in the genome, resulting in mutation and, if two or more copies are present, act as sites for homologous recombination which can lead to deletions, duplication and inversions of sections of the genome (Topic E3).

Mutagens

Chemical and physical agents that bind to DNA can increase the mutagenesis rate. Chemical mutagens can act in a number of ways:

1. **Base analogs** such as 5-bromouracil and 2-aminopurine are incorporated into the DNA during normal DNA replication, but undergo tautomeric shifts similar to the naturally occurring bases, but at much higher frequencies, leading to mutation.
2. **Intercalating agents** are flat, three-ringed compounds similar in shape to a base-pair. They can slot into the DNA, leading to a distortion of the helix which leads to slippage during DNA replication and the insertion and deletion of bases. Ethidium bromide and acridine orange are representatives of these agents
3. **DNA-modifying chemicals** are compounds that alter a base in the existing DNA, leading to mis-pairing at the next round of replication. Examples include hypoxanthine which converts cytosine into hydroxylaminocytosine which pairs with adenine, leading to the transition of G–C to A–T.

4. Ionizing radiation can be a potent cause of DNA damage. **UV light** is frequently used in the laboratory to induce mutations and its mechanism of action is the best understood. The principle damage caused by UV light is the formation of pyrimidine dimers, the most common being **thymine dimers**, between adjacent bases which causes a distortion of the DNA helix. This in itself is not mutagenic but mutations occur when the cell tries to repair the damage using an error-prone repair system called the SOS repair system discussed in Topic E4.

5. *In vitro* mutagenesis is frequently used now to change sequences precisely in the DNA when the DNA sequence is known. A copy of a mutated sequence is chemically synthesized and used to replace the wild-type sequence in the genome.

E3 RECOMBINATION AND TRANSPOSITION

Key Notes

General recombination

General (or homologous) recombination is mediated by RecA protein which causes the exchange of genetic information between DNAs with very similar sequences. Recombination proceeds via the production of a cross-shaped heteroduplex molecule, sometimes called a chi form which is resolved by endonuclease action to produce either recombinant DNA molecules or molecules in which a single-stranded region has been exchanged.

Site-specific recombination

Non-RecA protein-dependent recombination between specific sequences, generally resulting in phage genome integration into bacterial chromosomes, is mediated by phage-encoded proteins and host-encoded integration host factor (IHF).

Transposition

Transposable elements move between or within DNA molecules, mediated by a transposase enzyme encoded by the element. There are three types of transposable element: the simplest are insertion sequences; more complex are transposons which carry additional genes, often antibiotic resistance genes; and, thirdly, transposable phage, such as Mu, use transposition as a mechanism for replication.

Related topics

Mutations (E1)
Mutagenesis (E2)
DNA repair mechanisms (E4)

F plasmids and conjugation (E6)
Replication of bacteriophage (E8)

General recombination

General (or homologous) recombination is the exchange of genetic information between two double-stranded DNA molecules with identical or very similar sequences. In diploid cells this type of recombination is responsible for the exchange of genetic information between homologous chromosomes during meiosis (Topic G3). In haploid organisms, such as bacteria, there are also occasions when homologous regions of DNA may be present in the cell such as during chromosome replication or when plasmids or phage are present, allowing the opportunity for homologous recombination. In bacteria, genetic exchange is mediated by a protein called RecA which has a number of pivotal roles in the cell (see also DNA repair, Topic E4). An outline of the mechanism which leads to this strand exchange is shown in *Fig. 1*. It is sometimes called the **Holliday model** for reciprocal general recombination after the person who first proposed the mechanism. The two DNA molecules line up adjacent to each other and a nick is introduced into one strand of each. Strand invasion, mediated by **RecA protein**, involves complementary base-pairing between the nicked

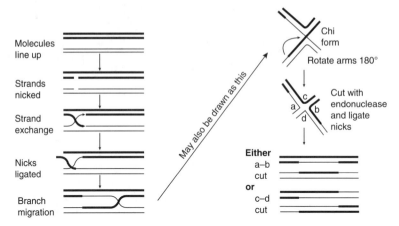

Fig. 1.   The Holliday model for general (homologous) recombination.

strand of one molecule and the intact strand of the other to form a cross-over. The single-stranded nicks are sealed by DNA ligase, joining the two molecules of DNA into a **heteroduplex** called the **Holliday structure**. The cross-over can move down the molecule, called **branch migration**, creating either short or long lengths of hybrid DNA duplexes. The resolution of the two molecules to produce recombinants is achieved by endonucleases cutting the DNA molecule. The nature of the products depends on where the cutting of the strands occur, as the hybrid molecule can also be drawn as a cross-like structure called a **chi form**. If cuts are made at points a and b, two DNA molecules will be resolved in which only a single strand of DNA has been exchanged. Any change in the DNA molecule will therefore be small as the two original molecules were very similar. If the cuts occur at points c and d, the consequences are much greater in that effectively the two DNA molecules are joined together. The effect of recombination on DNA molecules depends on whether they are linear or circular and whether recombination occurs at one or two sites (*Fig. 2*). One recombination event (cross-over) would lead to the joining of two linear molecules (*Fig. 2a*) or two circular molecules (*Fig. 2b*). Two recombination events at two different sites would lead to the integration of a new stretch of DNA into a DNA molecule (*Fig. 2c*).

Site-specific recombination

This type of recombination occurs in a number of specialist situations such as the integration of phage, such as λ, into the genome of bacteria (Topic E7). RecA protein is not required, the integration is mediated by a phage encoded enzyme and a bacterial enzyme called the integration host factor (IHF). These proteins interact with short homologous sequences, called *att* sites, found on the bacterial chromosome and on the phage molecule, forming a Holliday structure. Genetic exchange at these sites leads to integration of the phage into the genome (see *Fig. 3* in Topic E7). In a similar fashion excision of phage occurs as a reverse of that process.

Transposition

Transposable elements are mobile pieces of DNA that can move within a DNA molecule or between two DNA molecules at a low frequency (between 10^{-7} and 10^{-2}/generation). This movement does not require homology between the

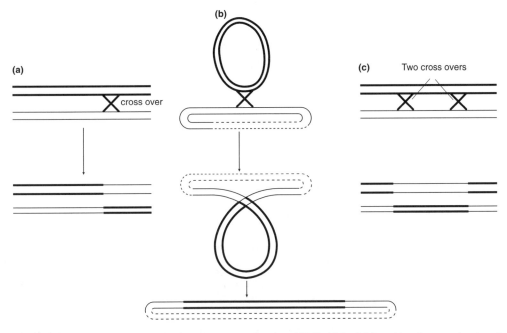

Fig. 2.   The results of recombination between two molecules of DNA: (a) the joining of two linear molecules of DNA; (b) the integration of a DNA sequence; (c) the joining of two circular molecules.

sequences or the presence of RecA protein. The ability to move is mediated by an enzyme called **transposase** which is carried on the transposable element. These elements are found in all kinds of organisms and have probably had a major role in the evolution of genomes by carrying genes between DNA molecules.

The simplest kind of transposable elements are insertion sequences (IS elements) found in bacteria. These are around 1 to 2 kb in size and consist of a stretch of DNA which encodes the transposase enzyme with a pair of inverted repeats, 9–41 bases in size, at either end (*Fig. 3*). These sequences can insert at random in the genome by a mechanism which leads to the duplication of a short sequence within the target site. There are a number of copies of the different IS elements in the genome and plasmids of bacteria, for example IS2 is present as five copies on the *E. coli* chromosome and as one copy on its F plasmid (Topic E6).

A second group of transposable agents found in bacteria are called **transposons** which carry other genes as well as the transposase enzyme. Frequently, these genes are antibiotic resistance genes (Topic F7), which may explain why these genes appear to spread rapidly through populations of bacteria leading to the appearance of strains that are resistant to a wide range of antibiotics. Transposons can be split into two groups:

● **Composite transposons** that are made up of a central region containing the additional genes between two copies of an IS element. Examples of these are Tn*10* which carries tetracycline resistance genes and Tn*1681* which codes for a heat stable enterotoxin.

● **Tn-3 type transposons** do not have IS elements but are capable of transposition by a mechanism which involves duplication of the element so that one

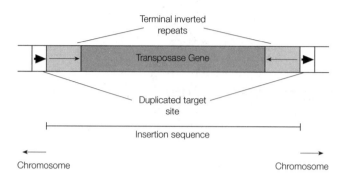

Fig. 3.   The structure of an E. coli *insertion sequence showing the duplication of a chromosomal target sequence which occurs during transposition.*

copy remains on the original site and a new copy is inserted at another site. An example of this is Tn-3 itself which carries penicillin resistance genes.

The third group of transposable elements, found in bacteria, are a group of bacteriophage typified by Mu and D108 which replicate by transposition (see Topic E8).

Finally, transposable elements are mutagenic. If they insert into the middle of a gene they will inactivate its function and therefore cause a mutation (Topics E1 and E2). They are often present on a genome as more than one copy and therefore can act as sites for homologous recombination. This can lead to various types of chromosome rearrangements such as duplications, inversions and deletions, which may cause large changes to the cells' genotype.

E4 DNA REPAIR MECHANISMS

Key Notes

Overview	As DNA damage is potentially lethal to the cell, a large number of DNA-repair systems have evolved to repair damage by mutagens. The best understood mechanisms are those that repair UV damage in *E. coli* of which there are at least four different systems.
Photoreactivation	Photolyase converts the UV-induced thymine dimers back into monomers in the presence of light.
Excision (dark) repair	The products of the *uvr A,B,C* and *D* genes form a complex that recognizes and removes damaged nucleotides from the DNA. DNA polymerase I fills in the missing nucleotides and DNA ligase joins the gap. This mechanism of repair is used for other forms of DNA damage as well as UV damage.
Recombinational (post-replication) repair	If replication occurs before the DNA has been repaired, there will be gaps in the DNA due to the inability of DNA polymerase to insert bases opposite a pyrimidine dimer. These gaps can be repaired by recombination with the sister DNA molecule that has an intact DNA sequence.
SOS-repair systems	A large number of genes are induced as a result of DNA damage. Among these emergency repair systems is an error-prone repair system which causes DNA polymerase III to insert any base opposite a pyrimidine dimer, allowing replication to continue. The SOS system is induced by RecA protein, activated by interaction with damaged DNA, which, in turn, inactivates a repressor of transcription, LexA protein.
Related topics	Structure and organization of DNA (C1) Control of gene expression (C5) DNA replication (C2) Mutagenesis (E2)

Overview

Errors during DNA replication and DNA damage due to exposure to mutagens happen all the time and microorganisms have evolved very complex systems to recognize and repair the damage. In the best situation the repair directly undoes the original damage, causing no change to the base sequence, but sometimes the amount of damage is so great that DNA synthesis cannot continue. In these situations an **SOS-repair system** is induced that allows quick and extensive repairs but at the expense of accuracy, leading to the accumulation of mutations. The repair of UV damage in *E. coli* is the best understood and will be the main subject covered in this topic. UV light induces the formation of a cyclobutane pyrimidine dimer and a consequent distortion of the DNA helix (Topic E2). These dimers are unable to form normal base-pairs so DNA synthesis stalls when DNA polymerase III reaches these adducts (Topic C2). A normal *E. coli* cell can cope with approximately 30 hits per molecule, and at least four different repair systems have evolved to cope with this damage and overcome the problem.

Photoreactivation

In a light-dependent repair mechanism, the enzyme photolyase, with the addition of flavin adenine dinucleotide (FAD), splits the cylobutane ring between the adjacent pyrimidines to restore the original monomers. This is the front-line defense against UV damage.

Excision (dark) repair

Excision repair is often called dark repair to distinguish it from photoreactivation. It is a general system used to repair a number of different types of DNA damage such as that caused by pyrimidine dimers and mis-matched base pairs. The products of the *uvr A, B* and *C* genes combine to form an endonuclease which recognizes distortions of the DNA helix caused by changes in the bases. The UvrABC endonuclease nicks the DNA on either side of the lesion (*Fig. 1*) and DNA polymerase I removes and replaces the bases. DNA ligase fills in the gap (Topic C2).

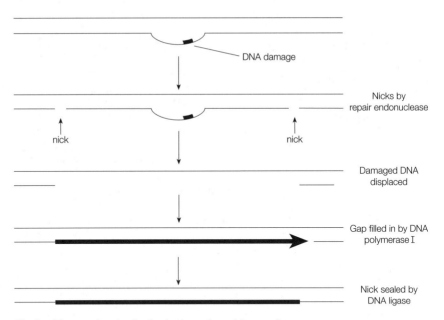

Fig. 1.   Diagram showing the basic stages in excision repair.

Recombinational (post-replication) repair

Thymine dimers cannot be replicated but, instead of stalling at these sites, DNA polymerase can jump over the lesion and resume DNA synthesis further down the DNA template. This leaves a gap in the newly synthesized DNA strand opposite the dimer. This cannot be repaired by excision repair as this requires an intact strand to be used as the template. Instead the gap can be filled by RecA-mediated recombination with the sister DNA helix, which contains the intact sequence that should be opposite the dimer (*Fig. 2a*). Although recombination does not repair the damage it does create the situation where both new DNA molecules can be repaired by excision repair.

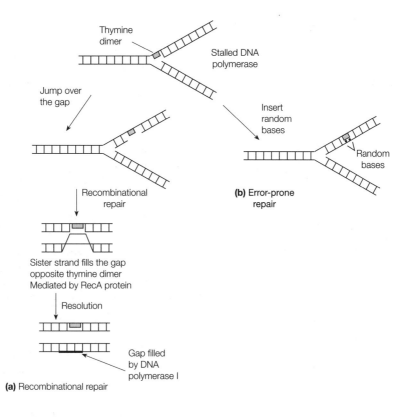

Fig. 2.   (a) Recombinational and (b) error-prone repair of UV-damaged DNA during replication.

SOS-repair systems

There are a number of genes and operons in the cell that are coordinately regulated by RecA protein and a repressor of transcription, LexA. These are involved in dealing with DNA damage and are called SOS-repair systems (see Topic C5). These include an error-prone repair system which interacts with DNA polymerase allowing it to continue replication of DNA past a pyrimidine dimer (*Fig. 2b*). The 3′ → 5′ proof-reading ability of the enzyme is inhibited so that random bases can be inserted opposite the dimer without accurate base-pairing (see Topic C2). This mechanism is the main reason why UV light is mutagenic because most of the other systems repair DNA accurately.

The SOS-repair systems are induced by RecA protein, activated by a conformation change, caused by the presence of damaged DNA. The activated RecA proteins causes LexA protein to be proteolytically cleaved so that it no longer acts as a repressor of transcription. The genes in the system may then be expressed as long as the DNA damage is present in the cell. Once the damage is removed, RecA protein is no longer activated as an inducer of LexA proteolysis, so LexA protein accumulates in the cell and inhibits transcription of the SOS-repair genes.

E5 PLASMIDS

Key Notes

Structure of plasmids	Plasmids are extra-chromosomal molecules of DNA that replicate independently of the bacterial chromosome. They are normally covalently closed, circular, supercoiled molecules but there are also instances of linear plasmids in both yeast and a few bacterial species. They range in size from about 2.2 to 210 kb.
Function of plasmids	Plasmids generally carry genes encoding functions that may be useful to the cell in certain circumstances but are not essential for normal cellular activities. Examples include antibiotic resistance genes, virulence genes and genes that allow bacteria to utilize unusual carbon sources such as catechol. Some plasmids are called cryptic as they do not carry any detectable function.
Replication of plasmids	Control of plasmid replication may be stringent, resulting in one or two copies of the plasmid per cell or relaxed, leading to multiple copies, in some cases as high as 40 copies per cell. Large plasmids tend to have low copy numbers.
Plasmid incompatibility	Plasmids that are similar to each other cannot survive in the same cell. This allows a useful method for classification of plasmids based on this incompatibility, called Inc groups. Plasmids in the same Inc group cannot be maintained in the same cell over a number of generations.
Conjugative plasmids	A number of plasmids have the ability to transfer themselves to other bacteria. These are called fertility (F) plasmids or resistance (R) plasmids, if they also carry genes for antibiotic resistance.

Related topics
Structure and organization of DNA (C1)
DNA replication (C2)

F plasmids and conjugation (E6)
Control of bacterial infection (F7)

Structure of plasmids

Bacterial cells often contain one or more types of extra-chromosomal DNA called plasmids. These are generally covalently closed, circular, supercoiled molecules of DNA that are capable of replication independently from the chromosome. To date a very large number of plasmids have been isolated from virtually every species of bacteria and a number of yeasts. Linear plasmids have also recently been found in *Borrelia*, *Streptomyces* and yeasts. The sizes of plasmids range from very small, an example being pUC8, an artificially constructed cloning vector, which is 2.1 kb (1.8 MDa in weight), to very large. The Ti plasmid of *Agrobacterium tumefaciens* is 213 kb (142 MDa) in size.

Table 1. *The types of genes that may be carried on plasmids and some examples of natural plasmids*

Types of genes carried by plasmids	Plasmid	Host
Fertility – ability to self-transmit	F	*E. coli*
	RP4	*Pseudomonas* sp.
Resistance to compounds toxic to bacteria		
Resistance to antibiotics	RP4	*Pseudomonas aeruginosa*
	RP1	*E. coli*
Resistance to heavy metals	FP2	*P. aeruginosa*
Production of compounds toxic to other organisms		
Colicins (antibacterial compounds)	ColE1	*E. coli*
Antibiotic production	SCP1	*Streptomyces coelicolor*
Degradative enzymes that allow the breakdown of unusual carbon sources		
Toluene degradation	pWWO	*P. putida*
Camphor degradation	CAM	*Pseudomonas* sp.
Virulence factors – compounds that enhance a microbe's ability to infect and colonize a host		
Tumor production in plants	Ti	*Agrobacterium tumefaciens*
Adhesion to mucous membranes	K88, K99, CFA	*E. coli*
Enterotoxin production	Ent (P307)	*E. coli*
	pZA10	*Staphylococcus aureus*
Symbiosis	Sym	*Rhizobium* sp.
Self defense		
Restriction and modification enzymes	R1	*E. coli*
UV light resistance	Col1b	*E. coli*

Function of plasmids

Plasmids frequently carry genes that are useful to the bacteria but are not essential for the normal growth and division of the cell. Examples of the different types of genes carried by plasmids are shown in *Table 1*. A number of plasmids have been isolated from bacteria which do not appear to have any function. These plasmids are frequently called **cryptic** plasmids.

Replication of plasmids

Very small plasmids use the host's replication machinery to make copies of themselves (Topic C2). All they need is an **origin of replication** to which DNA polymerase and the DNA synthetic machinery can bind in order for the plasmid to be replicated. Larger plasmids may carry additional genes whose gene products are necessary for that plasmid's replication. Some plasmids, called **episomes,** can integrate into the chromosome and are replicated when the chromosome is copied.

The number of copies of a plasmid in the cell varies depending on the mechanism that controls plasmid replication. Some plasmids are under a **relaxed** form of control which allows multiple rounds of replication of the plasmid, leading to the presence of a **high copy number** of the plasmid being present in the cell at one time (up to 40 copies). It is generally small plasmids, especially some of the genetically manipulated cloning vectors now available, that exist at these high numbers. Larger plasmids tend to be **stringently** controlled, ensuring that only one to three copies of the plasmid are present in the cell at one time. At cell division large plasmids are distributed to the daughter cells by attachment to membrane sites by a mechanism similar to chromosome segregation (Topic D8) but multi-copy plasmids are distributed to daughter cells by random segregation.

Plasmid incompatibility

Similar plasmids cannot survive in the same cell unless a very strong selection pressure is maintained to ensure that they do. They are described as **incompatible**. The reason for this is still not fully understood but it is thought that incompatible plasmids might share a common mechanism to control replication, which means that after a few rounds of replication one of the plasmid types is lost from the cell. This incompatibility between plasmids has led to a useful classification system in which plasmids that cannot coexist with each other are placed in the same incompatibility group. There are at least 30 different incompatibility groups found in *E. coli*. On the other hand, plasmids from different incompatibility groups can exist in the same cell and up to seven different types of plasmid have been found in one *E. coli* cell.

Conjugative plasmids

An important group of plasmids are the conjugative plasmids, which have been found in a number of bacteria. These are plasmids which are capable of transferring themselves between bacteria by a mechanism called **conjugation**. At their simplest they are represented by the **F-factor** from *E. coli*, which was isolated originally as it caused the transfer of genes between bacteria (Topic E6). Other conjugative plasmids such as the R plasmids carry additional genes such as antibiotic resistance genes. The conjugative plasmids are generally large plasmids whose replication is stringently controlled with the exception of a plasmid, RK6, which is maintained as a relaxed plasmid. Many conjugative plasmids only transfer themselves between bacteria of the same species, but one important group is capable of transferring itself among a wide range of Gram-negative bacteria. These are called **promiscuous** plasmids and are typified by the antibiotic resistance plasmid, RP4, from *Pseudomonas putida*. All R plasmids, but particularly the promiscuous ones, cause concern in a medical context because of their potential to spread antibiotic resistance genes through a wide range of clinically important bacteria (Topic F7).

E6 F PLASMIDS AND CONJUGATION

Key Notes

Conjugation	Conjugation is the mechanism by which genetic material is transferred between two bacterial cells by plasmids. A cell containing a conjugative plasmid (F+, fertility+) forms a mating pair with a cell that does not contain a conjugative-plasmid (F−) by means of an F-pilus on the surface of the cell. The pilus contracts, pulling the two cells into contact, and the DNA is transferred from the plasmid-containing cell, the donor, to the recipient. The *tra* genes, carried on the F plasmid, contain all the information for the conjugative process.
DNA transfer	The DNA that is transferred is single stranded and is synthesized by rolling-circle replication. The plasmid is nicked in one strand at a site called *oriT*, origin of transfer. The intact strand acts as a template for DNA synthesis while the nicked strand is displaced and transferred into the recipient cell. Once inside the recipient cell, a complementary strand of DNA is synthesized to form a new F factor.
Integration and Hfr strains	F factors can integrate into specific sites in the *E. coli* chromosome by homologous recombination between insertion sequences found on the chromosome and the plasmid. Even though integrated, the plasmid is still capable of initiating the rolling-circle mechanism of replication at *oriT*, followed by the transfer of single-stranded DNA into a recipient. The whole chromosome therefore acts like a large F plasmid, allowing the transfer of chromosomal genes to another bacterium. Genes closest to the plasmid are transferred first. In the recipient, the donor chromosomal DNA is integrated into the recipient chromosome by recombination. Strains which have an F plasmid integrated into the chromosome are called Hfr (high-frequency recombination) strains.
F' factors	F plasmids integrated in the chromosome normally excise themselves as precisely as they entered to give an intact F plasmid. However, in some cases, the excision is not precise and a piece of the chromosome is integrated into the F plasmid. These are called F' factors.
Related topics	DNA replication (C2) Plasmids (E5) Recombination and transposition (E3)

Conjugation

An important mechanism by which DNA can be transferred between cells is **conjugation**, mediated by some plasmids, which requires direct contact between cells. Conjugative plasmids, typified by the F plasmid of *E. coli*, have the ability

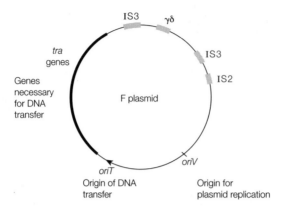

Fig. 1.   A simplified map of an F plasmid.

to transfer themselves between cells and, in some cases, also to transfer pieces of chromosomal DNA. F plasmids and their relatives carry a group of genes called the *tra* genes which encode all the proteins required to form a mating pair with another bacterium that does not contain an F plasmid. DNA is then transferred from the plasmid-containing cell, called the donor or F$^+$ cell, to the recipient, F$^-$ cell. A map of the F plasmid of *E.coli* is shown in *Fig. 1*. The plasmid is large, 95 kb in size, and normally exists as one or two copies in the cell. Vegetative replication of the plasmid at cell division is by normal semi-conservative, bidirectional DNA synthesis from *oriV* followed by segregation of the plasmid into the daughter cells (see Topics C2 and E5). The F plasmid also contains a number of insertion sequences (Topic E3) distributed throughout the molecule (*Fig. 1*).

Cells containing F plasmids (F$^+$) produce specialist F-pili on their surface which are encoded by the *tra* genes. These are proteinaceous, filamentous structures that attach to the surface of F$^-$ cells to form a mating pair (*Fig. 2a*). They will not form mating pairs with F$^+$ due to a phenomenon called **surface exclusion** which is mediated by a pair of *tra*-gene encoded proteins in the outer membrane of F$^+$ cells. The F-pilus retracts on contact, pulling the two cells into close contact.

DNA transfer

A single-stranded copy of the F plasmid is transferred into the recipient cell. It is produced by rolling-circle replication (see also Topic C2). A single-strand nick is made in the DNA at a site called *oriT* (origin of transfer) which is located beside the *tra* genes (*Fig. 1*). The 3' end of the nicked DNA is used as a primer to start synthesis of a new strand of DNA, using the intact strand as a template by rolling-circle replication (see *Fig. 6*, Topic C2). The nicked strand is displaced and transferred through a pore into the recipient cell (*Fig. 2b*). Synthesis of a complementary strand of DNA occurs in the recipient cell. The order of transfer of the F plasmid is such that the last genes to be transferred are *tra* genes. The plasmid circularizes in the recipient to give an F$^+$ cell and the mating-pair separates (*Fig 2c*).

Integration and Hfr strains

As mentioned previously, the F plasmid has a number of insertion sequences scattered about the molecule as does the *E. coli* chromosome. Occasionally, recombination between these homologous sequences leads to the integration of

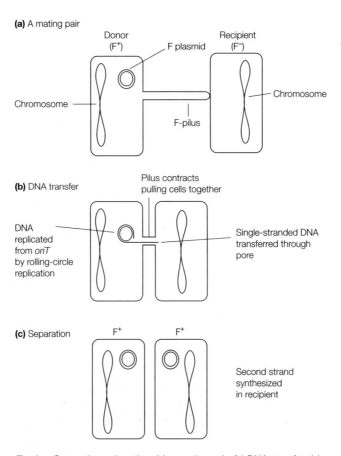

Fig. 2.   Stages in conjugation: (a) a mating pair; (b) DNA transfer; (c) separation.

the F plasmid into the genome at specific sites (*Fig. 3*). Strains that have an F plasmid integrated into the chromosome are called Hfr strains (standing for high-frequency recombination for the reasons described below). The F plasmid, once integrated, is still capable of forming mating pairs and initiating transfer

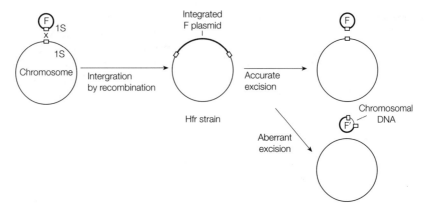

Fig. 3.   Integration and excision of F plasmids from the chromosomes showing the production of F' factors by aberrant excision.

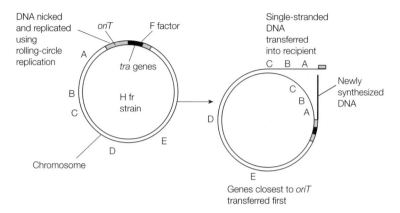

Fig. 4.   Transfer of chromosomal genes to a recipient by Hfr strains.

replication at *oriT*. However, in this case, the chromosome plus plasmid acts like a large F factor so F-plasmid DNA followed by the *E. coli* chromosome is transferred into the recipient (*Fig. 4*). Chromosomal genes closest to the *oriT* are transferred first. The order of transfer can be clockwise or anticlockwise, depending on the orientation in which the F plasmid is inserted. The number of genes transferred depends on how long the mating pair is held together but, in theory, it takes 100 min to transfer the whole *E. coli* chromosome. The last genes to be transferred are the *tra* genes. In practice, the mating pairs break at random and, normally, only part of the DNA is transferred. The incoming single-stranded, chromosomal DNA cannot recircularize to form a plasmid or a second chromosome. However, it can integrate into the recipient chromosome by homologous recombination (Topic E3), thus allowing the transfer of genes from donor to recipient. This is why the strains are named high-frequency recombination strains as they allow the transfer of chromosomal genes from one strain to another at high frequency. Hfr strains are used in genetic mapping for determining the position of genes on the chromosome of *E. coli* as the order of genes can be determined by the time at which they enter a recipient during mating.

F' factors

F factors excise themselves from the chromosome by a process which is the reverse of integration. Normally this is an accurate process but occasionally an error is made and the plasmid picks up a piece of the adjacent chromosome to form an F' factor (*Fig. 3*). These F' factors can be used to transfer chromosomal genes into a recipient cell and are used by bacterial geneticists particularly for creating partial diploids of genes in a cell for investigating dominance/recessive relationships between alleles of the same gene.

E7 BACTERIOPHAGE

Key Notes

Introduction

Bacteriophage, normally called phage, are viruses that infect bacteria. They are obligate intracellular parasites that are capable of existence as phage particles outside the bacterial cell but can only reproduce inside the cell. They consist of a nucleic acid genome surrounded by a protein coat called a capsid. Phage possess the ability to infect a bacterium and redirect the cell to synthesize phage components.

Structure of bacteriophage

Phage, like viruses, have a number of shapes. They may be icosahedrons, almost spherical shapes consisting of 20 triangular faces, or filamentous or complex structures consisting of icosahedral heads with helical tails. The genome can be DNA or RNA, single stranded (ss) or double stranded (ds), circular or linear. Some phage are very small such as the *E.coli* ss RNA phage, MS2, whose genome contains information for only four proteins, while others are large like T4 whose genome encodes around 135 different proteins.

A typical phage life cycle

A typical life cycle of a phage starts by the adsorption of the phage on to a specific receptor on the bacterial cell surface followed by penetration of the genetic material into the host. Bacteria contain restriction-modification systems that are designed to degrade foreign DNA, so many infections are unsuccessful. Enzymes encoded by the phage genome are synthesized, which is followed by nucleic acid replication. Finally, phage capsid proteins are made and assembled into a new phage coat, at the same time packaging a copy of the genome. The phage are then released, normally by lysis, into the surrounding medium.

Lysogenic life cycle

Some phage, a typical example being phage λ, instead of starting the lytic cycle on entry into the cell, produce repressor proteins that stop phage replication. Instead, the phage enters a state called lysogeny in which its genome is replicated at the same time as the host chromosome and is passed from one generation to the next. The phage may be induced from this lysogenic state spontaneously and enter into the lytic cycle. In order to replicate with the host, most lysogenic (sometimes called temperate) phage integrate into the chromosome to form a prophage but some, like P1, exist in the cytoplasm like a plasmid.

Growth and assays for phage

The number of infectious phage particles can be measured using a plaque assay to count the number of plaque-forming units (pfus) in a sample. A plaque is a clear area in a turbid lawn of bacteria, where phage released by a bacterium have gone on to lyse bacteria in the surrounding area. A plaque is therefore equivalent to one phage in the original sample.

Related topics

Control of gene expression (C5)
Recombination and transposition (E3)
Replication of bacteriophage (E8)

Transduction (E9)
Virus structure (K1)

Introduction

Bacteriophage, normally shortened to **phage**, is the name given to viruses that infect bacteria. They are genetic entities consisting of a nucleic acid genome and a protein coat, called a **capsid**, but they are not cellular. Although capable of existence as particles outside a cell, the phage genetic material must enter a cell in order for it to replicate. Bacteriophage are therefore obligate, intracellular parasites of bacterial cells. There are many types of bacteriophage, which specifically infect different types of bacteria. This specificity of infection is based on the presence of receptors for that phage on the surface of the bacterium (the **host**). Once inside the host, the phage genome directs the synthesis of components for new phage particles using the host's biosynthetic machinery. These new phage particles are then assembled and released. This is called the **lytic** cycle.

Bacteriophage are easy to handle and have made good model systems for studying viruses in general. One does not need expensive tissue culture, or work with whole animals and plants, to study phage organization, replication and assembly. Phage have always been important tools for microbial geneticists for moving genes around and for gene mapping by transduction (Topic E9) but, most recently, phage, in particular phage λ, have proved to be extremely useful tools for cloning genes and creating gene libraries. It is important to note that most of our knowledge is about *E. coli* phage, but phage do infect most other types of bacteria.

Structure of bacteriophage

Bacteriophage are made up of a nucleic acid genome surrounded by a protein coat called a capsid whose role is to protect the genetic material and to aid in the infection of a new host. Some phage may also carry additional enzymes with the nucleic acid inside the capsid. The coat is made up of protein subunits arranged in highly ordered structures which give the phage a distinctive shape. The number of different types of protein that go to make up the capsid may range from one in simple phage like MS2 to >25 in a complex phage like T4.

There are three morphological shapes associated with bacteriophage structure (*Fig. 1*):

Icosahedron
This is an almost spherical shape consisting of 20 triangular faces. Icosahedrons are very common shapes in nature as they are very efficient ways to create an enclosed shell from subunits.

Filamentous
These are long protein tubes formed by capsid proteins assembled into a helical structure.

Complex
Some phage consist of icosahedral heads attached to helical tails. In the case of a group of phages called the T-even phages the head is actually an elongated icosahedron. The tails can be contractile or non-contractile, sheathed or non-sheathed and they may also have base plates and tail fibers associated with them. The role of the tail is to assist the injection of the genetic material into the cell.

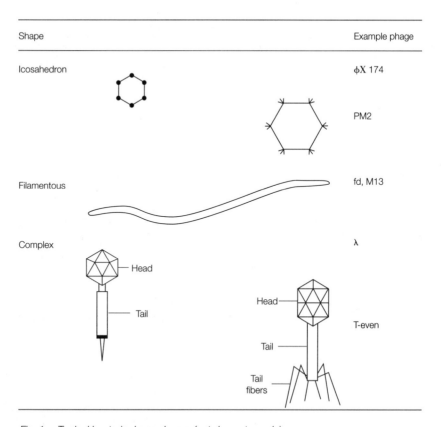

Shape		Example phage
Icosahedron		φX 174
		PM2
Filamentous		fd, M13
Complex	Head / Tail	λ
	Head / Tail / Tail fibers	T-even

Fig. 1. Typical bacteriophage shapes (not drawn to scale).

Bacteriophage range in size from about 25 nm (icosahedron diameter) for MS2 to 110 nm × 85 nm (head size) for T4 to which is attached a 25 nm × 110 nm tail. The filamentous phage such as M13 have measurements in the region of 6 nm wide by 860 nm long. A few bacteriophage, such as φ6, have membrane envelopes but, in general, they are not as common as in animal viruses (Topic K1).

The genetic material inside the phage capsid may be RNA or DNA, which can be ss or ds, linear or circular (*Table 1*). The mechanism by which these different types of nucleic acid molecules are replicated is described in Topic E8.

A typical phage life cycle

Although there are differences between the life cycles of bacteriophage they all tend to follow a similar pattern, which is outlined below and in *Fig. 2*.

- **Adsorption (attachment).** Phage infection of a host starts by attachment to specific receptors on the surface of the cell. These receptors vary in nature, being proteins or polysaccharides, which may be present on the surface all the time or only produced under certain conditions. For example, phage T4 binds to lipoproteins in the outer membrane of *E. coli* whereas phage λ binds to a maltose-transport protein only present in the outer membrane when bacteria are grown in maltose-containing medium.
- **Penetration.** Generally, only the genetic material is injected into the host, leaving the empty protein coat on the surface of the cell. Lysozyme associated with the phage head is used to dissolve the peptidoglycan (Topic D3). The mechanism of penetration varies from phage to phage depending on its structure.

Table 1. The nature of phage genomes

Nucleic acid type		Structure	Example
DNA	Single-stranded	Circular	φX174
			M13
	Double-stranded	Linear	T phage
		Circular	PM2
RNA	Single-stranded	Linear	MS2
	Double-stranded	Linear	φ6

In the case of T4, which has a contractile tail, a highly complex mechanism has evolved. After attachment by its long tail fibers and contact of the base plate with the cell wall, the tail sheath contracts and the DNA is injected into the cell. In many cases the incoming DNA may be degraded by endonucleases, which are part of the host's **restriction-modification** defense systems, but occasionally DNA molecules will survive to set up an infection. These defense endonucleases, called **restriction enzymes**, bind to specific sequences in foreign DNA and cut the DNA at these sites. The microbe's own DNA is modified, normally by methylation, to prevent it being degraded by these restriction enzymes. Some phage such as the T-even phages have evolved mechanisms to modify their own DNA by glycosylation or methylation to prevent them being degraded by restriction enzymes on injection.

- **Nucleic acid replication**. Once inside the host the nucleic acid is transcribed and translated (or just translated in the case of RNA viruses) to produce enzymes that direct the synthesis of new phage nucleic acids. These are sometimes called **early proteins**. Many phage switch off the synthesis of host proteins and degrade the host genome, thus ensuring that all the cellular biosynthetic machinery is directed to the production of phage components. After a short time there is usually a switch to the production of large amounts of phage structural proteins, scaffold proteins required for phage assembly and proteins required for lysis and phage release. These are referred to as **late proteins.**

- **Phage assembly**. Once sufficient phage capsid components and nucleic acid has been synthesized the new phage particles are assembled spontaneously while, simultaneously, packaging the nucleic acid into the capsid.

- **Release**. Many phage are released by lysing the bacterial cell wall which is why this life cycle is often called the lytic cycle and the phage are referred to as virulent phage. Enzymes soften the cell wall and the phage are released as bursts of between 50–1000 phage per cell. The time from infection to release is about 22 min at 37°C for T4. Other phage, such as the filamentous phage M13, release phage through the cell wall without damaging the cell so phage particles can be released over a long period of time. The bacteria continue to grow but at a reduced rate.

Lysogenic life cycle

A number of phage, typified by phage λ, have an alternative to a lytic life cycle inside the host. On entering the host these phage do not enter the lytic cycle. Instead they enter into a state called **lysogeny** in which they either integrate into the host chromosome to form a **prophage** (e.g. phage λ) or exist in the cytoplasm like a plasmid (e.g. phage P1). Consequently, these lysogenic (sometimes called **temperate**) phage replicate at the same time as the host

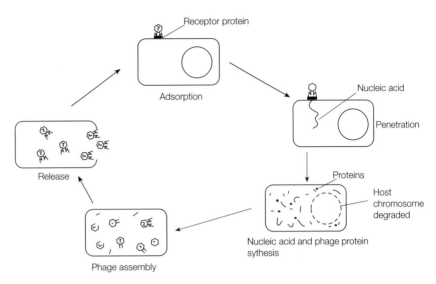

Fig. 2.   The life cycle of a typical lytic phage.

chromosome and are passed on to daughter cells (*Fig. 3*). The protein in λ that is responsible for the phage entering into lysogeny is a repressor, produced by the *cI* gene, which prevents expression of all the phage genes necessary for phage replication and assembly. Instead, the phage genome (which is linear) circularizes due to the presence of complementary ss ends called **cos** sites. An integrase protein, the *int* gene product, mediates site-specific recombination between a site on the phage genome, *attP*, and a homologous site, *attB*, on the *E. coli* genome, which leads to phage DNA integration into the chromosome (*Fig. 2* in Topic E3). The lysogenic state can be interrupted spontaneously at any time, but especially when there is induction of the SOS response by DNA-damaging agents such as UV light, as activated RecA protein inactivates the λCI repressor protein (see Topic E4). The repressor proteins become inactivated and transcription of the genes for phage production is induced. The phage is excised from the chromosome by a mechanism that is the reverse of integration mediated by the product of the *xis* gene, and λ then enters the lytic cycle.

Lysogenic phage are of particular interest because they can carry functional genes which can change the phenotype of the host. This is called lysogenic conversion and the properties the lysogenic phage confer often contribute to the ability of a bacterium to survive in a host (Topic F5). Epsilon phage of *Salmonella* alter the O-antigen portion of the bacterial LPS, thus changing the antigenic properties of the bacterium. A number of bacterial toxin genes are also carried on temperate phage including diphtheria toxin carried by phage β of *Corynebacteria diphtheriae* (Topic F6). Examples of phage conversion in this category are being discovered regularly; cholera toxin has just been added to this list.

Finally, the bacteriophage λ has proved to be an extremely useful phage for gene cloning. The central region of its genome contains genes that are essential for lysogeny but not required for lytic growth. This region can therefore be replaced with up to 20 kb of foreign DNA without interfering with phage replication in the lytic cycle, making it an ideal system for producing large amounts of foreign DNA.

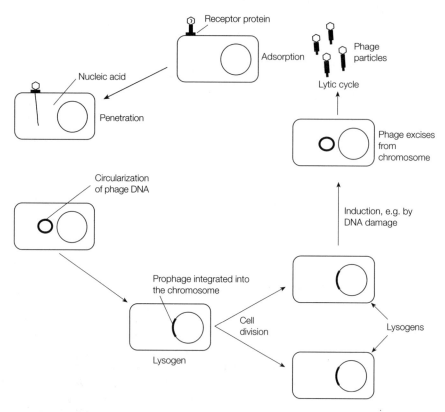

Fig. 3. The life cycle of a lysogenic phage.

Growth and assays for phage

Bacteriophage are grown in the laboratory in either liquid or solid culture. Bacteria and phage are mixed together at a ratio where the number of phage is considerably less than the number of bacteria. The mixture is incubated, allowing the bacteria to replicate and become infected with phage. The infected bacteria then lyse, releasing large numbers of phage. Consequently, the number of phage increases at a much greater rate than the bacteria until at a critical point the number of phage present is enough to infect all the bacteria in the culture. All the bacterial cells therefore lyse virtually at the same time, which can be seen as a complete reduction in the turbidity of the broth culture (see Topic D9). Alternatively, a soft agar is added to the bacteria and phage mixture, and this is poured on to the surface of an agar plate. The bacteria grow and the phage infect as described above until, eventually, the bacteria are all infected, causing **confluent lysis** of the bacterial lawn.

The plate method of growth is also used to measure the number of infectious phage particles in a preparation. The phage preparation is diluted, mixed with cells and, after a short time to allow adsorption of the phage on to the cells, soft agar is added and the mixture is poured on to the surface of an agar plate. The infection of a bacterium by a phage particle causes the production of a large number of phage, which can infect adjacent cells. When these cells lyse, yet more phage are released which infect even more cells. The uninfected bacteria continue to grow to form a bacterial lawn. After overnight growth, the foci where a bacterium was originally infected by a phage is seen as a clear area of lysis called a **plaque** in a turbid lawn of bacteria. The number of plaques in the lawn indicates the number of infectious phage in the original preparation. This number is measured as plaque-forming units (pfus).

E8 REPLICATION OF BACTERIOPHAGE

Key Notes

RNA phage	Most RNA phage contain ss linear molecules. The RNA is a positive strand that acts as mRNA as well as the genetic material. A phage-encoded RNA replicase synthesizes a complementary copy (negative strand) of the RNA which acts as a template for the production of new RNA genomes and mRNA. Only one ds RNA phage is known but little is known about how it replicates.
Single-stranded DNA phage	These may be icosahedral phage, such as φX174, or filamentous, such as M13. On entry into the cell, the DNA is converted into a ds replicative intermediate called an RF form. Multiple copies of the RF form are produced by bidirectional replication. These then act as templates for RNA synthesis and for the synthesis of new ss genomes by rolling-circle replication.
Double-stranded DNA phage	Linear DNA from T-phage is synthesized by bidirectional, semi-conservative replication. The problem of synthesis of the 5′-ends created by this mechanism is solved by the formation of concatamers of phage DNA by base-pairing between terminal direct repeats. Phage-length genomes are then cut by endonucleases. λ phage circularizes in the host and replicates by the rolling-circle method.
Mu – a novel mechanism of replication	The ds *E. coli* phage Mu integrates randomly into the chromosome where it can either exist as a lysogen or initiate replication by replicative transposition to multiple sites in the chromosome.
Related topics	DNA replication (C2) Bacteriophage (E7) Recombination and transposition (E3) Transduction (E9)

RNA phage

All ss RNA phage contain positive-strand RNA which means that it can act as mRNA as soon as it enters the cell. They are very small phage, containing a minimum of genetic information. MS2 is a typical example; it infects *E. coli* by binding to the F-pilus, which is present on the surface of cells containing an F plasmid (Topic E6). The 3569-nucleotide genome has been sequenced and encodes four proteins: a **replicase** enzyme, a coat protein, a maturation protein and a lysis protein. In order to fit these into the genome the proteins are over-lapping (*Fig. 1*), which is a common feature of phage genomes. The enzyme that replicates the RNA is an RNA-dependent RNA polymerase which is actually made up of the phage-encoded replicase and polypeptides produced by the host. The events during replication of MS2 are shown in *Fig. 2*. The

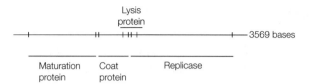

Fig. 1. Genetic map of MS2.

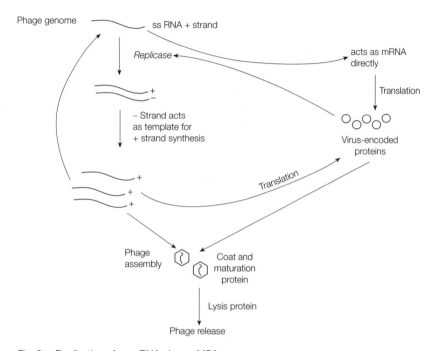

Fig. 2 Replication of a ss RNA phage, MS2.

replicase makes a copy of the RNA strand called a **negative strand**. This then acts as a template for the production of new positive strands of RNA which can act as mRNA or, once enough phage components have been made, as new phage genomes. The RNA is extensively folded and packaged with the coat protein (180 copies) and the maturation protein (one copy). The lysis protein then causes the cell to burst, releasing the new phage particles.

Single-stranded DNA phage

These are typified by the icosahedral phage φX174 and the filamentous phage M13, both of which have circular genomes. Although the DNA in these phage is replicated by similar mechanisms, their life cycles are very different, as φX174 is a lytic phage whereas M13 can leave its host without killing the cell. In both cases, however, on entry into the cell the ss DNA must be converted into a ds form for replication to occur. This ds intermediate is called the RF form, which is then replicated to produce a number of copies. These act as templates for the synthesis of mRNA and new phage proteins and also for the production of new ss molecules of DNA by rolling-circle replication (see *Fig. 6*, Topic C2). The new phage-length strands of ss DNA are cut off and ligated into a circle, packaged and released from the cell either by lysis (φX174) or by extrusion through the membrane (M13). Because of the nature of its replication, M13 has been used as a cloning vector for the production of ss DNA to be used for DNA sequencing, as DNA can be inserted into the RF form and then subsequently purified in a ss form from the progeny phage.

Double-stranded DNA phage

Replication of the T-phage group, which contain linear DNA molecules, proceeds bidirectionally from a single origin of replication. The problem for linear molecules is how to replicate the last bit of DNA at the 5′-end of each strand which is left ss due to the nature of DNA replication (Topic C2). This is overcome by a number of strategies. In T7 there is a direct repeat of 160 bases at the ends of the molecule (*Fig. 3*). This allows base-pairing between the ss DNA portions on two separate DNA molecules to form a **dimer** twice as long as the original strand. The non-replicated portions can then be filled in by DNA polymerase and ligase. Other DNA molecules will be joined until eventually a long **concatamer** is formed. A phage endonuclease cuts the concatamer at a specific site to produce phage lengths of DNA which can be packaged into heads. In the

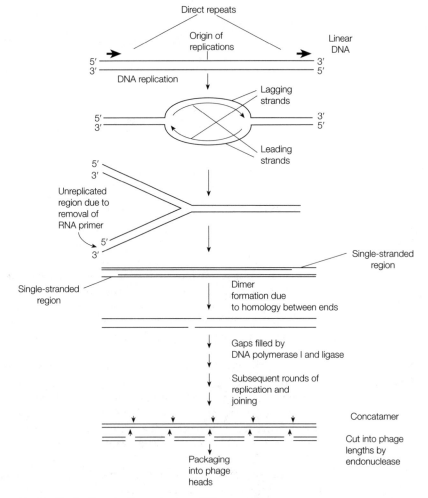

Fig. 3.   Replication of a linear, ds phage, T7.

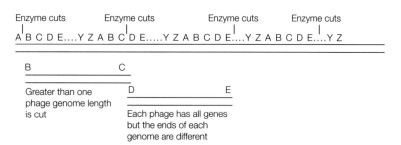

Fig. 4. *The cutting of replicated T4 DNA for packaging into phage heads. Each head contains more than one genome.*

case of T4 a slightly different strategy is adopted in that the enzyme cuts the concatamer into lengths that will fit into the phage head but not necessarily at the same point in the DNA. The molecule packaged into the head has actually more than a genome's worth of DNA, and although all the genes will be present, the ends of the molecules will be different between different phage particles (*Fig. 4*). This phenomenon is called **circular permutation**.

In phage λ, DNA synthesis of the linear DNA proceeds via a circular intermediate formed by base-pairing between *cos* ss overhangs on the DNA (Topic E7) and new ds molecules are produced using the rolling-circle method of replication (see *Fig. 6*, Topic C2).

Mu – a novel mechanism of replication

Mu is an *E. coli* phage which has a very unusual mechanism of replication. It is a lysogenic phage, in that it can integrate into the chromosome of the host. However, unlike λ, it inserts randomly into the genome, causing mutations if it inserts into the middle of a gene. It is, therefore, a useful tool for geneticists. Mu replicates by **replicative transposition** (Topic E3). On induction, the phage DNA transposes to multiple sites on the host genome, producing many copies of the DNA. Synthesis of mRNA and phage-encoded proteins occurs and eventually new phage particles are assembled and released by lysis.

E9 TRANSDUCTION

Key Notes

Transduction	Transduction is the movement of host DNA from one bacterial cell (donor) to another cell (recipient) by a bacteriophage.
Generalized transduction	Generalized transduction is the result of a rare error during phage assembly, leading to the packaging of a piece of host genome into the phage head instead of the phage DNA. Such transducing phage are capable of infecting new hosts and injecting in the DNA but, as this DNA is not phage DNA, it will be lost unless the DNA is integrated into the host genome by recombination.
Specialized transduction	This is a feature of some lysogenic phage which spend some time integrated into sites in the host chromosome. At induction, instead of the phage genome accurately excising from the chromosome, it excises a piece of adjacent chromosome as well. Phage particles containing these genes will infect recipient cells where the whole phage may integrate into the chromosome to form a lysogen, or the chromosomal DNA may be integrated by recombination.
Related topics	Recombination and transposition (E3) Bacteriophage (E7)

Transduction

A mechanism by which DNA can be moved from one cell to another is through the agency of a phage particle; this is called **transduction**. Phage can pick up pieces of chromosomal or plasmid DNA from one bacterium, called the **donor**, and transfer it to a **recipient** bacteria where it may be integrated into the recipient genome by recombination (Topic E3). Transduction has been found in a wide range of bacteria and phage and is thought to play a significant role in the transfer of genetic information in nature. Transduction is also an important tool for geneticists as it allows the transfer of genes between bacteria and the mapping of bacterial genomes; genes that are close together will be transferred to a recipient cell together. Many of the original uses of transducing phage have been replaced by gene cloning, but they are still useful tools today for the creation of bacterial strains and the analysis of genomes.

Generalized transduction

Generalized transduction is the transfer of any piece of host DNA to a recipient by a phage particle. This is due to the fact that some phage, at the point of phage assembly, will accidentally package a phage-size piece of host DNA into the phage head instead of the phage genome (*Fig. 1*). These transducing phage, on release from the host, are capable of attaching to another cell and injecting the DNA into this recipient. The DNA, however, is not phage DNA

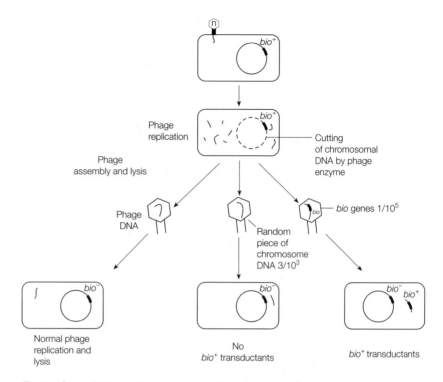

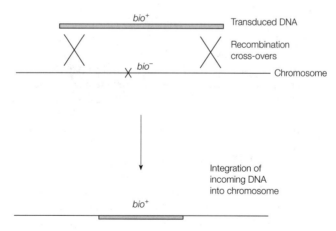

Fig. 1.   Generalized transduction showing the transfer of bio *genes to a recipient.*

Fig. 2.   Integration of transducing DNA into the host chromosome by recombination.

and is not capable of replicating. It will therefore only survive if it is integrated into the host chromosome by recombination.

Frequently, especially if the donor and recipient are closely related, the new DNA will not make a significant difference to the recipient, but if the donor and recipient have different genotypes, a gene or genes of the recipient cell may be converted to that of the donor (*Fig. 2*). Transduction is easy to carry out in the laboratory. A phage preparation is added to a culture of the donor strain which is allowed to incubate until complete lysis has been achieved (Topic E7). The phage **lysate** containing a few transducing phage is then used to infect a recipient culture of bacterium and this is plated out on agar plates. Bacteria that receive the wild-type phage will be lysed, but bacteria that receive transducing phage will be able to grow as colonies. The transfer of genes of interest may be selected by using appropriate media (Topics D7 and E1). The frequency at which a particular gene may be transferred is about one cell in every 10^6 to 10^8. Phage P22 in *Salmonella typhimurium* and phage P1 in *E. coli* are two systems frequently used in the laboratory.

Specialized transduction

Specialized transduction is a rare occurrence associated with a number of lysogenic phage that integrate into the host chromosome. Normally when a prophage is induced it excises from the chromosome in a very precise fashion to form a complete phage genome (Topic E7). Very rarely an excision event occurs which is inaccurate; a piece of chromosome adjacent to the phage is removed and a piece of the phage is left in the chromosome (*Fig. 3*). The excised phage is replicated normally, packaged into phage heads and released to infect another cell. The nature of the genes lost from the phage particle is dependent on which end of the genome is lost. In the example shown in *Fig. 3* where λ is integrated at its normal *att* site, on excision, it may pick up *gal* or *bio* genes to form λ*gal* or λp*bio*. The λ*gal* phage have lost essential genes for making mature phage so are incapable of replicating and forming plaques in a new host unless these functions are provided by a helper λ phage. The λp*bio* phage are plaque-forming, hence the 'p' designation, without any additional functions being provided, as the genes that are lost are not essential for phage replication. The fate of the DNA in the recipient is various as shown in *Fig. 4*. The transducing phage may enter the lytic cycle and produce many copies of the phage and the genes it carries or it may integrate into the host chromosome. Integration may be by site-specific recombination at the *att* site to form a lysogen (*Fig. 4a*) in which case there are now two copies of the transduced gene(s) (partial diploid), or recombination may occur between genes on the phage and in the recipient chromosome (*Fig. 4b*). In contrast to generalized transduction, the particles produced by specialized transduction are capable of moving genes at a high frequency of transduction in a specific manner, as all phage in a preparation carry the same piece of DNA. Using a variety of genetic techniques specialized transducing phage have been produced for most areas of the *E. coli* chromosome which were used in the past for the production of large amounts of DNA for a particular gene, the transfer of genes between strains and for the creation of partial diploid strains in order to study gene function.

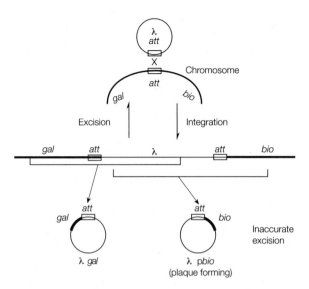

Fig. 3.   *Production of specialized transducing phage.*

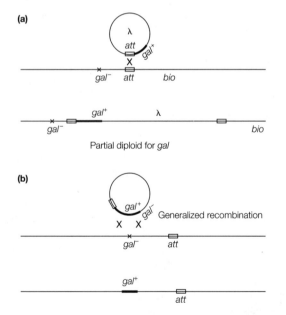

Fig. 4.   *Some of the fates of λ specialized transducing phage on infections of a recipient: (a) formation of λ lysogen; (b) replacement of recipient chromosomal DNA with DNA carried on λ by generalized recombination.*

E10 TRANSFORMATION

Key Notes

Overview	Transformation is the result of the uptake of free DNA from the surrounding medium and the incorporation of that DNA into the genome. A number of bacteria such as *Bacillus, Streptococcus, Neisseria* and *Haemophilus* are capable of natural transformation. In eukaryotes the ability of cells to take up DNA is called transfection.
Competence	The ability of a cell to take up DNA is dependent on the cell being in a particular state called competence. Competence is related to the presence of receptors for DNA being present on the surface of the cell. Other bacteria, such as *E. coli,* can be induced to competence by chemical treatment under cold conditions.
Uptake and fate of DNA	Depending on the species, DNA binding and transfer into the cell may be of specific or non-specific sequences. The DNA bound is normally ds but in most cases it is made ss as it enters the cell. *Haemophilus influenzae* takes up ds DNA. The incoming DNA is integrated into the chromosome by recombination.
Related topics	Structure and organization of DNA (C1)

Overview

Our knowledge that DNA is the genetic material has come from experiments based on the fact that some species of bacteria are capable of naturally taking up DNA from their surrounding medium, and incorporating it into their genomes, by a process called **transformation**. Fred Griffith in the late 1920s first showed that a rough, non-capsulated, avirulent strain of *Streptococcus pneumoniae* could be converted into a smooth, capsulated, virulent strain by the addition of heat-killed smooth cells. Oswald Avery and his group in a series of experiments in the 1930s showed that the material that carried this transforming ability was DNA and not the proteins of the cell. They were fortunate in their choice of experimental microorganism as only a few species are naturally able to take up DNA, including *Streptococcus, Bacillus, Neisseria, Haemophilus* and some species of archaebacteria.

Uptake of DNA by eukaryotic cells is usually referred to as **transfection**.

Competence

Not all species within a genera will necessarily be capable of transformation nor all cells within a population. The ability of a bacterial cell to take up DNA is associated with the development of a **competent** state in which DNA receptors on the cell surface and other transformation-specific proteins are present. For example, in *Neisseria gonorrhoeae*, only piliated cells are competent for transformation. Competence is generally dependent on growth conditions. Competence may occur in only a few cells in a culture (*B. subtilis*) or in nearly all cells (*S. pneumoniae*) depending on the genera. Competence may last for a few minutes (*S. pneumoniae*) or for several hours (*B. subtilis*).

Competence can be induced in species that are not naturally transformable, for example, *E. coli*, using chemical treatments such as ice-cold $CaCl_2$ treatment followed by a brief heat shock. This competent state is not the same as natural competence.

Uptake and fate of DNA

The nature of the DNA taken up is dependent on the genera. Most bacteria will take up any DNA molecule, but *Haemophilus* requires the presence of a 10 bp specific sequence on the DNA. *Haemophilus*, a Gram-negative bacterium, binds and takes up ds DNA in contrast to the Gram-positive bacteria, *Bacillus* and *Streptococcus*, which bind ds DNA but degrade one strand as the other strand is imported into the cell. The incoming linear DNA is protected from degradation by competence-specific proteins and is integrated into the genome by recombination involving RecA protein (*Fig. 1*). During artificially induced competence in *E. coli*, circular plasmid DNA are preferentially taken up and in this case remain independent of the chromosome.

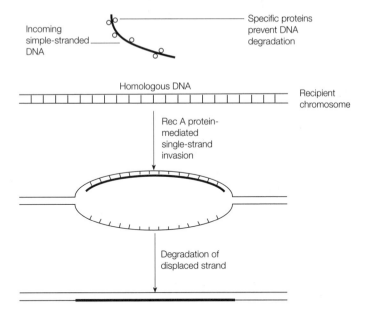

Fig. 1. Integration of ss DNA into the chromosome by recombination.

Transformation is probably an important mechanism by which genes are transferred naturally in the environment. Transformation has also been used in genetic mapping as genes that are close together are more likely to be taken up on the same fragment of DNA and incorporated together into the chromosome of a recipient cell.

F1 BACTERIA IN THE ENVIRONMENT

Key Notes

Bacteria in the environment

Bacteria are found in environments ranging from hydrothermal vents where temperatures reach 100°C to polar regions. Bacteria perform many important tasks such as nutrient recycling and many form important symbiotic associations with plants, increasing soil fertility and plant growth.

Nutrient recycling

Microorganisms play a major role in biogeochemical cycling of carbon, nitrogen and sulfur. Carbon is present in many forms including cellulose, lignin and hydrocarbons. Degradation of these substrates is controlled by factors such as the structure of the individual components within the molecule, the environmental conditions and the microbial community present. The carbon cycle is divided into aerobic and anaerobic processes. Complex organic material is broken down by fermentation or under aerobic conditions by respiration, releasing methane (CH_4) and CO_2. The fixation of CO_2 by aerobic chemolithoautotrophs generates new biomass. The nitrogen cycle involves several distinct processes. Nitrification is the aerobic process of nitrite and nitrate formation from ammonium ions. The process of denitrification is a dissimilatory process (loss of nitrogen from the immediate environment) that produces nitrogen gas and nitrous oxide. Nitrogen assimilation occurs when inorganic nitrogen is used as a nutrient. Nitrogen fixation (incorporation of gaseous nitrogen products into biomass) can be carried out by aerobic or anaerobic bacteria. In the sulfur cycle, sulfide can be used by a variety of photosynthetic and chemolithoautotrophic organisms. Sulfate can undergo sulfate reduction by *Desulfovibrio*. Dissimilatory reduction occurs when sulfate is used as an external electron acceptor (anaerobic respiration) to form sulfide. Sulfate can be used directly for amino acid and protein biosynthesis.

Bacterial associations with plants

The relationship between microbial communities and plant roots is complex. The release of substrates by plants increases the soil microbial population in the surrounding region, and nitrogen fixation performed by bacteria increases ammonium ion availability for the plant. This area is called the rhizosphere and is very important in soils with low fertility. *Rhizobium*, a nitrogen fixing organism, forms symbiotic associations with legumes. This association requires that the bacteria infect plant root cells. Infections frequently stimulate root cell division, causing the formation of root nodules.

Related topics

Heterotrophic pathways (B1)
Electron transport and oxidative phosphorylation (B2)

Autotrophic metabolism (B3)
Biosynthetic pathways (B4)

Bacteria in the environment

Bacteria are found in most environments ranging from hydrothermal vents where temperatures reach 100°C to polar regions. Bacteria, along with the other microorganisms, are essential components of every ecosystem where they perform many important tasks such as nutrient recycling. Microorganisms also form important symbiotic associations with plants, increasing soil fertility and plant growth.

Nutrient recycling

The microbial community plays a major role in **biogeochemical cycling**. As this term suggests, both biological and chemical processes are involved in the cycling.

Microorganisms are essential in the transformation of carbon, nitrogen and sulfur and metals such as iron. Significant gaseous components occur in the carbon and nitrogen cycles and, to a lesser extent, in the sulfur cycle; thus, bacteria can often fix atmospheric carbon and nitrogen compounds (see Topic B3) , even when they are lacking in their environment.

Carbon cycle

Carbon can be present in a variety of forms ranging from simple compounds such as CH_4 to more complex macromolecules such as starch, cellulose, lignin, hydrocarbons and microbial biomass. A variety of factors control the rate of breakdown of these substrates; these include the structure of the individual components within the molecule (e.g. the major component in lignin is phenyl-propane), the environmental conditions and the microbial community present. Typically, carbohydrate-rich substrates are degraded first with the more

(a)

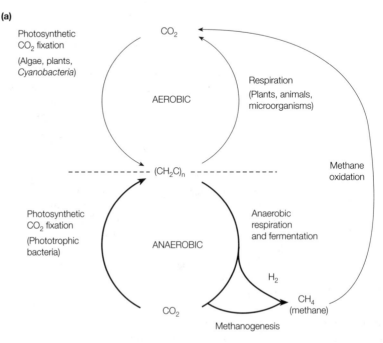

Fig. 1.   Biogeochemical cycles in the environment: (a) carbon cycle; (b) niotrogen cycle; (c) sulfur cycle. Processes that occur under aerobic conditions are shown by (→). Anaerobic reactions are shown by (➤). Those processes that can occur under aerobic or anaerobic conditions are shown by (- - ➤).

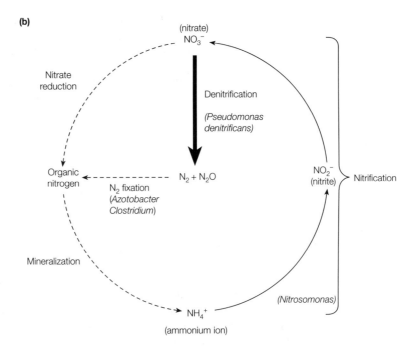

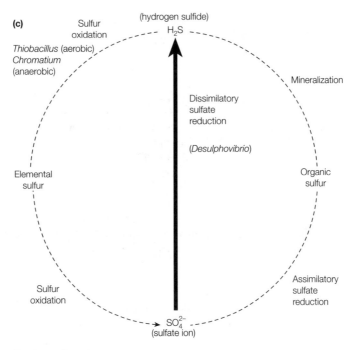

Fig. 1. continued.

complex molecules such as lignin being the most recalcitrant (resistant to degradation). The carbon cycle can be divided into those processes that occur under anaerobic conditions and those that require oxygen (*Fig. 1a*). Organic material is broken down by fermentation or under aerobic conditions by respiration (see Topic B1) releasing CO_2 or forming reduced products such as CH_4. The fixation of CO_2 for the production of biomass occurs through the activities of *Chromatium, Chlorobium* and aerobic chemolithoautotrophs (see Topic B3).

Nitrogen cycle

The nitrogen cycle involves several distinct processes (*Fig. 1b*). **Nitrification** is the aerobic process of ammonium ion oxidation to nitrite and subsequent nitrite oxidation to nitrate. *Nitrosomonas* and *Nitrosococcus*, for example, play significant roles in the first step, and *Nitrobacter* and related chemolithoautotrophic organisms carry out the second step. **Heterotrophic nitrification** by bacteria and fungi contributes significantly to these processes in more acidic environments. The process of **denitrification** is a dissimilatory process (see Topic B2) performed by heterotrophs such as *Pseudomonas denitrificans*. The major products of denitrification include nitrogen gas and nitrous oxide. **Nitrogen assimilation** occurs when inorganic nitrogen is used as a nutrient and incorporated into new microbial biomass (see Topic B4). **Nitrogen fixation** can be carried out by aerobic or anaerobic bacteria. Under aerobic conditions, bacteria such as *Azotobacter* and *Azospirillum* contribute to this process (see Topic B4). Under anaerobic conditions, the most important nitrogen fixers are members of the genus *Clostridium*. Nitrogen fixation can also occur through the activities of bacteria that develop symbiotic associations with plants.

Sulfur cycle

Microorganisms contribute greatly to the sulfur cycle, a simplified version of which is shown in *Fig. 1c*. Photosynthetic microorganisms transform sulfur by using sulfide as an electron source (see Topic B2). In the absence of light, sulfide can be used by *Thiobacillus* and other lithoautotrophs (see Topic B3). In contrast, sulfate can undergo **sulfate reduction** when organic reductants are present; *Desulfovibrio* can derive energy by using sulfate as an oxidant under these conditions. **Dissimilatory reduction** occurs when sulfate is used as an external electron acceptor (anaerobic respiration) to form sulfide (see Topic B2). In comparison, the reduction of sulfate for amino acid and protein biosynthesis is described as **assimilatory reduction**. When pH and oxidation–reduction conditions are favorable, several key transformations in the sulfur cycle also occur as a result of chemical reactions.

Bacterial associations with plants

Microorganisms of various kinds are associated with the leaves, stems, flowers, seeds and roots of plants. The microbial community influences plants in many direct and indirect ways. The plants release a wide range of potential microbial substrates, inhibitors and stimulants. The soil microbial populations respond to the release of organic materials in the immediate vicinity of the plant root by increasing their numbers and changing the characteristics of the microbial community; this region is called the **rhizosphere** and is very important in soils with low fertility. In the rhizosphere, nitrogen fixation performed by bacteria such as *Azotobacter* and *Azospirillum* increase ammonium ion availability for the plant. *Rhizobium*, a prominent member of the rhizosphere community can fix nitrogen and forms symbiotic associations with legumes. This association requires that the bacteria infect plant root cells and move to adjacent root cells via an **infection thread** (a cellulose structure synthesized by the plant). Where root cell division is stimulated by infection a **root nodule** arises.

F2 BACTERIA IN INDUSTRY

Key Notes

Traditional biotechnology	Bacteria are used in industry to make products such as butanol, acetic acid and antibiotics. Biotechnology is the collective term for these processes. Recent advances in large-scale culture techniques and strain improvement have enhanced the efficiency and reliability of these microbial processes.
Modern biotechnology	Genetic engineering of bacteria allows a much greater range of products to be made by bacteria. Human hormones, blood proteins and proteins for vaccine development can now be made in bulk using these modified bacteria.
Biodeterioration	The ability of bacteria to degrade materials such as paints, textiles, concrete and oils (biodeterioration) can cause significant problems in a wide variety of industrial environments. The microbial degradation of oils can, however, be used in the control of oil spills.
Fermented foods	Fermentation is a useful way of preserving food and it is also a means of creating new food flavors and odors. Yoghurt and cheese are produced from milk by lactic acid fermentation. In cheese production the fermentation results in coagulation of milk solids with the formation of a curd. The treated cheese curd can be ripened with or without additional microorganisms. Cheese curd inoculation is used in the manufacture of blue cheeses. A variety of meat products, especially sausage, can be fermented.
Food spoilage	Bacteria play a significant role in food spoilage where they produce detrimental changes in the food such as the formation of putrid odors. Foods with easily utilizable carbohydrates, proteins and fats are ideal environments for spoilage by microorganisms. More importantly, food can act as a good medium for the transmission of bacterial pathogens.
Related topics	Heterotrophic pathways (B1) Mutagenesis (E2) Control of gene expression (C5) Recombination and transposition (E3)

Traditional biotechnology

Bacteria have been used for thousands of years in the manufacture of a wide range of products (**biotechnology**); however, much has changed over the last 30 years. Most industrial microbial processes are now carried out in **fermenters** (large-volume stirred tank systems) where processes can be controlled to maximize production and reduce failure from contamination. Advances in the technology of strain improvement have further enhanced these processes. The formation of vinegar is a good example of how processes have changed. Vinegar is produced by the aerobic conversion of ethanol to acetic acid by *Acetobacter* and *Glucobacter*. Vinegar can be made from any alcoholic liquor such as wine or cider. The traditional **open-vat** system is still used in some countries and requires that the alcoholic liquor is placed in shallow vats that allow considerable exposure to air. Large-scale production systems currently used include the **bubble method**. This is a continuous fermentation system where liquor is added to the fermenter at a rate equivalent to removal of vinegar. This system has a high efficiency with 90–97% of the alcohol converted into acetic acid. Bacteria are currently used in a wide range of processes ranging from the synthesis of antibiotics to the formation of biofuels such as methane (*Table 1*).

Table 1. Major bacterial products of industrial interest

Products	Bacteria used in the processes
Industrial products	
Acetone and butanol	*Clostridium acetobutylicum*
2,3-Butanediol	*Enterobacter*
Enzymes, e.g. proteases	*Bacillus*
Amino acids	*Corynebacterium glutamicum*
Vitamins	*Blakesela*
Organic acids	*Acetobacter*
Medical Products	
Antibiotics	*Streptomyces, Bacillus*
Steroids	*Arthrobacter*
Insulin, interferon, human growth hormone	*E. coli*
Biofuels	
Hydrogen	*Photosynthetic bacteria*
Methane	*Methanobacterium*

Modern biotechnology

Since the late 1970s advances in genetic engineering have produced a revolution in biotechnology and now an increasing number of products are being made by genetically engineered organisms. The greatest interest is in the production of mammalian proteins and peptides by microorganisms, since many of these materials have high pharmaceutical value and are expensive or difficult to produce by other methods. If the gene or genes that code for production of a mammalian protein can be cloned into a microorganism, and good expression of the gene(s) obtained, then a biotechnological process for making this protein can likely be developed. Such materials produced by modified organisms include hormones, blood proteins, immune modulators (interferons) and proteins for vaccine development.

Biodeterioration

Bacteria can also pose significant problems for industry. The microbial breakdown of paints, textiles, concrete, metals and oils is termed **biodeterioration.**

These breakdown processes are more common in humid environments as water is usually the limiting factor in microbial biodeterioriation. Many of these processes, such as the degradation of concrete by *Thiobacillus*, renders the material unusable. Microbial degradation of oils by organisms such as *Pseudomonas* can be used in the control of oil spills. Fungi are also capable of biodeterioration (see Topics H6 and H7).

Fermented foods   Fermentation has been a major way of preserving food for thousands of years. Fermentation is a means by which microbial growth causes chemical and/or textural changes which form a product that can be stored for extended periods. The fermentation process is also used to create new food flavors and odors.

Dairy products

Yoghurt is produced by *S. thermophilus* and *L. bulgaricus*. With these organisms growing in concert, acid is produced by *Streptococcus* (see Topic B1), and aroma components are formed by *Lactobacillus*. Freshly prepared yoghurt contains 10^9 g^{-1} of bacteria. Cheese is another microbial fermentation product. All cheese results from a lactic acid fermentation of milk. The process requires the growth of a **starter culture** in milk which produces flavor changes and acid production, resulting in coagulation of milk solids and formation of a curd. **Renin**, an enzyme from calf stomachs (which can also be produced by genetically engineered microorganisms), is used to assist in curd formation. The curd is heated and pressed to remove the watery part of the milk or **whey**, salted, and then ripened. The cheese curd can be ripened with or without additional microorganisms. Cheese curd inoculation is used in the manufacture of Roquefort and blue cheese.

Meat and fish

A variety of meat products, especially sausage, can be fermented: country-cured hams, summer sausage, salami and izushi. *Pediococcus cerevisiae* and *L. plantarum* are most often involved in sausage fermentations. Izushi is based on the fermentation of fresh fish, rice and vegetables by *Lactobacillus*.

Food spoilage   Bacteria also play a significant role in food **spoilage** (detrimental changes to the food). Foods with easily utilizable carbohydrates, proteins and fats are ideal environments for spoilage by microorganisms. For example, milk undergoes a

Table 2.   *Major diseases and poisonings associated with food*

Bacterium	Disease	Foods involved
Food infections		
Enteropathogenic *E. coli* (0157)	*E. coli* enteritis	Cheese, meats
Listeria monocytogenes	Listeriosis	Dairy products
Salmonella typhimurium	Salmonellosis	Meats, poultry,
S. enteritidis		eggs, dairy products
Vibrio parahaemolyticus	*V. parahaemolyticus* gastro-enteritis	Seafood
Food poisoning (caused by toxin release)		
Clostridium perfringens	Perfringens food poisoning	Meats and poultry
C. botulinum	Botulism	Fish, meats and canned foods
Bacillus cereus	*B. cereus* food poisoning	Meats, rice, cereals, potatoes

series of steps which include increased lactic acid production by *S.lactis,* growth of more acid-tolerant organisms and finally protein digestion by bacteria that results in the formation of a putrid odor and bitter flavor. In canned foods, spoilage is typically caused by fermentative bacteria such as *Clostridia* which alter the texture and taste of the food as a result of protein degradation. The production of CO_2 and H_2S during spoilage causes cans to swell. More importantly the presence of bacterial pathogens in food presents a significant problem (*Table* 2 and Topic F3).

F3 BACTERIAL DISEASE – AN OVERVIEW

Key Notes

Pathogens	Infectious disease is the change in the structure or function of a host caused by a microorganism. Microorganisms that are able to cause disease in their host are described as pathogens. Virulence is a quantitative measurement of a pathogen's disease-causing capacity. The features of bacteria that confer the ability to infect a host and avoid the immune system are called virulence factors. The microbes that are normally found colonizing a host and do not cause damage are called commensals.
The nature of bacterial disease	Bacterial diseases may be chronic, such as tuberculosis, or acute, typified by Staphylococcal food poisoning. The symptoms of the disease depend on the site of infection, the ability of the organism to combat the immune system and the toxic products it produces. Some bacteria such as *Streptoccocus pyogenes* are capable of causing many different diseases form impetigo to rheumatic fever. Others may show a range of different symptoms during the course of a disease, such as *Treponema pallidum*, the causative agent of syphilis.
Koch's postulates	Koch's postulates are a number of criteria that have been used in the past to prove that a bacterium is responsible for a particular disease.

1. The bacteria should be found in all cases and at all sites of the disease.
2. The bacteria should be isolated from the infected person and maintained in pure culture.
3. The pure cultured microbe should cause symptoms of the disease on inoculation into a susceptible individual.
4. The bacteria should be reisolated from the intentionally infected host.

Related topics	Human host defense mechanisms (F4) Detrimental effects of protozoa:
	 Entry and colonization of human parasitic relationships (J7)
	   hosts (F5) Virus infection (K8)
	 Detrimental effects of fungi in their
	   environment (H7)

Pathogens
Although millions of microorganisms consisting of a wide range of species live in and on the human body, their presence is normally beneficial to the host and, in some cases, is even essential. These microbes are referred to as the **normal flora** or **commensals**. The healthy human body has a wide range of defense mechanisms, the immune system, which protects itself from microbes capable of causing damage. However, **pathogenic** bacteria, which are capable

of causing **disease** – that is, a change in the normal structure or function of the host's body – as well as infecting their host, do exist. This change to the host, manifested as a set of **symptoms,** may be due to the effect of microbial products such as toxins or the result of the host's immune reaction to the presence of the bacteria (Topic F4). Pain, fever, redness and swelling are common symptoms of bacterial disease. There are many definitions associated with the bacterial ability to cause disease, which are listed in *Table 1*.

Table 1. Some definitions used in describing pathogenicity

Pathogens	Bacteria that are capable of causing disease
Colonization	The establishment and multiplication of bacteria on a site in the host
Infection	The establishment and multiplication of a pathogen on a site in the host
Commensals or resident microflora	Bacteria that are normally found colonizing sites in the body
Asymptomatic carriage	Infection with disease-causing bacteria without the production of any symptoms
Systemic	Infection where bacteria are spread throughout the body
Sepsis	The presence of bacteria or their products in the blood or tissues
Septicemia	The multiplication of bacteria in the blood
Toxemia	Disease caused by the presence of a toxin in the blood

For a bacterium to be pathogenic it must be able to spread between hosts, invade and remain within the host, acquire nutrients, and avoid or damage the host immune system so that the bacterium can become established in the host. The features of bacteria that allow them to do this are called **virulence factors** (see Topic F5). Many virulence factors are carried on plasmids (Topic E5) or phage (Topic E7).

The nature of the host is an important factor in pathogenicity. Bacteria that are capable of infecting normal, healthy hosts may be considered as true pathogens but, if the host is damaged, other **opportunistic pathogens** may be able to infect and cause damage. *Pseudomonas aeruginosa,* for example, is unable to infect healthy skin but will infect and cause serious damage to burnt skin. People who are immunocompromised as a result of genetic defects, medical treatment or infection with certain viruses such as HIV are particularly susceptible to infection and may be infected by virtually any microbe.

The nature of bacterial disease

Bacillus anthracis was the first bacterium proven to be the causative agent for a disease, anthrax, in the 1870s. This work was done by Robert Koch. Subsequently, the **etiologies** (causes) of many common and sometimes fatal diseases were shown to be a result of bacterial infection. Bacterial diseases range from chronic infections such as tuberculosis caused by *Mycobacteria tuberculosis* and other related mycobacterial species which are capable of survival within macrophages in the lung (Topic F5), to acute infections such as Staphylococcal food poisoning which is caused by production of an enterotoxin that induces vomiting and diarrhea (Topic F6).

The type of disease caused by a bacterium depends on the site at which it infects (skin, respiratory tract, gastrointestinal tract, genito-urinary tract, tissues or blood stream), which normally reflects the mechanism by which it is transmitted (Topic F5), its ability to combat the immune system (Topic F4) and the toxic products that it produces. In many cases of bacterial disease such as diphtheria, an infection of the upper respiratory tract by *Corynebacteria diphtheriae*; tetanus, an infection of deep wounds where anerobic conditions prevail by *Clostridium tetani*; and cholera, an infection of the intestinal tract by *Vibrio cholera*, the symptoms are due to the production of very powerful toxins capable of causing extensive cell damage at distances from the site of infection. Other diseases such as gonorrhea, a sexually transmitted disease caused by *Neisseria gonorrhoeae*, and Legionnaires disease, a form of pneumonia caused by inhalation of an airborne bacterium, *Legionella pneumophila*, result from the ability of these microbes to survive within cells (Topic F5) and avoid the immune defense mechanisms. Normally, a number of different virulence mechanisms will be associated with the ability of a particular microbe to cause disease.

Some bacteria are capable of causing just one kind of disease such as cholera or whooping cough (caused by *Bordatella pertussis*). Other bacteria may cause a range of different diseases, depending on the site at which they act, the virulence factors they carry and the stage of the disease process. An example is *Streptococcus pyogenes*, which is a major cause of sore throats. However, if an erythrogenic exotoxin is produced by *S. pyogenes*, the disease caused is the more serious one of scarlet fever, which is characterized by a diffuse rash. In a small number of cases *S. pyogenes* infections may result in rheumatic fever, a creeping arthritis, which may also be associated with heart valve damage, or acute glomulerulonephritis (kidney damage). These diseases are thought to be the result of autoimmune responses due to the presence of antigens on the bacteria which are similar to host antigens. *S. pyogenes* is also capable of causing a skin disease, impetigo, and a number of new strains of the organism are able to cause a potentially fatal septicemia as a result of the toxin, streptococcal pyrogenic exotoxin. Another example of the variety of symptoms caused by one bacterium is shown by the spirochaetes, *Borrelia burgdorfei* (causes Lyme's disease, transmitted by deer ticks) and *Treponema pallidum* (syphilis). These have three stages of disease, if the original infection is not treated, reflecting the movement of the microbe in the body with time.

This section has attempted to describe just a few of the many bacterial diseases that are about. Others will be discussed in the context of virulence factors in later sections.

Koch's postulates For some diseases it is very easy to identify the causative agent and establish how it causes disease. The symptoms of cholera, for example, can be clearly explained by the presence of the organism *Vibrio cholera* in the gut and the production of a toxin which causes the secretion of copious amounts of water from the gut (Topic F6). For other diseases, the cause and effect relationship may not be as clear. For example, *Helicobacter pylori* has only recently been accepted as the cause of many gastric ulcers as it was believed for a long time that bacteria could not survive in the stomach.

Traditionally, the relationship between a bacterium and a disease has been established if the bacterium fulfills a number of criteria laid down by Robert Koch in the late 1800s:

1. The bacteria should be found in all cases and at all sites of the disease.
2. The bacteria should be isolated from the infected person and maintained in pure culture.

3. The pure cultured microbe should cause symptoms of the disease on inoculation into a susceptible individual.
4. The bacteria should be reisolated from the intentionally infected host.

In practice, in many cases it has proved difficult to fulfill all of these criteria.
Pure culture of bacteria in the laboratory has often been a problem. Although
we are fairly certain that *Mycobacterium leprae* is the cause of leprosy, it has yet
to be cultured in laboratory medium, and has so far been cultured only in an
animal model. In the case of an intestinal disease called Whipple's disease, the
bacterial agent responsible has only been discovered using modern polymerase
chain reaction (PCR)-based DNA detection methods. Criteria 3 and 4 of Koch's
postulates are also difficult to meet if no animal model for the disease is available or there is a lack of human volunteers.

A modern-day modification of these postulates is now being used to establish if a particular virulence factor is necessary for the production of an infection.
Potential virulence factors are first identified, then removed by mutation or
replaced by cloning and the microbe is then tested for virulence. In this way a
clear understanding of the disease process of a number of microorganisms has
been obtained.

F4 HUMAN HOST DEFENSE MECHANISMS

Key Notes

Overview	The host immune defense system consists of two components: a non-specific constitutive defense aimed at preventing infection and an inducible response which responds to the presence of a specific microbe or cellular component in the host.
Defense of surfaces	The body is protected by a number of defense mechanisms that are present at all times and act immediately on contact with any foreign body. These include physical barriers and mechanical removal, chemicals such as lysozyme and lactoferrin, the normal microbial flora on mucosal membrane surfaces, the protein complement system and phagocytosis.
Phagocytosis	Phagocytic polymorphonucleocytes (PMNs) and macrophages ingest and destroy foreign molecules or cells. Their activity is enhanced (opsonized) by the presence of complement proteins or antibodies on the surface of the microbe.
Complement	Complement is a set of proteins circulating in the blood stream which, on activation by the presence of microbes or antibody–antigen complexes, attract phagocytes, enhance phagocytic ability and form a membrane–attack complex (MAC), which can kill Gram-negative bacteria.
The inflammatory response	Infection results in a complex series of events, mediated by proteins, called the inflammatory response, which helps to prevent the spread of the microbes and aids in their destruction. This results in the classic symptoms of infection: pain, redness, swelling and fever.
Specific induced immune response	Antibodies (humoral immunity), produced by B lymphocytes, and receptors on T lymphocytes (cell-mediated immunity) recognize and bind to specific foreign molecules (antigens), resulting in a variety of responses depending on the nature of the target. On a first encounter, the response takes between 5 and 7 days to reach full potential but, there is also a memory component in this system, which allows a much more rapid response on a second exposure.
Antibodies	Antibodies are proteins called immunoglobulins consisting of two light and two heavy chains. One end of the molecule carries the antigen-binding site and at the opposite end is the Fc portion which mediates complement activation and enhances phagocytosis.
Cell-mediated immunity	T cells have a major role in the stimulation of B cells to produce antibodies and in the killing of host cells that have intracellular microorganisms.
Related topics	Bacterial disease – an overview (F3) Bacterial toxins and human disease (F6) Entry and colonization of human hosts (F5)

Overview

The function of the immune system of the host is, initially, to prevent invasion of the host by microorganisms (bacteria, fungi, protozoa and viruses). However, if infection occurs, the system must recognize the presence of a foreign invader and act to neutralize or eradicate it. There are two components in this system:

- a non-specific, constitutive set of defenses which act immediately against most microbes;
- a specific, inducible reaction, mediated by antibodies and T cells, which is initiated by the presence of a specific organism or cellular component (antigen).

Defense of surfaces

Bacteria can enter the body through any site that is open to the environment. All these sites are therefore protected by a number of non-specific defense mechanisms (*Table 1*). The main external barrier, the **skin**, is a dry, thick layer of dead, keratinized cells virtually impermeable to all microorganisms. Natural openings in the skin such as pores, hair follicles and sweat glands are protected by the secretion of toxic chemicals such as fatty acids and **lysozyme**. Generally, the only way bacteria can penetrate this surface is if it is damaged by wounds or burning. With the exception of the eye, all other surfaces to which bacteria can gain access, such as the gastrointestinal tract, the respiratory tract and the genito-urinary tract are covered in **mucosal membranes**. The cells in these membranes are protected by thick layers of **mucus**, a mixture of proteins and polysaccharides, which prevents penetration of bacteria into the cell surface. Ciliated cells are also present, which help in the removal of bacteria by the constant movement of the mucus. Similarly, all mucous membranes are constantly flushed by fluids which:

- prevent bacteria becoming established at one site by moving them through the system;
- contain antibacterial compounds which kill or inhibit bacterial growth.

Table 1. *Components of the non-specific immune response*

Defense	Site	Function
Mechanical barriers	Skin, mucosal membranes, mucin	Prevents penetration of bacteria
Flushing	Tears, urine, mucus, saliva, liquid movement through gut, coughing and sneezing	Removal of microorganisms and prevention of attachment
pH	Fatty acids on skin, stomach acidity, vaginal acidity	Prevention of bacterial growth
Lysozyme	Most sites and tissues	Breaks down bacterial peptido-glycan
Lactoperoxidase	Mucosal membranes and lysosomes	Kills bacteria by producing toxic superoxide radicals
Lactoferrin/transferrin	Mucosal surfaces, blood and tissues	Chelates iron essential for bacterial growth
Commensal microflora	Skin and most mucosal surfaces, exception the lungs	Competes for colonization sites Colicins may inhibit bacterial growth
Phagocytosis	PMNs in blood/tissue; macrophages in tissues and lungs	Engulfs bacteria into phagosomes in which they are destroyed
Complement proteins	Blood/tissue. Activated by presence of microorganisms or antigen–antibody interactions	Attracts phagocytes. Enhances phagocytosis (opsonization) Kills Gram-negative bacteria

Finally, a major part of the non-specific defense system of the host is the complex set of residential commensals present on the skin and on most mucosal membranes. These microorganisms compete with the invading pathogenic bacteria for nutrients and attachment sites. They may also contribute to the host defense against disease by the production of antibacterial compounds, such as **colicins.** A good example of the role of commensals has been shown in mature, premenopausal women where lactic acid produced by *Lactobacillus* species keeps the pH of the vagina around 5 inhibiting the growth of potential pathogens.

Phagocytosis

Circulating in the blood and in the tissues are cells called phagocytes that are capable of ingesting and killing bacteria. Two main types of cell are important in the control of bacterial disease: short-lived **polymorphonuclear leukocytes** (PMNs, neutrophils) found mainly in the bloodstream, and **macrophages** found in the tissues and in the lungs. Macrophages can live for weeks and have a number of roles in the immune system. They are first produced from stem-cells as monocytes which circulate in the bloodstream but once attracted to a site of infection or entry into a tissue they differentiate into macrophages. The steps of ingestion and killing by phagocytic cells are shown in *Fig. 1*. The phagocyte ingests the bacterium, by endocytosis, into a membrane-surrounded bag called a phagosome. This in turn fuses with lysosomes that contain a range of antibacterial compounds which kill and destroy the bacterium. Phagocytosis can be enhanced by the presence of antibody and complement (see later) on the surface of the bacterium. This is called **opsonization.**

As well as phagocytosis, macrophages release proteins called **cytokines** that stimulate the **inflammatory response** whose role is to prevent bacterial multi-plication and spread. They also interact with the specific immune system to stimulate antibody production and activate T cells.

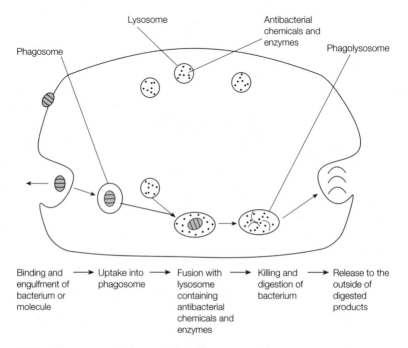

Fig. 1. The process of phagocytosis.

Complement

As well as lysozyme, iron-chelating agents and other antibacterial agents (*Table 1*), serum contains a number of proteins that make up the **complement** system. These proteins on their own are not toxic but are activated by the presence of components of microorganisms in the host. There are two pathways of activation: the classical pathway which is initiated by **antibody–antigen complexes** and the alternate pathway which is triggered by **microbial surface molecules** such as cell walls. Activation causes a cascade of proteolytic reactions which leads to the production of a number of products that activate the immune system:

- C3b binds to bacteria enhancing their phagocytosis;
- C3a and C5a trigger the inflammatory response and activate phagocytes;
- a membrane attack complex which forms pores in the membranes of Gram-negative bacteria, leading to cell death.

The inflammatory response

The infiltration of phagocytes to the site of infection, molecules released by macrophages, the complement cascade and tissue damage result in a complicated series of events that leads to the classic symptoms of **inflammation**: heat, redness, swelling and pain. The function of this is to prevent the multiplication and spread of the microbes and aid in their eventual destruction.

Specific induced immune response

In the presence of a specific foreign molecule, which might be a protein or a carbohydrate, as a single molecule or attached to a cell, the host can induce a highly efficient defense response. The immune response may be in the form of **antibodies**, produced by B lymphocytes, sometimes referred to as humoral immunity, or cell-mediated by T lymphocytes. Antibodies and the receptors on the surface of T cells recognize and bind to specific foreign molecules called **antigens**, resulting in a variety of responses depending on the nature of the target. This type of immune response takes between 5 and 7 days to reach full potential on a first encounter with a pathogen. However, an important feature of this response is the production of a **memory** component, which allows a much more rapid response on second exposure to the antigen.

Antibodies

Antibodies are proteins called immunoglobulins (Igs), made up of two heavy and two light chains, which create two functionally important areas, as shown in *Fig. 2*: two sites that recognize and bind the antigen (Fab) and the Fc portion at the other end of the molecule which mediates the function of the antibody. The result of a specific antibody–antigen interaction depends on the nature of the antigen and antibody and may include:

- enhanced phagocytosis;
- complement activation;
- neutralization of toxins;
- inactivation of proteins;
- inhibition of binding of toxins or bacteria to surfaces.

However, the final aim is to neutralize and remove the foreign molecule or infective agent from the host. There are several types of antibody molecule, which differ in structure and function. The normal circulating antibodies responsible for the immune response to bacteria are IgG and IgM. The antibody that protects mucosal membranes is secretory IgA.

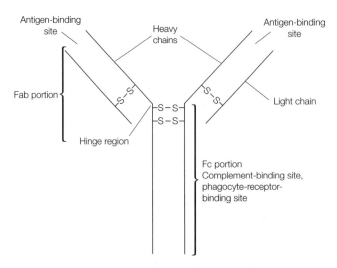

Fig. 2.   Basic structure of an antibody: immunoglobulin G.

Cell-mediated immunity

T cells carry antibody-like receptors on their surface and are induced to function by the presence of a foreign antigen in the blood. T cells play several roles in the specific immune defense system, depending on the nature of the antigen involved in the initial activation of the response:

- T helper cells are important for the stimulation of B cells to produce antibodies in response to protein antigens.
- Cytotoxic T cells kill cells that display foreign antigens, such as cells infected with *Mycobacterium tuberculosis*.
- Suppressor T cells are involved in down-regulating the immune response.

F5 ENTRY AND COLONIZATION OF HUMAN HOSTS

Key Notes

Overview

In order to set up an infection a bacterium must reach a site at which it can survive, accumulate nutrients to allow it to replicate, and overcome the host immune defense mechanisms.

Transmission between hosts

Bacteria are transmitted by aerosols, in water or food, by direct contact or via animal vectors. The sites in the host that they reach depend on the transfer mechanism and the microbe itself. Some microbes may remain on mucosal surfaces, others penetrate through into the tissues and others are injected directly into the tissues or bloodstream.

Colonization of surfaces

Adhesin molecules on the surface of cells bind to cell surfaces and prevent the mechanical flushing of the bacteria from the host. Adhesins may be polysaccharide or protein, pili or capsules. Motility and mucinases, which break down mucous, also assist in colonization.

Invasion of cells

Bacterial proteins called invasins induce non-phagocytic cells to take up bacteria. Some bacteria use this mechanism to penetrate through epithelial layers and to spread within the host.

Acquisition of nutrients

In order to survive within the host, bacteria must be able to acquire nutrients. Toxins that lyse cells and extracellular enzymes assist the bacteria to do this. Iron-chelating proteins are also important virulence factors as they allow the bacteria to compete with the host for iron.

Spread of bacteria

Extracellular enzymes and toxins which destroy host tissue allow the dissemination of the microbe in the host.

Avoidance of host defense mechanisms

Many mechanisms contribute to the ability of bacterial pathogens to withstand the onslaught of the host defense systems. Some particularly successful strategies include: destruction of phagocytes and complement; survival of bacteria within phagocytes; disguises that prevent the host from recognizing the bacteria as foreign.

Related topics

Bacterial cell structure (D2)
Bacterial cell envelope (D3)

Human host defense mechanisms (F4)
Bacterial toxins and human disease (F6)

Overview

What has a bacterium to do to cause infection (*Fig. 1*)? It must reach an appropriate site in the host at which it can survive and multiply. To do this it must overcome the host immune defense mechanisms as described in Topic F4. The properties or virulence factors that a bacterium requires to infect therefore depend on two factors:

● the mechanism by which the bacteria enter the host;
● the nature of the site of colonization, whether it is a surface such as the mucosal membranes or within the tissues of the host.

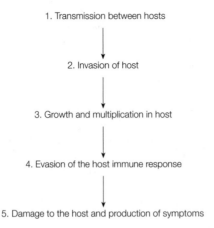

1. Transmission between hosts

2. Invasion of host

3. Growth and multiplication in host

4. Evasion of the host immune response

5. Damage to the host and production of symptoms

Fig. 1.   Stages in the disease process.

Bacteria that survive on mucosal surfaces, such as *Vibrio cholerae* in the gut, will encounter a different range of host defense mechanisms to bacteria, like *Yersinia pestis*, that are capable of survival in the bloodstream or tissues of the host.

Transmission between hosts

Organisms may be transmitted between hosts in a number of ways, as shown in *Table 1*. The route of transfer depends on the nature of the bacteria. Bacteria that can survive in the environment may be spread in food, water and air whereas others, such as *Neisseria gonorrhoeae*, cannot survive outside the human host and must be spread by direct contact. The nature of transmission also dictates the final site of infection: bacteria that are spread in the air or aerosols will gain access to the naso-pharynx and the lungs; those spread by food or water will gain access to the gut. Many bacteria will remain on the mucosal membranes such as *V. cholerae* and may only cause damage by the production of extracellular toxins or enzymes (Topic F6). Other bacteria such as *Salmonella typhi* enter into the host tissues by penetrating through the mucous membranes. Some bacteria enter directly into the host bloodstream, often through animal bites, such as *Y. pestis* carried by the rat flea and, finally, virtually any bacteria may enter the host through wounds and burns.

Colonization of surfaces

In order to survive in a host, the immediate priority for most bacteria is to stay there. Nearly all surfaces in the body are protected by some form of flushing movement that removes any invading bacteria that cannot stick to the surface of the epithelial cells. Most pathogenic bacteria carry adhesins on the surface of their cells to allow the attachment of the bacterium to host cell surfaces.

Table 1. Some examples of how bacteria are transmitted between hosts

Nature of transmission	Organism	Disease	Site of infection
Air/aerosols	Corynebacteria diphtheriae	Diphtheria	Throat but toxin passes into bloodstream
	Legionella pneumophila	Legionnaire's disease	Alveolar macrophages in the lung
	Bordatella pertussis	Whooping cough	Ciliated epithelial cells in the upper respiratory tract
Food/water-borne	Vibrio cholera	Cholera	Gut
	Salmonella typhi	Typhoid	Gut → invasion through mucosa → systemic infection
	Shigella dysenteriae	Dysentry	Gut→ invasion of mucosa
Direct contact	Neisseria gonorrhoeae	Gonorrhea	Genito-urinary tract → invasion of epithelial cells → systemic infection (1% of cases)
	Mycobacteria leprae	Leprosy	Skin
Vector-borne	Borrelia sp. (vector – deer tick)	Lyme's disease	Local to bite → systemic → neurons
	Yersinia pestis	Bubonic plague	Blood and lymph glands

Many adhesins are proteins and are often in the form of pili (fimbriae), long cylindrical-shaped structures which stick up from the surface of the cell (Fig. 2; see also Topic D2). The structure of the pili is thought to help overcome the electrostatic repulsion resulting from the negative charge on both types of cell. However, membrane proteins and other surface proteins can also act as adhesins (Table 2), sometimes in the absence of pili. The receptors to which the adhesins bind on the surface of the host cells are often specific glycoproteins or glycolipids which may not be present in all cell types. For this reason, microbes infecting different animals or different sites may require different adhesins. A good example of this are the enterotoxigenic strains of E. coli, which cause a diarrheal disease by the production of toxins that induce secretion of fluid from the epithelial cells (Topic F6) after colonization of the gut. E. coli strains pathogenic to pigs carry a piliated adhesin, K88, that binds to pig epithelium but not human. Disease in humans is caused by strains that carry different types of adhesin such as CFA/I.

Exopolysaccharides may also act as adhesins, an important example being those produced by Streptococcus mutans in the formation of dental plaque. Exopolysaccharides have also been recently implicated in the formation of microcolonies of Pseudomonas aeruginosa in the lungs of cystic fibrosis patients.

In addition to the problem of adhesion, most surfaces are covered with mucous so bacteria must first penetrate this to reach the epithelial cell surface. A number of virulence factors may contribute to the ability of bacteria to overcome this barrier (Table 3):

- motility caused by the presence of flagella will help the bacteria to move through the mucous;
- mucinases break down mucous.

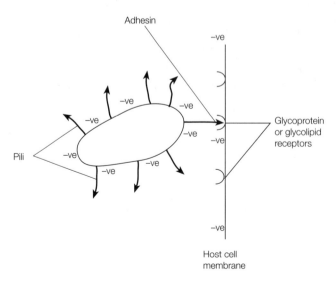

Fig. 2.   The role of pili in adherence to cell membranes.

Table 2.   Examples of the nature of some adhesins and the sites to which they bind

Adhesin	Bacterium	Binding site
K88 pili	*E. coli*	Pig intestinal mucosa
CFA/I, CFA/II pili	*E. coli*	Human intestinal mucosa
PAP pili	*E. coli*	Human urinary tract
Outer membrane protein II	*N. gonorrhoeae*	Urogenital tract
Protein F surface protein	*Strep. pyogenes*	Fibronectin
Exopolysaccharides	*Strep. mutans*	Teeth

Table 3.   Virulence factors that assist the bacterium to colonize a host

Virulence factors	Function
Flagella and chemotaxis	Movement to an appropriate site
Adhesins	Attachment to cell surfaces; prevents bacteria from being washed out
Mucinases	Break down mucus, help bacteria gain access to cell surface
Ciliostatic proteins	Inhibit action of cilia, preventing removal of bacteria from surface
Enzymes such as DNases, hyaluronidases, proteases, collagenases	Nutrient acquisition; dissemination of the microbe
Toxins	Nutrient acquisition; dissemination of the microbe; destruction of cells of the immune system
Siderophores	Iron scavenging
Invasins	Trigger non-phagocytic cells to take up bacteria

Finally, ciliostatic proteins may inhibit the movement of cilia on the surface of cells in the naso-pharynx, therefore reducing movement.

Invasion of cells   Some bacteria such as *Shigella dysenteriae, Listeria monocytogenes, Y. entercolicitca* and *S. typhi* can penetrate through the epithelial layer to set up either localized or systemic infections. Bacteria may penetrate through specialist M cells in the gut. These are naturally phagocytic cells which have a role in the gut-associated immune system. Other microorganisms that penetrate through the epithelium have the ability to induce naturally non-phagocytic cells to become phagocytic. Proteins produced by the bacteria called **invasins** bind to receptors in the host cell surface, **integrins**, and induce rearrangements of actin molecules, part of the cell cytoskeleton, under the membrane. The bacteria are then engulfed into the cells by an invagination of the host cell membrane and may pass through the cell in these membrane vesicles as is the case with *S. typhi*, or, similar to *Sh. dysenteriae*, escape from the vesicle into cytoplasm. The nature of the subsequent disease depends on other virulence factors present on the bacteria. *Sh. dysentriae* spreads to adjacent cells and causes a localized infection with extensive tissue damage due to multiplication of the bacteria in the cells and to the production of a toxin (Topic F6). In contrast, *S. typhi*, which causes typhoid fever, enters the bloodstream and spreads throughout the body. The disease symptoms of high fever and anorexia are as a result of the presence of the bacteria in the blood and the effect of LPS on the immune system (Topic F6).

Acquisition of   In order to replicate, bacteria must be able to acquire nutrients from the host.
nutrients   Bacteria produce many toxins (Topic F6) and enzymes that lyse cells, causing varying amounts of damage ranging from local tissue damage to the death of the host. There is much debate as to the role of these compounds and why they are produced by bacteria, but it may be simply that they are virulence factors which allow the release of nutrients from the cell. Another group of virulence factors that contribute to bacterial growth are siderophores, iron-chelating agents, which can remove iron from the transferrin and lactoferrin (*Table 3*). Iron is essential for bacterial growth, and mechanisms that keep the free concentration of iron below the level necessary for bacterial growth are an important aspect of the host defense system (Topic F4).

Spread of bacteria   The spread of bacteria from the sites of infection is aided by the production of a large number of extracellular enzymes, some of which are listed in *Table 3*. Toxins and enzymes that break down cells (Topic F6) may also play a role in the dissemination of the organism.

Avoidance of host   A large number of virulence factors have been found that help the bacteria to
defense   avoid the host defense mechanisms. A list of different types is given in *Table 4*
mechanisms   but this is by no means all of them. As the initial line of defense for the host is to destroy the bacteria by the action of phagocytes, potentiated by components of the complement system (Topic F4), avoidance of phagocytosis and complement killing are important features of successful pathogens. Phagocytosis may be avoided by killing the phagocyte (cytolytic toxins), inhibiting their movement (toxins that inhibit cellular function) or inhibiting phagocytosis (capsules). One of the most interesting strategies that has evolved in a number of bacteria is the ability to survive within the phagocytes. This can lead to serious problems as

Table 4. Examples of virulence factors that protect bacteria from the host's immune response

Target	Virulence factors	Function
Phagocytes	Cytolytic toxins	Destruction of phagocytic cells
	Toxins that inhibit cellular function	Inhibition of chemotaxis/mobilization of phagocytes
	Capsules and cell surface proteins such as Protein M of *Strep. pyogenes*	Inhibition of phagocytosis
	Mycobacterial cell-envelope sulfatides	Prevention of phagosome–lysosome fusion
	Listeriolysin O (*L. monocytogenes*)	Breaks down the phagosome membrane allowing the bacteria to escape into the cytoplasm
	Catalase, superoxide dismutase, cell-envelope components	Protection of cell from intracellular killing by phagocytes
Complement	Proteases	Destruction of components of complement
	Capsules and long LPS side chains	Prevent activation of complement or hinder access of complement to surface
Antibodies	IgA proteases	Inactivation of IgA on mucosal surfaces
	Protein A (*Staph. aureus*)	Binding of Fc portion of IgG disguises bacterium by a coat of IgG
	Capsules containing sialic acid	Sialic acid is a host polysaccharide therefore not recognized as foreign
	Antigenic variation	Evasion of the antibody response

bacteria such as *Mycobacterium tuberculosis* and *L. monocytogenes* can persist for a long time in the host without being eradicated. Three different mechanisms can help the bacteria in this survival:

● prevention of fusion of the phagosome with the lysosome;
● escape from the phagosome into the cytoplasm before fusion with the lysosome;
● production by the bacteria of enzymes and other proteins that overcome the effect of the intracellular killing components of the phagosome.

Bacteria interact with complement in a number of ways in order to destroy or minimize its antibacterial activity:

● proteases may destroy the individual components of the proteolytic cascade;
● potential complement-activating components may be hidden by the presence of capsules or long O-antigen side chains on LPS (Topic D3);
● bacteria shed molecules that bind to complement, therefore acting as decoys.

Bacteria have evolved a number of strategies to avoid the effect of the host's specific immune response. Bacteria such as *N. gonorrhoeae* that survive on mucosal surfaces often produce specific proteases which destroy IgA, the secretory immunoglobulin. Other bacteria try to prevent the induction of the specific immune reaction by looking like part of the host. They can do this by either producing polysaccharide capsules that resemble host polysaccharides such as sialic acid (*E. coli*) or by coating themselves with host proteins such as fibronectin. Protein A produced by *Staphylococcus aureus* is a very interesting example of this phenomenon in that it binds to the Fc portion of immunoglobulins (Topic F4) therefore creating a coat of antibodies on its surface but with the antigen-binding part sticking up. This is seen as normal host protein by the host defense system and, therefore, acts as a disguise. Finally, antigenic variation is a very neat way of avoiding the specific immune response. In this case, the bacteria change the antigenicity of some component of their cell surface such as pili or outer membrane proteins. This means that the host's specific antibody response, which is against a component that is no longer there, is ineffective so the bacteria survive.

F6 BACTERIAL TOXINS AND HUMAN DISEASE

Key Notes

Endotoxins	Endotoxin is the Lipid A portion of lipopolysaccharide (LPS) found on the surface of Gram-negative bacteria. It causes septic shock by induction of the host immune system when present in the bloodstream.
Exotoxins	The word exotoxin may be used to describe any extracellular protein produced by bacteria which induces a toxic effect on host cells. Exotoxins may be split into three groups based on their sites of action: membrane-damaging toxins, toxins that act on sites inside the cell, and superantigens, toxins that over-induce the immune system.
Membrane-damaging toxins	Damage to the host cell membranes may be as a result of enzyme activity that destroys components of the cell membrane, such as sphingomyelinases and phospholipases, or by holes made in the membrane by pore-forming proteins such as *Staphylococcus aureus* α-toxin.
Intracellularly acting toxins	Some of the most toxic compounds known to man such as botulinum toxin and diphtheria toxin are bi-partite toxins consisting of two components: a binding (B) part that binds to receptors on the host cell surface and an active (A) component that is the actual toxic part. Once inside the cell many of the known bi-partite toxins act by ADP-ribosylating a molecule in the host cell; for example, the eukaryotic elongation factor EF-2 (diphtheria toxin) and the Gs protein, which regulates the activity of adenylate cyclase in eukaryotic membranes (cholera toxin). Other activities associated with toxins include ribonuclease (Shiga toxin) and metallopeptidases (botulinal and tetanus toxin).
Related topics	Bacterial cell envelope (D3) Entry and colonization of human Human host defense mechanisms (F4) hosts (F5)

Endotoxins

Traditionally, bacterial toxins have been split into two groups on the basis of whether they are associated with the bacterial cell, **endotoxins**, or released into the medium by growing bacteria, **exotoxins**. Endotoxin activity is due to the lipid A portion of LPS, which is an integral part of the outer membrane of Gram-negative bacteria (Topic D3). The endotoxin is normally released from Gram-negative cells in the bloodstream after lysis as a result of the action of the host immune system or by the action of some antibiotics. Once released, the LPS activates the complement cascade by the alternative pathway (Topic F4) and induces the release of cytokines and other defense mediators from macrophages and other cells. These bioactive compounds are a normal part of the host defense mechanisms and help to rid the body of infection, but when induced by LPS, the amounts of these are much higher than normal and cause toxicity to the host itself. This is called **septic** shock which has the symptoms of fever, circulatory collapse and organ failure, and eventually death. Components of Gram-positive cells, including lipoteichoic acids and peptido-glycan, can also induce septic shock.

Exotoxins

Traditionally, the word exotoxin has been used to describe bacterial proteins produced by both Gram-negative and Gram-positive bacteria that are toxic to host cells. Generally, the word exotoxin is shortened to toxin, when describing bacterial proteins, as not all exotoxins are necessarily secreted into the surrounding medium but may be held on the surface of the cell or in the periplasmic space. The exotoxins are generally heat labile; although there are a number of heat-stable toxins such as the *E. coli* ST enterotoxin. They are water soluble, so can act at a site distant from the original site of infection. In some cases, such as botulism, there is no need for colonization of the host by the bacterium at all, the toxin alone is enough to cause the disease (toxemia).

The nomenclature of toxins is varied:

- Some toxins, especially those that are the clear cause of disease symptoms, are named after the bacteria that produces them such as cholera toxin (produced by *Vibrio cholerae*) or diphtheria toxin (*Corynebacterium diphtheriae*).
- Others are named after their sites of action such as neurotoxins (acting on nerve cells), enterotoxins (acting on the gut) or luekotoxins (active against white blood cells). Toxins that are active against a number of cell types may just be described as cytotoxins.
- Some toxins are just given a letter or number such as *Pseudomonas aeruginosa* toxin A or *Staphylococcus aureus* α-toxin.
- Toxins may be named on the basis of their activities such as phospholipase C (produced by *Listeria monocytogenes*) and adenylate cyclase produced by *Bordatella pertussis*.

Some bacteria, especially the Gram-positives, may produce a number of different toxins, others may only produce one. Toxin production is generally very carefully controlled at the level of transcription in bacteria (Topic C5) and, in some cases, only a subset of bacteria within a strain will actually secrete a toxin. Most toxins are either membrane-damaging toxins or toxins that act inside the cell. A small third group, called **superantigens,** interact with T-helper cells, causing over-induction of the immune system and symptoms similar to shock.

The role of toxins in the disease process is a matter of some debate. However, as discussed in Topic F5 they may play a number of roles including:

- the release of nutrients from cells;
- destruction of cells associated with the immune defense;
- escape from the phagolysosome by intracellular bacteria.

Membrane-damaging toxins

These cytolytic toxins cause lysis by destroying the integrity of the plasma membranes of the host cell. Usually they are active against a wide range of cell types and can be divided into two groups on the basis of their mechanisms of action: enzymic action and pore forming. A large number of different Gram-positive bacteria produce enzymes such as phospholipases and sphingomyelinases, which break down the lipid components on the membrane. This destabilizes the membrane, eventually leading to cell lysis. The second group of toxins includes the thiol-activated cytolysins, streptolysin O (produced by *Streptococcus pyogenes*) and listeriolysin O (produced by *L. monocytogenes*) and *Staph. aureus* α-toxin. These toxins form pores in the host cell membranes which results in the leakage of nutrients and essential ions from the cell and eventually cell lysis.

Intracellularly acting toxins

The most potent group of toxins known to man are the bacterial exotoxins that act on intracellular targets. These include proteins such as botulinum toxin which inhibits the release of neurotransmitters in the peripheral nervous system, cholera toxin that acts in the gut to induce watery diarrhea and diphtheria toxin which inhibits protein synthesis (*Table 1*). Although these toxins cause a very disparate set of symptoms at different sites in the body they do tend to have a common **bipartite** structure. One part of the molecule is responsible for binding to specific receptors in the host cell membrane, called the B-portion while a separate part of the molecule, the A-portion carries the toxic enzymic activity. These toxins are, therefore, frequently called **A–B type toxins**. The structure of two of these toxins is shown in *Fig. 1*. Diphtheria toxin is one of the simplest of the A–B toxins consisting of just one A and one B portion. These are synthesized together as a single molecule, then cleaved to give two peptides

Table 1. *Some examples of bacterial exotoxins that act inside the cell*

Bacteria	Toxin	Site of action	Mode of action	Symptom/role in disease
Vibrio cholerae *E. coli*	Cholera toxin LT toxin	Intestine	ADP-ribosylation of Gs (stimulatory) protein which regulates adenylate cyclase	Water secretion into the intestine resulting in watery diarrhea
Corynebacterium diphtheriae *Pseudomonas aeruginosa*	Diphtheria toxin Exotoxin A	Many cell types	ADP-ribosylation of EF-2 leads to inhibition of protein synthesis	Cell death, general organ damage
Bordatella pertussis	Pertussis toxin	Many cell types	ADP-ribosylation of Gi (inhibitory) protein which regulates adenylate cyclase	Adherence to cells, cell damage, fluid secretion
Shigella dysenteriae *E. coli*	Shiga toxin Vero (Shiga-like) toxin	Many cell types	RNA glycosidase enzyme modifies 28S rRNA in 60S ribosome subunit. Inhibits protein synthesis	Cell death
Clostridum tetani	Tetanus toxin	Neurons in central nervous system	Metallopeptidase – inhibits release of neurotransmitters	Spastic paralysis
Cl. botulinum	Botulinal toxin	Peripheral neurons	Metallopeptidase – inhibits release of neurotransmitters	Flaccid paralysis

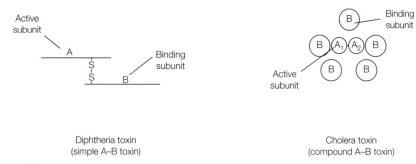

Fig. 1.   Structure of A – B type exotoxins.

joined by a disulfide bridge. In contrast, cholera toxin consists of five B subunits surrounding an active A_1–A_2 molecule, which is synthesized separately.

The B portion is responsible for the cell specificity of most of these toxins. It will bind only to cells that have the correct receptor. Hence, in the case of botulinum toxin, the receptor glycoprotein or glycolipid is only found in the peripheral neurones whereas diphtheria toxin binds to growth factor precursor found on many cell types. After binding to a receptor, the toxin–receptor complex is internalized in the cell either by endocytosis, followed by translocation of the A portion from the endocytic vacuole into the cytoplasm, or by translocation of the A portion across the host plasma membrane, directly into the cytoplasm.

In many cases the enzymic activity of the A–B type toxins is ADP-ribosylation of a target site in the cell. The A-portion catalyzes the removal of the ADP-ribosyl group from NAD and the attachment of that group to a cellular protein. The target proteins and the consequences of the ADP-ribosylation are various, as shown in *Table 1*, and the varied nature of these explains why toxins cause a wide range of different symptoms. Diphtheria toxin, for example, inhibits protein synthesis, with one molecule of the toxin enough to kill the cell. In contrast, cholera toxin modifies Gs protein, a regulatory protein of adenylate cyclase, causing it to be permanently switched to the production of cAMP. In the gastrointestinal tract, where the toxin acts, the most significant effect of the rise in cAMP is the production of an ion imbalance, leading to massive water loss from the cell. This is seen as copious watery diarrhea, causing dehydration and death, if not treated.

Other enzymic activities are also associated with the A portion of A–B type toxins. Shiga toxin, produced by *Shigella dysenteriae*, is a ribonuclease which inactivates the 60S subunit of the ribosome. The neurotoxins, botulinum toxin, which causes flaccid paralysis, and tetanus toxin, which induces spastic paralysis, are both zinc-requiring endopeptidases. The different symptoms caused by each toxin are probably as a result of the different cell specificities of the toxins. Botulinum toxin acts on the peripheral nervous system whereas tetanus toxin interferes primarily with the central nervous system.

F7 CONTROL OF BACTERIAL INFECTION

Key Notes

Disease in the population	The most effective tools in the control of bacterial diseases have been public health policies, which have removed pathogenic bacteria from water and food and reduced their populations in the environment. Similarly, better housing and nutrition has raised the health of individuals, thus making them more resistant to infection.
Vaccination	Whole microbes (live, attenuated or killed), fractions of microbes or inactivated microbial products, such as toxoids, may be used to induce a specific immune response in a host. This provides protection from infection with that microbe in the long term. Vaccines are not available for all bacterial infections.
Antibiotics	Antibiotics are molecules, produced by microbes, which inhibit the growth, or kill, other microbes. The antibiotics normally used to treat bacterial infections are those that are selectively toxic to bacterial cells and do not harm the host. These antibiotics target sites such as peptidoglycan, ribosomes and nucleic acid synthesis which are significantly different between prokaryotic and eukaryotic cells. The effectiveness of antibiotic therapy is endangered by the development of bacterial antibiotic resistance and, particularly, by the spread of antibiotic-resistance genes between bacteria.
Related topics	Translation (C7) Human host defense mechanisms Bacterial cell envelope (D3) (F4) Bacterial cell-wall synthesis (D4) Virus vaccines (K10) Plasmids (E5) Antiviral chemotherapy (K11)

Disease in the population

Deaths from bacterial diseases have become a rare occurrence in the developed world in the last 50 years as a result of a number of measures designed to reduce the spread of (and even eradicate) pathogenic bacteria in our environment and to improve the health of the population. These are shown in *Table 1*. The importance of sanitation, good housing and nutrition in reducing the incidence of bacterial disease cannot be emphasized enough. Even before the introduction of vaccines and antibiotics, the reduction of diseases such as tuberculosis had occurred owing to public health measures which prevented the spread of the causative microorganism, *Mycobacterium tuberculosis*, between individuals. In the developing countries, where there is a lack of sanitation and a large incidence of malnutrition, bacterial diseases are still a major cause of death.

Table 1. Strategies used to control bacterial infections

Prevention of the spread of bacteria within the population	Improvements in water treatment and sewage disposal prevents the spread of water-borne diseases such as cholera
	Introduction of food hygiene regulations stops the spread of food-borne diseases such as typhoid
	Reduction in overcrowding prevents the spread of aerosol-borne diseases such as tuberculosis
	Control of the vector (e.g. rats or mosquitos) prevents the spread of animal-borne diseases
	Vaccination reduces the incidence of the organism in the environment
	Education improves the general standard of hygiene in the population
Improvement in the health of the individual	Better nutrition improves an individual's resistance to disease
	Vaccination protects an individual against infection with specific bacteria
	Antibiotics can cure an individual of a bacterial infection

Vaccination

Vaccination has proved to be a useful way of protecting the individual, and the population, from a number of bacterial diseases. Vaccination, or active immunization, is the artificial introduction of antigens from a microbe into an individual, in a controlled way, leading to the stimulation of the immune system without the symptoms of the full-blown disease. This leads to the production of memory cells within the host (see Topic F4) so that on a second encounter with the microbe the immune system can generate a rapid antibody response thereby preventing infection. Vaccines may be whole cells or cellular fractions (e.g. cell-surface components such as capsule polysaccharides) or inactivated bacterial toxins (toxoids). Both live and dead bacteria are used in vaccines. Live vaccines are more useful as they mimic more effectively the natural disease process, causing long-lasting immunity. However, the bacteria must be attenuated in some way so that they do not cause disease in the host. Examples of some of the bacterial vaccines used today are given in *Table 2*. Although vaccines have been particularly useful in the control of diseases such as diphtheria, tetanus (both are toxoid vaccines) and tuberculosis (a live, attenuated vaccine), there are many bacteria for which it has proven difficult to produce a safe, effective vaccine. Finally, passive immunization, the introduction of antibodies to a particular microbe, is used occasionally to protect against infection.

Antibiotics

Molecules that inhibit the growth of or kill bacteria are called antibacterial agents. If these compounds are isolated from microbes they are called **antibiotics**. There are many millions of antibiotics, produced mainly by soil bacteria and fungi, that are active against bacteria but only a few can be used to control human bacterial disease. These few have the property of being toxic to bacteria but have no significant effect on the human host. The reason for this so-called **selective toxicity** is that the sites at which these antibiotics act are either unique to bacteria such as peptidoglycans (Topics D3 and D4) or very different between prokaryotes and eukaryotes such as ribosomes (Topic C7) and nucleic acid

Table 2.   *Examples of vaccines available for protection against bacterial diseases*

Disease	Vaccine components
Diphtheria	Inactivated toxin (toxoid)
Tetanus	Inactivated toxin (toxoid)
Tuberculosis	Attenuated *Mycobacterium bovis* (BCG)
Whooping cough	Subcellular fractions and pertussis toxoid
Haemophilus influenzae meningitis	Capsular polysaccharide – linked to a protein carrier
Meningococcal meningitis	Capsular polysaccharides
Typhoid	Killed cells of *Salmonella typhi*
	Live oral attenuated strain Ty21A
	Polysaccharides

synthesis (Topics C2 and C4). *Table 3* shows the sites in the bacterial cell at which a number of clinically important antibiotics act.

Some of the factors that affect antibiotic therapy are listed below:

● The antibiotic must reach the site of infection in the host.
● The antibiotic has to reach its target site in the cell. This is easier for antibiotics, such as penicillin, which act on peptidoglycan than for those, like tetracycline, which must penetrate through the plasma membrane to reach their target sites, the ribosomes.
● Gram-negative bacteria are often intrinsically resistant to the action of antibiotics due to the presence of the outer membrane which acts as an additional barrier for the antibiotic to cross and protects the peptidoglycan (Topics D2 and D3).
● Broad-spectrum antibiotics are effective against a wide range of different Gram-positive and Gram-negative bacteria whereas other antibiotics may have only a narrow range.
● All the pathogenic bacteria must be eradicated from the host by either inhibiting the growth of the microbes (**bacteriostatic** antibiotics), which can then be removed by the immune systems, or by killing them directly (**bactericidal** antibiotics).

Table 3.   *Target sites for antibiotics in bacterial cells*

Target site	Mode of action	Example
Cell wall (peptidoglycan biosynthesis,Topic D4)	Inhibition of cross-linking	Penicillins and cephalosporins (β-lactam antibiotics)
	Inhibition of polymerization	Glycopeptide antibiotics (e.g. vancomycin)
Protein synthesis (Topic C7)	Inhibition of translocation of ribosome	Aminoglycoside antibiotics (e.g. Gentamicin)
	Inhibition of binding of aminoacyl tRNAs	Tetracycline
Nucleic acid synthesis (Topic C2)	Inhibition of tetrahydrofolic acid synthesis	Sulfonamides and trimethoprim
	Inhibition of DNA gyrase	Quinolone antibiotics (e.g. ciprofloxacin)

Antibiotics have proved to be of great benefit to humankind and have ensured that people no longer need die from diseases such as streptococcal scarlet fever or as a result of wound infections. However, there is a problem. Bacteria can become resistant to the action of antibiotics. The mechanisms by which they do this include the production of enzymes that break down the antibiotic, reduction in permeability to the antibiotic and alterations to the target site as shown in *Table 4*. Antibiotic resistance may arise by mutation (Topic E1) but more often the genes for antibiotic resistance are transferred between bacteria by conjugation, transduction and transformation (Topics E6, E9 and E10). Antibiotic resistance carried by plasmids has caused particular concern as these plasmids may carry genes that confer resistance to many different antibiotics at the same time (Topic E5). Multiply-resistant bacteria are therefore becoming a problem, particularly in the hospital environment, and the fear is that it will not be long before there is a bacterial strain that is untreatable by all known antibiotics.

Table 4.   Common mechanisms of antibiotic resistance in bacteria

Mechanism of resistance	Examples	
Antibiotic inactivation	β-Lactamase	Penicillin resistance
	Chloramphenicol acetyl transferase (CAT)	Chloramphenicol resistance
	Aminoglycoside modifying enzymes	Aminoglycoside resistance
Reduction in permeability	Reduced uptake	Natural resistance of many Gram-negative bacteria due to presence of outer membrane
	Antibiotic efflux	Tetracycline resistance
Alteration of target site	Change in target site so that it is no longer sensitive to drug	Sulfonamide resistance
	New target site produced that is not sensitive to the antibiotic	Methicillin resistance
	Overproduction of target site	Trimethoprim resistance

G1 TAXONOMY

Key Notes

Current taxonomic status	The objective of current taxonomic schemes is to create monophyletic groups of microorganisms which are assumed to have a single ancestor. This is achieved by studying characters of an individual and comparing them with those of other, closely related, organisms. Using such information phylogenetic trees can be constructed that can indicate probable evolutionary sequences of protistan microbes.
Amitochondrial protists	At the base of the phylogenetic tree are the protists that do not have mitochondria. These are anaerobic organisms that may inhabit sediments or animal intestines or they can be animal parasites. It is from this group that other lineages of microorganisms have evolved.
Mitochondrial protists	Mitochondrial protists possess mitochondria with either discoid, tubular or lamellar cristae, depending on their position on the phylogenetic tree. They are aerobic organisms. They can be free-living, symbiotic or parasitic on other species.
Sequence of evolutionary events	A number of phenomena prevent us from knowing what the exact events of evolution were. Both primitive and derived features can be found in some protists, and loss of features, parallel evolution and transfer of genetic material can all contribute to the difficulty in creating a definitive tree.
Related topics	Fungal cell structure and growth (H1) Algal cell structure and growth (I1) Related phyla: the water and slime Taxonomy of protozoa (J1) molds (H5)

Current taxonomic status

Establishing relationships within the different members of the protistan microbes relies on studying **characters**, which are features or attributes of an individual organism that can be used to compare it with another organism. These features can be morphological, anatomical, ultrastructural, biochemical or based on sequences of nucleic acids. The objective of such a study is to create **monophyletic** groups which are assumed to have a single ancestor, usually extinct; a similar approach is used in the creation of classification systems for bacteria (Topic B1). A **cladistic** approach would not assume features from an ancestor, but would merely define a monophyletic group on the basis of shared characters. New information based on the presence and type of mitochondria, and the DNA sequencing of ribosomal RNA, place the plant, animal and fungal members of the eukaryotic microbes into a complex **phylogenetic tree** (*Fig. 1*).

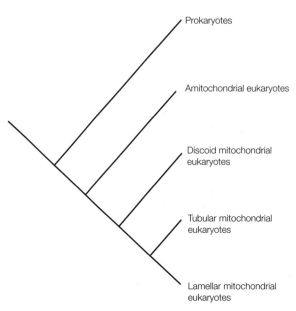

Fig 1.   Tentative phylogenetic tree of the probable evolutionary sequence of protistan microbes.

Amitochondrial protists

At the base of the tree are amitochondrial protists, which separated from other linages before mitochondria were acquired (see Topic J1). These organisms most strongly resemble the colorless, **anaerobic** cell from which, by symbiotic acquisition of mitochondria and in some cases chloroplasts, the rest of the linages developed.

Mitochondrial protists

The next group in the protistan linage are those organisms that have mitochondria with **discoid christae**. These organisms are **aerobes**. The euglenids, ameboflagellates and acrasid slime molds share these characteristics (see Topics H5, I1 and J1). Many members of this group have two flagellae, and are freeliving while others are parasites within animals and man.

Organisms having mitochondria with **tubular** cristae appear next on the phylogenetic tree, and within this group we see the myxomycete and dictyosteloid slime molds, dinoflagellates, ciliates, apicomplexans, diatoms, oomycetes, radiolaria and heliozoans (see Topics H5, I1 and J1).

At the top of the tree are organisms that have mitochondria with **lamellar** cristae, structures that are seen in the most complex protistan organisms, the green algae, chytrids, higher fungi, plants and animals.

Sequence of evolutionary events

Exact orders of evolution are difficult to determine, because of morphological inconsistencies within groups. For instance, dinoflagellates have a number of less specialized features, but they also have more derived features including being biflagellate and having scales, a characteristic which brings them close to the ciliates and sporozoa (see Topic I3).

Even sequencing of nucleic acid to provide phylogenetic **gene trees** may not answer all the questions. For instance, sequencing the small subunit nuclear encoded rRNA reveals that one can follow the evolution of an individual gene, but not necessarily the whole organism, as different regions of rRNA genes evolve at different rates. Phylogenetic patterns may also be confused because of transfer of genetic information between linages as a consequence of **endosymbiosis** and other mechanisms. **Parallel** evolution within these organisms is also likely to occur, and superficial similarities within distinct linages may arise because of loss of features, which is likely to be a significant factor within very simple organisms.

G2 EUKARYOTIC CELL STRUCTURE

Key Notes

Eukaryotes	Eukaryote cells have complex, membrane-bound, subcellular organelles which compartmentalize cell functions.
Plasma membrane	The plasma membrane is a semi-permeable barrier between the outside and inside of the cell, and it is involved in cell–cell recognition, endo- and exocytosis and adhesion to surfaces. Transport systems in the membrane allow it to import materials selectively into the cell.
Cytoplasm	The cytoplasm is 70–85% water, but also contains proteins, sugars and salts in solution. The organelles of the cell are suspended in the cytoplasm. Both fungal and algal cells have single membrane-bound vacuoles in their cells.
Cytoskeleton	The cytoskeleton of the cell is made up of microtubules, intermediate fibers and microfilaments, which maintain the shape of the cell.
Nucleus and ribosomes	The nucleus is a double membrane-bound organelle which contains the chromosomal DNA of the cell. Inside the nucleus is the nucleolus, which is the site of ribosomal RNA synthesis. Ribosomes are made of two subunits of RNA plus proteins, and they are the site of DNA translation and protein synthesis.
Endoplasmic reticulum	The endoplasmic reticulum (ER) is a complex of membrane tubes and plates which is continuous in places with the nuclear membrane. The ER can be smooth or it may be termed rough where ribosomes are attached to it. The main function of this organelle is the synthesis and transport of proteins and lipids.
Golgi	The Golgi body is a series of flattened, membrane-bound fenestrated sacs and vesicles. Vesicles secreted from the ER fuse with the Golgi, and their contents are then further processed by resident biochemical processes. Processed materials are then secreted from the Golgi in vesicles which fuse with other organelles or with the plasma membrane.
Lysosomes and peroxisomes	Lysosomes and peroxisomes are membrane-bound sacs secreted from the Golgi. Lysosomes contain acid hydrolases involved in intracellular digestion. Peroxisomes contain amino and fatty acid-degrading enzymes and the enzyme catalase, which detoxifies hydrogen peroxide released by degradative processes.
Mitochondria and hydrogenosomes	Mitochondria are the site of respiration and oxidative phosphorylation in aerobic organisms. They are bound by a double membrane, the inner one being in-folded to form plates or tubes called cristae. ATP production is

located in particles attached to the cristae. Hydrogenosomes are organelles found in some amitochondrial, anaerobic protozoans. Their function is energy production.

Chloroplasts

Chloroplasts are double membrane-bound organelles which contain the photosynthetic pigment chlorophyll. Within the chloroplast are stacks of flattened sacs called thylakoids where the photosynthetic systems are located.

Cell walls

Cell walls are found in the algae (cellulose-based) and the fungi (chitin-based). They delimit the outside of the cell from the environment and are important in maintaining cell rigidity and controlling excess of water influx due to osmosis.

Flagella

Flagella are microtubule-containing extensions of the cell membrane. They provide the cell with motility by their flexuous bending.

Related topics

The microbial world (A1)
Heterotrophic pathways (B1)
Electron transport and oxidative
     phosphorylation (B2)
Autotrophic metabolism (B3)

DNA replication (C2)
Bacterial cell structure (D2)
Taxonomy (G1)
Cell division and ploidy (G3)
Cellular structure of protozoa (J2)

Eukaryotes

Eukaryotic cells are compartmentalized by **membranes**. The cell contains several different types of membrane-bound organelle in which different biochemical and physiological processes can occur in a regulated way (*Fig. 1*). Membranes also transport information, metabolic intermediates and end-products from the site of biosynthesis to the site of use.

Plasma membrane

The plasma membrane of eukaryotes is a **semi-permeable** barrier that forms the boundary between the outside and the inside of the cell. It is similar to that of the prokaryotes (see Topic D3), except that it contains **sterols**, flat molecules that give the membrane a greater rigidity and which stabilize the eukaryotic cell. There are transport systems in the membrane that selectively import materials into the cell, and it is also involved in **endo-** and **exo-cytosis**, where food particles are engulfed and waste products are expelled from the cell in membrane bound vesicles. The plasma membrane is involved in key interactive processes between cells, like **cell–cell recognition** systems, as well as adhesion of the cell to solid surfaces.

Cytoplasm

The cytoplasm is a dilute solution (70–85% water) of proteins, sugars and salts in which all other organelles are suspended. It has **sol–gel** properties, which means it can be liquid or semi-solid depending on its molecular organization. Vacuoles act as storage sites for nutrients and waste products in a very weak solution. The high water content of the vacuole maintains a high cell turgor pressure.

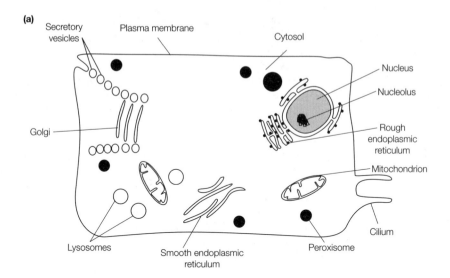

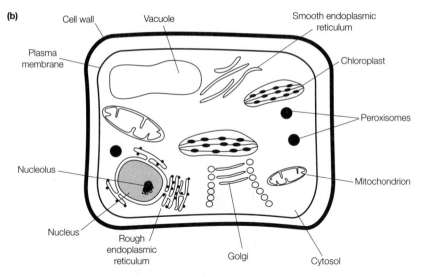

Fig. 1.   Eukaryote cell structure: (a) structure of a typical animal cell; (b) structure of a typical plant cell.

Cytoskeleton

The eukaryotic cell is further stabilized by a cytoskeleton made up of **tubulin** containing **microtubules** (25 nm diameter), and **microfilaments** (4–7 nm diameter) and **intermediate fibers** (8–10 nm diameter) made of an actin much like that of contractile muscle. The cytoskeleton is a dynamic structure, providing support for the cell but also the machinery for ameboid movement, cytoplasmic streaming and nuclear and cell division (see Topic G3).

Nucleus and ribosomes

The nucleus contains the DNA of the microbe. In eukaryotic microbes the nucleus usually contains more than one **chromosome**, in which the DNA is protected by **histone** proteins. In diploid organisms these chromosomes are paired (see Topic G3). The nucleus is surrounded by the **nuclear membrane**, a double membrane which is perforated by many pores where the two membranes fuse. It is via these pores that the nucleus remains in constant control of the rest of the cell machinery via **mRNA** and **ribosomes**. In places the nuclear membrane is also continuous with the endoplasmic reticulum. Within the nucleus is the **nucleolus**, an RNA-rich area where ribosomes are synthesized (see Topics C4 and C7 for DNA transcription and translation). Eukaryotic ribosomes are essentially very similar to those of prokaryotes (see Topic C7) but they are slightly larger, their two subunits are of 60S and 40S, making a dimer of 80S. Their function is the same as that in prokaryotes (see Topic C7).

Endoplasmic reticulum

The outer membrane of the nucleus is in places continuous with a complex, three-dimensional array of membrane tubes and sheets, the endoplasmic reticulum (ER). Tubular ER can be studded with ribosomes, and described as **rough ER** (RER), where **ribosomal translation** and **protein modification** takes place (see Topics C2 and C4). These proteins can either be secreted into the lumen of the ER or inserted into the membrane. Plates of **smooth ER** are associated with **lipid synthesis** and **protein** and **lipid transport** across cells.

Golgi

The Golgi is composed of stacks of a flattened series of membrane-bound sacs or **cisternae**, surrounded by a complex of tubes and vesicles. There is a definite **polarity** across the stack, the *cis* or forming face receiving vesicles from the ER, the contents of these vesicles then being processed by the Golgi, to be budded from the sides or the *trans* (maturing) face of the organelle. The Golgi apparatus processes and packages materials for secretion into other subcellular organelles or from the cell membrane. Golgi in fungi are less well developed than in algae, and tend to have fewer or single cisternae. They are sometimes termed **dictyosomes.**

Lysosomes and peroxisomes

The Golgi body generates these single-membrane-bound organelles which contain enzymes (**acid hydrolases** in the lysosome, **aminases, amidases** and **lipases** plus **catalase** in the peroxisomes) needed in the digestion of many different macromolecules. The internal pH of the lysosome is **acidic** (pH 3.5–5) to enable the enzymes to work at the optimum pH, and this pH is maintained by proton pumps present on the membrane. The breakdown of amino and fatty acids by the peroxisomes generates **hydrogen peroxide**, a potentially cytotoxic by-product. The enzyme **catalase**, also present in the peroxisome, degrades the peroxide into water and oxygen, protecting the cell.

Mitochondria and hydrogenosomes

Mitochondria are double-membrane-bound organelles where the processes of **respiration** and **oxidative phosphorylation** occur (see Topics B1 and B2). They are approximately 2–3 μm long and 1 μm in diameter. Their numbers in a cell vary. They contain a small, circular DNA molecule which encodes some of the mitochondrial proteins, and 70S ribosomes (see Topics C4 and C7). The inner membranes of the mitochondria contain an **ATP/ADP transporter** that moves the ATP, which is synthesized in the organelle, outwards into the cytoplasm. ATP production is located on particles attached to the cristae, the inner infolded mitochondrial membrane (*Fig. 2*). Their structures differ slightly between the three protistan groups. Not all protozoa have mitochondria (see Topic J2) and in these cells metabolism is essentially anaerobic (see Topic J4). In the aerobic, very primitive eukaryotes' mitochondrial **cristae** are discoid (see Topic G1).

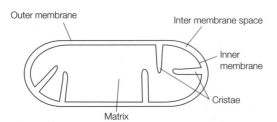

Fig. 2.   Structure of a mitochondrion. From Hames B.D. et al., Instant Notes in Biochemistry.
© *BIOS Scientific Publishers Limited, 1997.*

Mitochondria of the fungi are large and highly lobed with flat, plate-like cristae, while those of the algae have much more inflated cristae.

Hydrogenosomes are found in anaerobic protozoa, and they are membrane-bound organelles containing **electron-transport pathways** in which **hydrogenase** enzymes transfer electrons to terminal electron acceptors which generate molecular hydrogen (see Topic J4).

Chloroplasts

Chloroplasts are **chlorophyll**-containing organelles that can use light energy to fix carbon dioxide into carbohydrates (**photosynthesis**) (see Topics B3 and I2). They are bound by double membranes and contain flattened membrane sacs called **thylakoids** where the light reaction of photosynthesis takes place (*Fig. 3*). In algae these organelles are large, almost filling the cell. The **pyrenoid** is a proteinaceous region within the chloroplast where polysaccharide biosynthesis takes place.

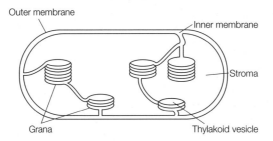

Fig. 3.   Structure of a chloroplast. From Hames B.D. et al., Instant Notes in Biochemistry.
© *BIOS Scientific Publishers Limited, 1997.*

Cell walls

The protoplasts of plant and fungal cells are in most cases surrounded by rigid cells walls. In the fungi the cell wall is composed of a microcrystalline polymer of **chitin** (repeating units of β 1–4 linked NAG) and amorphous β-glucans, while plant cell walls are composed of **cellulose** (repeating units of β1–4 linked glucose) and **hemicelluloses** (see Topics H1 and I1).

Flagella

Flagella are membrane-bound extensions of the cell, which contain microtubules (see Topic J2). The microtubules are arranged as a bundle of nine doublets around the periphery of the flagellum, with a pair of single microtubules running within them. This structure is called the **axoneme**. Flagella provide cells with motility, because they flex and bend when supplied with ATP.

G3 CELL DIVISION AND PLOIDY

Key Notes

Cell cycle

The cell cycle describes all the events that occur in a cell from the end of one cell division to the end of the next. There are four phases to the cell cycle, which include cell growth (G_1), DNA synthesis (S), a second gap or growth phase (G_2) and finally nuclear division (N).

Mitosis and asexual cell division

Mitosis is nuclear division which results in progeny nuclei that are identical to the parent. It is usually followed by cytokinesis, cell division, which produces cells that have the same phenotype as the parent. In some eukaryotic microbes multiple nuclear divisions may occur without cytokinesis, giving rise to large, multinucleate cells.

Meiosis and sexual cell division

Meiosis is a nuclear division where there is a halving of chromosome numbers from a diploid to a haploid state. In organisms with an extended diploid phase of their life cycle, meiosis produces haploid gametes, and it is immediately followed by gamete fusion and formation of a new diploid organism. In organisms with an extended haploid stage, diploid formation is immediately followed by meiosis and produces the new haploid organism.

Chromosomes

Eukaryotic microorganisms package the large amount of DNA they contain into chromosomes. Chromosomes contain a single, linear double strand of DNA tightly bound with histone proteins. There are usually between four and eight chromosomes per cell.

Histones

Histones are basic proteins that bind to DNA to condense and fold it. They are vital to the structure of DNA and have been highly conserved during evolution.

Related topics

Structure and organization of DNA (C1)
DNA replication (C2)
Reproduction in fungi (H3)

Algal cell structure and growth (I1)
Reproduction in the algae (I3)
Reproduction (J5)

Cell cycle

The cell cycle consists of an **interphase** during which growth of the cell (G_1) occurs, the cell increases in volume to maximum size and there is synthesis of cytoplasmic constituents and RNA (*Fig. 1*). Synthesis of DNA occurs next, during the **S** phase, as chromosomes are duplicated in preparation for nuclear division. During the final part of the cell cycle a second gap or growth phase occurs (G_2), when specific cell division-related proteins are synthesized. Nuclear division (**N**) then follows to complete the cell cycle (cf. the bacterial cell cycle, D8).

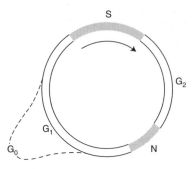

Fig. 1.   The eukaryotic cell cycle. The S phase is typically 6–8 h long, G_2 is a phase in which the cell prepares for mitosis and lasts for 2–6 h, and nuclear division (N) takes only about 1 h. The length of G_1 is very variable and depends on the cell type. Cells can enter G_0, a quiescent phase, instead of continuing with the cycle cell. From Hames B.D. et al., Instant Notes in Biochemistry. © BIOS Scientific Publishers Limited, 1997.

Mitosis and asexual cell division

Asexual cell division in unicellular eukaryotic organisms is synonymous with growth. Division is usually by **binary fission** after a single nuclear division. The parent cell divides, usually longitudinally, into two even-sized identical progeny cells. Division can be by **multiple fission** (see Topic J5), where many nuclear divisions occur, producing either a large multinucleate **coenocyte** or many uninucleate progeny cells. All cells produced from mitosis are genetically identical to their parent.

In both cases **somatic** cell division is preceded by the mitotic division of the cell nucleus. In mitosis the replicated DNA from the S phase of the life cycle is separated equally into two progeny cells. The events of mitosis can be separated into four stages for convenience, but each flows into the other as a continuous process.

The first phase in mitosis is the **prophase**. In this phase microtubules form from the the **microtubule organizing centers** (**MTOCs**). In fungi these structures are known as **spindle pole bodies** (**SPBs**), located close to the nuclear envelope. In the motile species of fungi, protozoa and some algae the MTOC is called a **centriole** and it becomes surrounded by microtubules in a process termed **aster formation** (*Fig. 2a*). The MTOCs begin to move towards opposite poles of the nucleus, and spindle microtubules appear between them. Single chromosomes which have been duplicated to form **chromatids**, are joined together at their **centromeres**. These centromeres are also attached to spindle microtubules. By late prophase–early **metaphase**, MTOCs are opposite each other and the spindle is complete, with chromosomes aligned across the center in a **metaphase plate**. In many species of fungi, chromosomes remain extremely indistinct during mitosis, and they do not appear to assemble across a metaphase plate.

The nuclear envelope may disappear between the prophase and the beginning of the metaphase in some eukaryotic microbes, but in most fungi the nuclear envelope remains intact throughout the process, the microtubules of the spindle penetrating through it.

In the next phase of mitosis, the **anaphase**, pairs of chromatids that were held together at the centromere begin to separate simultaneously, and spindle microtubules begin to pull them towards the two poles of the cell. By the end of the

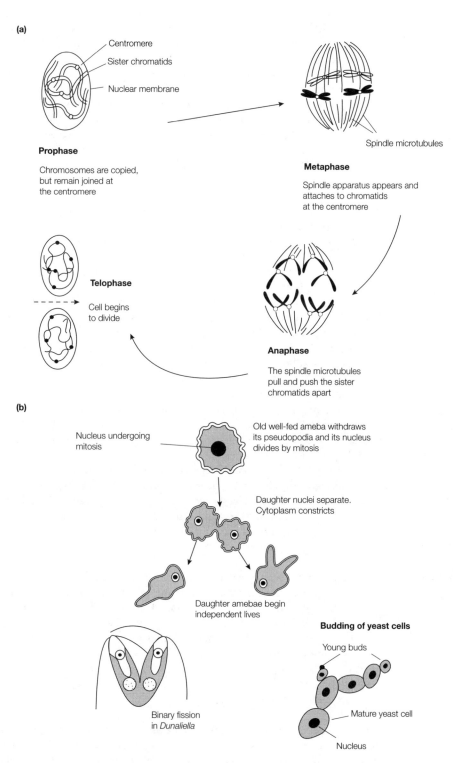

Fig. 2. Events of mitosis: (a) nuclear division; (b) cell division. Reproduced from Beckett, B.S., Biology, 1978, with kind permission from B.S. Beckett.

anaphase the chromatids have been pulled close to the MTOCs. In many species of the higher fungi there is an asynchronous chromosome separation during the anaphase.

In the final phase of mitosis, the **telophase**, the aster microtubules disappear, and the nuclear envelope reforms if it has disintegrated. In the two progeny nuclei the MTOC duplicates, and **cell division** commences with the division of the cytoplasm by an invaginating plasma membrane or the formation of a **cell plate** by Golgi-derived vesicles across the midline between the two nuclei (*Fig. 2b*). In some eukaryotic microbes separation can be by **budding**, producing a progeny cell that is much smaller than the parent.

Meiosis and sexual cell division

Most fungi, and some algae and protozoa, are **haploid** for most of their life cycle. They have only one set of chromosomes. The **diploid** stage (two sets of chromosomes) is often very transient and found only in resting structures such as spores. The life cycle ends with meiosis, which returns the new cell to its haploid state. In organisms with a dominant diploid vegetative phase, meiosis occurs just before cell division, producing haploid gametes which then fuse to reform the diploid.

At the end of the interphase and before meiosis begins the duplication of chromosomes to chromatids occurs as in mitosis. However, as the cell is diploid, **homologous** pairs of chromosomes associate during **prophase 1** (*Fig. 3*), and at this point it is possible for genetic recombination to occur (cf. Topic E3).

During metaphase 1 chromosomes assemble across the metaphase plate, and in anaphase 1 homologous chromosomes are separated. Telophase 1 is very transient, and the chromosomes rapidly move into the prophase and metaphase 2, where pairs of chromatids assemble across the metaphase plate. The chromatids separate from each other at anaphase 2, and in telophase 2 cytoplasm begins to separate around the four progeny nuclei, each containing a haploid complement of chromosomes.

In some circumstances **multiple sets** of chromosomes can exist in a cell and this is termed **polyploidy**. Nuclear division in polyploid organisms is complex and often results in the loss of single chromosomes, leading to odd numbers of chromosomes in some progeny cells. This is termed **aneuploidy**. Polyploid and aneuploid cells are usually unable to participate in meiosis because of their odd chromosome numbers.

Chromosomes

Compared with the prokaryotes, eukaryotic microbes contain much more DNA, and therefore have had to evolve structures which pack, store and present DNA during different parts of the cell cycle (see Topic C1). There are usually between four and eight chromosomes per cell, but there can be more or less. Chromosomes contain a single linear double strand of DNA, tightly condensed with **histone** proteins. The combined DNA and histone is often termed **chromatin**. The histone proteins bind to the DNA and create three levels of folding. During the prophase, chromatin is very condensed but, in the interphase, chromatin is described as dispersed.

Histones

Histones are very basic proteins because they contain many basic amino acids (lysine and arginine), with positively charged side chains. These side chains associate with the negatively charged phosphate groups on the DNA molecule. The histones are vital to the structure of DNA and have been **highly conserved** during evolution (see Topic C1).

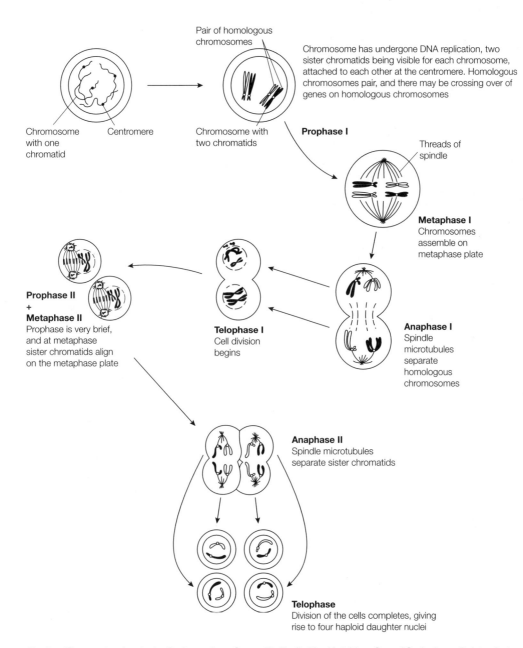

Pair of homologous chromosomes

Chromosome has undergone DNA replication, two sister chromatids being visible for each chromosome, attached to each other at the centromere. Homologous chromosomes pair, and there may be crossing over of genes on homologous chromosomes

Chromosome with one chromatid

Centromere

Chromosome with two chromatids

Prophase I

Threads of spindle

Metaphase I
Chromosomes assemble on metaphase plate

**Prophase II
+
Metaphase II**
Prophase is very brief, and at metaphase sister chromatids align on the metaphase plate

Telophase I
Cell division begins

Anaphase I
Spindle microtubules separate homologous chromosomes

Anaphase II
Spindle microtubules separate sister chromatids

Telophase
Division of the cells completes, giving rise to four haploid daughter nuclei

Fig. 3.   The events of meiosis. Redrawn from Gross, T., Faull, J.L., Kettridge, S. and Springham, D. Introductory Microbiology, *1995, with kind permission from Kluwer Academic Publishers.*

H1 FUNGAL CELL STRUCTURE AND GROWTH

Key Notes

Fungal structure	Fungi are heterotrophic, eukaryotic organisms with a filamentous, tubular structure, a single branch of which is called a hypha. A network of hyphae is called a mycelium. Hyphae are bound by firm, chitin-containing cell walls and contain most eukaryotic cell organelles.
Fungal taxonomy	There are four phyla within the fungi, divided from each other on the basis of differences in their mechanisms of sexual reproduction. The four phyla are the Zygomycotina and Chytridiomycotina (these two phyla are termed the Lower fungi) and the Ascomycotina and Basidiomycotina (termed the Higher fungi). A fifth group exists which contains fungi where sexual reproduction is not known but where asexual reproduction is seen. These fungi are placed in the phylum Deuteromycotina.
Cell-wall structure and growth	Fungal cell walls are formed from semi-crystalline chitin microfibrils embedded in an amorphous matrix of β-glucan. In the higher fungi, hyphae grow by tip growth followed by septation. Lower fungal hyphae grow by tip growth but remain aseptate.
Colonial growth	Colonial growth is characterized by the radial extension of mycelium over and through a substrate, creating a circular or spherical fungal colony.
Kinetics of growth	Fungal growth can be measured by measuring mycelial mass changes with time under excess of nutrient conditions. From this information the specific growth rate can be calculated. After a lag phase, a brief period of exponential growth follows as hyphal tips are initiated. As the new hypha extends, it grows at a linear rate until nutrient depletion causes a retardation phase, followed by a stationary phase.
Hyphal growth unit	Hyphal growth may also be measured by microscopy and by counting the total numbers of hyphal tips, and dividing that number by the total length of mycelium in the colony, the average length of hypha required to support a growing tip can be calculated. This is termed the hyphal growth unit.
Peripheral growth zone	The peripheral growth zone is the region of mycelium behind the tip, which permits radial extension at a rate equal to the specific growth rate.

Related topics   Bacterial cell structure (D2)   Algal cell structure and growth (I1)
                     Bacterial growth and cell cycle (D8) Cellular structure of protozoa (J2)

Fungal structure   Fungi are filamentous, non-photosynthetic, eukaryotic microorganisms that have a **heterotrophic** nutrition (see Topic B1). Their basic cellular unit is described as a **hypha** (*Fig 1*). This is a tubular cell which is surrounded by a rigid, **chitin**-containing cell wall. The hypha extends by tip growth, and multiplies by branching, creating a fine network called a **mycelium**. Hyphae contain nuclei, mitochondria, ribosomes, Golgi and membrane-bound vesicles within a plasma–membrane bound cytoplasm (see Topic G2). The subcellular structures are supported and organized by microtubules and endoplasmic reticulum. The cytoplasmic contents of the hypha tend to be concentrated towards the growing tip. Older parts of the hypha are heavily vacuolated and may be separated from the younger areas by cross walls called **septae**. Not all fungi are multicellular, some are **unicellular** and are termed **yeasts**. These grow by binary fission or budding, creating new individuals from the parent cell.

(a)

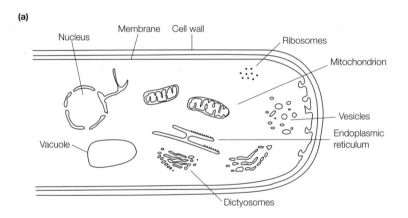

(b)

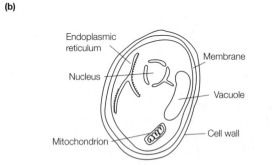

Fig. 1.   (a) Hyphal and (b) yeast structures. Redrawn from Grove, S.M., Bracker, C.E. and Morré, B.J., American Journal of Botany, vol. 57, pp. 245–66, 1970, with kind permission from the Botanical Society of America.

Fungal taxonomy   In the past the fungi were a **polyphyletic** group which contained microorganisms that had very different ancestors. Current thinking now prefers a **monophyletic** classification where all groups within a phylum are descendants of one ancestor (see Topic G1). Fungi are currently divided into four major phyla on the basis of their morphology and sexual reproduction.

Two of the phyla are described as **Lower** fungi, the **Zygomycotina** and the **Chytridiomycotina**. In the lower fungi the vegetative mycelium is non-septate, and complete septa are only found in reproductive structures. Asexual reproduction is by the formation of **sporangia**, and sexual reproduction by the formation of **zygospores**.

The other two phyla are described as the **Higher** fungi and have a more complex mycelium with elaborate, perforate septa. They are divided into the **Ascomycotina** and the **Basidiomycotina**. Members of the Ascomycotina produce asexual **conidiospores** and sexual **ascospores** in sac-shaped cells called **asci**. Fungi from the **Basidiomycotina** rarely produce asexual spores, and produce their sexual spores from club-shaped **basidia** in complex fruit bodies.

A fifth group exists in the higher fungi which contains all forms that are not associated with a sexual reproductive stage, and these are termed the **Deuteromycotina**.

Within each of the major phyla are several classes of fungi, indicated by names ending with -*etes*, for example Basidiomycetes, and within classes can be sub-classes or orders, indicated by names ending in -*ales*, within which there are genera and then species (see Topic D1). The important differences between fungi used to distinguish taxonomic groups are summarized in *Table 1*.

Table 1. Features of the main groups of fungi

Group	Perforate septae present or absent	Asexual sporulation	Sexual sporulation
Lower fungi			
Zygomycotina	Absent	Non-motile sporangiospores	Zygospore
Chytridiomycotina	Absent	Motile zoospores	Oospore
Higher fungi			
Ascomycotina	Present	Conidiospores	Ascospore
Basidiomycotina	Present	Rare	Basidiospore
Deuteromycotina	Present	Conidiospores	None

Cell-wall structure and growth

Fungal cell walls are rigid structures formed from layers of semi-crystalline chitin **microfibrils** that are embedded in an amorphous matix of β-**glucan**. Some protein may also be present. Growth occurs at the hyphal tip by the fusion of membrane-bound vesicles containing wall-softening enzymes, cell-wall monomers and cell-wall polymerizing enzymes derived from the Golgi with the hyphal tip membrane (*Fig. 2*). The fungal wall is softened, extended by turgor pressure and then rigidified.

Septae are cross walls that form within the mycelium. Growth in lower fungi is not accompanied by septum formation, and the mycelium is cenocytic. Septae only occur in the lower fungi to delimit reproductive structures from the parent mycelium, and they are complete (see Topic H3). In higher fungi, growth of the mycelium is accompanied by the formation of incomplete septae. Septae in the ascomycetes are perforate, and covered by endoplasmic reticulum membranes to limit movement of large organelles such as nuclei from compartment to compartment. This structure is called the **dolipore** septum. In dikaryotic

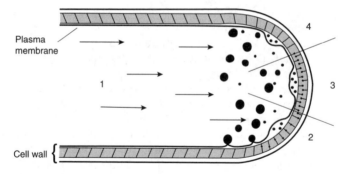

Fig. 2. *Hyphal tip growth. Step 1, vesicles migrate to the apical regions of the hyphae. Step 2, wall-lysing enzymes break fibrils in the existing wall, and turgor pressure causes the wall to expand. Step 3, amorphous wall polymers and precursors pass through the fibrillar layer. Step 4, wall-synthesizing enzymes rebuild the wall fibrils. Redrawn from Isaac, S., Fungal–plant Interactions, 1991, with kind permission from Kluwer Academic Publishers.*

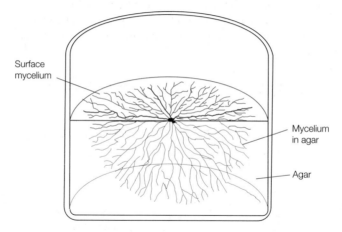

Fig. 3. *Colonial growth patterns of filamentous fungi. Reproduced from Ingold, C.T. and Hudson, H.J., The Biology of Fungi, 1993, with kind permission from Kluwer Academic Publishers.*

basidiomycetes, septum formation is co-ordinated with divisions of the two mating-type nuclei, maintaining the dikaryotic state by the formation of **clamp connections**. These septae resemble crozier formation in the formation of asci (see Topic H3).

Colonial growth   Hyphal tip growth allows fungi to extend into new regions from a point source or **inoculum**. Older parts of the hyphae are often emptied of contents as the cytoplasm is taken forwards with the growing tip. This creates the radiating colonial pattern seen on agar plates (*Fig. 3*), in ringworm infections of skin and fairy rings in grass lawns.

Kinetics of growth   When fungi are filamentous their growth rate cannot be established by cell counting using a hemocytometer or by turbidometric measurements (which can be used to measure bacterial and yeast growth) (see Topics D7 and D9). However, by measuring mass (M) changes with time (t) under excess nutrient conditions the **specific growth rate** (μ) for the culture can be calculated using the formula:

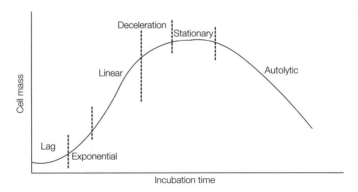

Fig. 4.   Phases of fungal growth.

$$\frac{\mathrm{d}M}{\mathrm{d}t} = \mu M$$

Fungal growth in a given medium follows the growth phases of lag, acceleration, exponential, linear, retardation, stationary and decline (*Fig. 4*; see Topic D8).

Exponential growth occurs only for a brief period as hyphae branches are initiated, and then the new hypha extends at a linear rate into uncolonized regions of substrate. Only hyphal tips contribute to extension growth. However, older hyphae can grow aerially or differentiate to produce sporing structures.

Hyphal growth unit

It is also possible to observe hyphal growth by microscopy, measuring tip growth and branching rates of mycelium. From these data the hyphal growth unit and the **peripheral growth zone** can be calculated. The hyphal growth unit (G), which is the average length of hypha that is required to support tip growth, is defined as the **ratio** between the total length of mycelium and the total number of tips:

$$\frac{\text{Total number of tips}}{\text{Total length of mycelium}} = \text{Hyphal growth unit}$$

In most fungi the hyphal growth unit includes the tip cell plus two or three sub-apical compartments. The ratio increases exponentially from germination, but stabilizes to give a constant figure for a particular strain under any given set of environmental conditions.

Peripheral growth zone

The peripheral growth zone is the region of mycelium behind the tip, needed to support maximum growth of the hyphal tip. This zone permits radial extension at a rate equal to that of the **specific growth rate** of unicells in liquid culture. The mean rate of hyphal extension (E) is a function of the hyphal growth unit and the specific growth rate

$$E = G\mu$$

The **radial extension rate** (K_r) is a function of the peripheral growth zone (ω) and the specific growth rate (μ):

$$K_r = \omega\mu$$

H2 FUNGAL NUTRITION

Key Notes

Carbon nutrition	Fungi require organic carbon compounds to satisfy their carbon and energy requirements. They obtain this carbon by saprotrophy, symbiosis or parasitism.
Carbon metabolism	Fungi normally utilize glycolysis and aerobic metabolism of carbohydrates. Some can use fermentative pathways under reduced oxygen levels. A few fungi are truly anaerobic.
Nitrogen nutrition	Fungi cannot fix gaseous nitrogen but can utilize nitrate, ammonia and some amino acids as nitrogen sources.
Macro-, micro-nutrients and growth factors	Most macro- and micro-nutrients that fungi require are present in excess in their environments. Phosphorus and iron may be in short supply, and fungi have specific mechanisms to obtain these nutrients. Some fungi may require external supplies of some vitamins, sterols and growth factors.
Water, pH and temperature	Fungi require water for nutrient uptake and are therefore restricted to damp environments. They occupy acidic environments between pH 4 and 6, and by their activity further acidify it. Most fungi are mesophilic, growing between 5° and 40°C, but some can tolerate high or low temperatures.
Secondary metabolism	Secondary metabolites, derived from many different metabolic pathways, are produced by fungi when vegetative growth becomes restricted by nutrient depletion or stress.

Related topics
Heterotrophic pathways (B1)	Algal nutrition and metabolism (I2)
Autotrophic metabolism (B3)	Physiology of protozoa (J3)
Biosynthetic pathways (B4)	

Carbon nutrition   Fungi are heterotrophic for carbon (see Topic B1). They need organic compounds to satisfy energy and carbon requirements. There are three main modes of nutrition: **saprotrophy**, where fungi utilize dead plant, animal or microbial remains; **parasitism**, where fungi utilize living tissues of plants and animals to the detriment of the host; and **symbiosis**, where fungi live with living tissues to the benefit of the host.

Carbohydrates must enter hyphae in a soluble form because the rigid cell wall prevents endocytosis. Soluble sugars cross the fungal wall by diffusion, followed by active uptake across the fungal membrane (see Topics D5 and G2). This type of nutrition is seen in the symbiotic and some parasitic fungi. For the saprophytic fungi most carbon in the environment is not in a soluble form but is present as a complex polymer like cellulose, chitin or lignin. These polymers have to be broken down enzymically before they can be utilized. Fungi release **degradative enzymes** into their environments. Different classes of enzyme can be produced, including the **cellulases**, **chitinases**, **proteases** and multi-component **lignin-degrading enzymes**, depending on the type of substrate the fungus is growing on. Regulation of these enzymes is by **substrate induction** and **end-product inhibition** (see Topic B4).

Carbon metabolism

Once within the hypha, carbon and energy metabolism is by the processes of **glycolysis** and the **carboxylic acid cycle** (see Topic B1). Fungi are usually aerobic, but some species, for example, the yeasts, are capable of living in low oxygen-tension environments and utilizing **fermentative** pathways of metabolism (see Topic B4). Recently, truly **anerobic** fungi have been discovered within animal rumen and in anaerobic sewage-sludge digestors. These fungi have an energy metabolism much like that of some parasitic protozoa (see Topic J4).

Nitrogen nutrition

Fungi are heterotrophic for nitrogen (see Topic B4). They cannot fix gaseous nitrogen, but they can utilize nitrate, ammonia and some amino acids by direct uptake across the hyphal membrane. Complex nitrogen sources, such as peptides and proteins, can be utilized after extracellular proteases have degraded them into amino acids.

Macro-, micro-nutrients and growth factors

Phosphorus, potassium, magnesium, calcium and sulfur are all macronutrients required by fungi. All but phosphorus are usually available to excess in the fungal environment. Phosphorus can sometimes be in short supply, particularly in soils, and fungi have the ability to produce extracellular **phosphatase** enzymes which allow them to access otherwise unavailable phosphate stores.

Micronutrients include copper, manganese, sodium, zinc and molybdenum, all of which are usually available to excess in the environment (see Topic D7). Iron is relatively insoluble and therefore not easily assimilated, but fungi can produce **siderophores** or organic acids, which can **chelate** or alter iron solubility and improve its availability (see Topic D5).

Some fungi may require pre-formed vitamins, for example, thiamin and biotin (see Topic D7). Other requirements can be for sterols, riboflavin, nicotinic acid and folic acid.

Water, pH and temperature

Fungi require water for nutrient uptake and they are therefore restricted to fairly moist environments such as host tissue if they are parasites or symbionts, or soils and damp substrates if they are saprophytes. Desiccation causes death unless the fungus is specialized, as they are in the lichens (see Topic H6). Some fungi are wholly aquatic (see Topic H5).

Fungi tend to occupy acidic environments, and by their metabolic activity (respiration and organic acid secretion) tend to further acidify it. They grow optimally at pH 4–6.

Most fungi are **mesophilic**, growing between 5° and 40°C. Some are **psychrophilic** and are able to grow at under 5°C, others are **thermotolerant** or **thermophilic** and can grow at over 50°C.

Secondary metabolism

Nutrient depletion, competition or other types of metabolic stress which limit fungal growth promote the formation and secretion of secondary metabolites. These compounds can be produced by many different metabolic pathways and include compounds termed **antibiotics** (active against bacteria, protozoa and other fungi) (see Topic F7), **plant hormones** (gibberellic acid and indoleacetic acid (IAA)) and **cytotoxic** and **cytostimulatory** compounds.

H3 REPRODUCTION IN FUNGI

Key Notes

Life cycles	All fungi undergo a period of vegetative growth where their mycelium exploits a substrate. This stage is followed by asexual and sexual reproduction, which differs in each of the four phyla.
Reproduction in Chytridiomycetes	Asexual reproduction in the Chytridiomycetes is usually by the formation of motile, uniflagellate zoospores within sporangia. Sexual reproduction is by the formation of oospores.
Reproduction in Zygomycetes	Zygomycete fungi reproduce asexually by the formation of non-motile sporangiospores within sporangia. Sexual reproduction is by the formation of zygospores.
Reproduction in Ascomycetes	Ascomycete fungi reproduce asexually by the formation of conidiospores from hyphal tips. Sexual reproduction is by the formation of ascospores.
Reproduction in Basidiomycetes	Basidiomycete fungi rarely reproduce asexually. Sexual reproduction is by the formation of basidiospores on the gills or pores of large fruit bodies.
Related topics	Bacterial growth and cell cycle (D8) Reproduction (J5) Reproduction in the algae (I3)

Life cycles

Each of the four fungal groups is characterized by differences in their life cycles. All fungi are characterized by having a period of vegetative growth where their biomass increases. The length of time and the amount of biomass needed before sporulation can occur varies. Almost all fungi reproduce by the production of **spores**, but a few have lost all sporing structures and are referred to as *mycelia sterilia*. Different types of spore are produced in different parts of the life cycle.

Reproduction in Chytridiomycetes

Chytrids are quite distinct from other fungi as they have extremely simple thalli and motile zoospores. Some species within this group can be so simple that they consist of a single vegetative cell within (**endobiotic**) or upon (**epibiotic**) a host cell, the whole of which is converted into a **sporangium**, a structure containing spores. These types are termed **holocarpic** forms.

Other members of this group have a more complex morphology, and have **rhizoids** and a simple mycelium. Asexual reproduction in the chytrids is by the production of motile **zoospores** in sporangia that are delimited from the vegetative mycelium by complete septae. The zoospores have a single, posterior flagellum. Sexual reproduction occurs in some members of the chytrids by the production of **diploid** spores after either somatic fusion of haploid cells, either two different mating-type mycelia, fusion of two motile **gametes**, or fusion of one motile gamete with a **non-motile egg** (*Fig. 1*). The resulting spore may

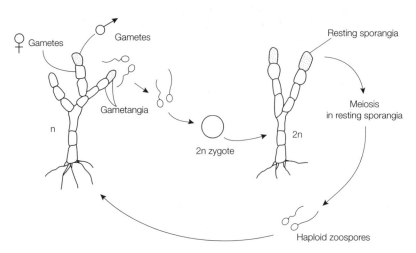

Fig. 1.   Chytrid life cycle.

undergo meiosis to produce a haploid mycelium or it may germinate to produce a diploid vegetative mycelium, which can undergo asexual reproduction by the production of diploid zoospores. The diploid mycelium can also produce resting sporangia in which meiosis occurs, generating haploid zoospores that germinate to produce haploid vegetative mycelium.

Reproduction in Zygomycetes

In the lower fungi, asexual reproduction begins with the production of aerial hyphae. The tip of an aerial hypha, now called a **sporangiophore**, is separated from the vegetative hyphae by a complete septum called a **columella** (*Fig. 2*). The cytoplasmic contents of the tip differentiate into a sporangium containing many asexual spores. The spores contain haploid nuclei derived from repeated mitotic divisions of a nucleus from the vegetative mycelium (see Topic G3). Dispersal of the spores is by wind or water (see Topic H4).

In sexual reproduction, two nuclei of different mating types fuse together within a specialized cell called a **zygospore** (*Fig. 2*). In some species the different mating-type nuclei may be within one mycelium (**homothalism**). In other species, two mycelia with different mating-type nuclei must fuse (**heterothalism**). In both cases, fusion occurs between modified hyphal tips called **progametangia**, which once fused are termed the zygospore. Within the developing zygospore meiosis occurs; usually three of the nuclear products degenerate, leaving only one nuclear type present in the germinating mycelium (see Topic G3).

Reproduction in Ascomycetes

The vegetative stage of the Ascomycete life cycle is accompanied or followed by asexual sporulation by the production of single spores called conidia from the tips of aerial hyphae called conidiophores (*Fig. 3*). The spores can be delimited by a complete transverse wall formation followed by spore differentiation (*Fig. 3a*) termed **thallic** spore formation, or more usually by the extrusion of the wall from the hyphal tip, termed **blastic** spore formation (*Fig. 3b*). These spores can be single celled and contain one haploid nucleus, or they can be multicellular and contain several haploid nuclei produced by mitosis (see Topic G3).

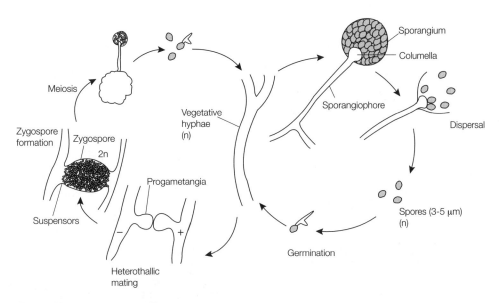

Fig. 2. Life cycle of a typical Zygomycete.

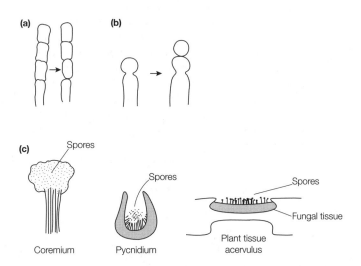

Fig. 3. Asexual reproduction in the Ascomycetes: (a) thallic spore formation; (b) blastic spore formation; (c) aggregations of conidiophores. Redrawn from Ingold, C.T. and Hudson, H.J., The Biology of Fungi, 1993, with kind permission from Kluwer Academic Publishers.

Spores can be produced from single, unprotected conidiophores or they can be produced from **aggregations** that are large enough to be seen with the naked eye (*Fig. 3c*). The conidiophores can aggregate into stalked structures where the spores produced are exposed at the top (**synnema** or **coremia**). Alternatively, varying amounts of sterile fungal tissue can protect the conidia, as in the flask-shaped **pycnidia**. Some species produce conidia in plant tissue, and the conidial

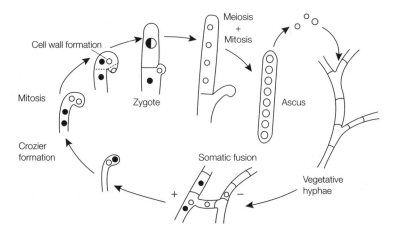

Fig. 4.   Sexual reproduction in the Ascomycetes.

aggregations erupt through the plant epidermis as a cup-shaped **acervulus** or a cushion-shaped **sporodochium**.

Sexual reproduction in this group occurs after **somatic** fusion of different mating-type mycelia. A transient diploid phase is rapidly followed by the formation of **ascospores** within sac-shaped **asci** differentiated from modified hyphal tips. In the initial stages of ascal development hooked hyphal tips form, called **croziers** or **shepherds' crooks** because of their shape. They have distinctive septae at their base which insure that two different mating-type nuclei are maintained in the terminal cell. Formation of the septae is coordinated with nuclear division (*Fig. 4*). In yeasts all these events occur within one cell, after fusion of two mating-type cells, the whole cell being converted into an ascus.

In more complex Ascomycetes many asci form together, creating a fertile tissue called a **hymenium**. In some groups the hymenium can be supported or even enclosed by large amounts of vegetative mycelium. The whole structure is called a **fruit body** or **sporocarp** and is used as a major taxonomic feature (*Fig. 5*). They can become large enough to be seen with the naked eye. Flask-shaped sexual reproductive bodies are called **perithecia**, cup-shaped bodies are called **apothecia** and closed bodies are called **cleistothecia**. These structures

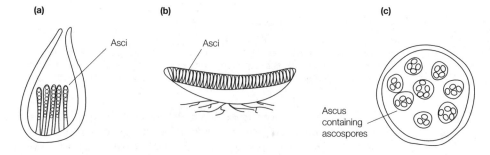

Fig. 5.   Structures of sexual sporocarps in the Ascomycetes: (a) perithecium; (b) apothecium; (c) cleistothecium. Redrawn from Ingold, C.T. and Hudson, H.J., The Biology of Fungi, 1993, with kind permission from Kluwer Academic Publishers.

have evolved to protect the asci and assist in spore dispersal, but the hymenium itself is unaffected by the presence of water.

Reproduction in Basidiomycetes

This group of fungi are characterized by the most complex and large structures found in the fungi. They are also distinctive in that they very rarely produce asexual spores. Much of the life cycle is spent as vegetative mycelium, exploiting complex substrates. A preliminary requisite for the onset of sexual reproduction is the acquisition of two mating types of nuclei by the fusion of compatible hyphae. Single representatives of the two mating-type nuclei are held within every hyphal compartment for extended periods of time. This is termed a **dikaryotic** state, and its maintenance requires elaborate septum formation during growth and nuclear division (see Topic H1).

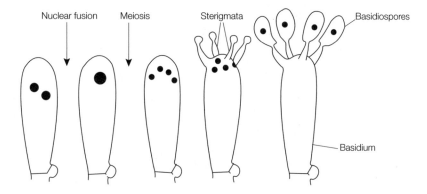

Fig. 6. Basidium formation. Redrawn from Ingold, C.T. and Hudson, H.J., The Biology of Fungi, 1993, with permission from Kluwer Academic Publishers.

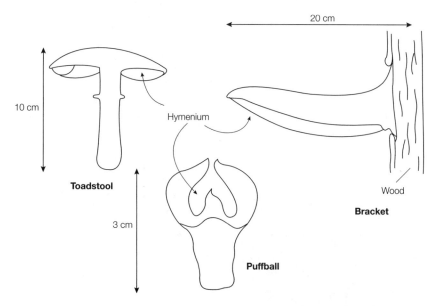

Fig. 7. Structure of sexual sporocarps in the Basidiomycetes. Redrawn from Ingold, C.T. and Hudson, H.J., The Biology of Fungi, 1993, with kind permission from Kluwer Academic Publishers.

Onset of sexual-spore formation is triggered by environmental conditions and begins with the formation of a **fruit body primordium**. Dikaryotic mycelium expands and differentiates to form the large fruit bodies we recognize as mushrooms and toadstools. Diploid formation and meiosis occur within a modified hyphal tip called a **basidium** *(Fig. 6)*.

Four spores are budded from the basidium. Basidia form together to create a hymenium which is highly sensitive to the presence of free water. The hymenium is distributed over sterile, dikaryotic supporting tissues which protect it from rain. The hymenium can be exposed on **gills** or **pores** beneath the fruit body, seen in the **toadstools** and **bracket fungi**, or enclosed within chambers as in the **puffballs** and **truffles** *(Fig. 7)*.

H4 FUNGAL SPORE DISCHARGE AND DISPERSAL

Key Notes

Fungal spores	Spores allow fungi to spread, to maintain genetic diversity and to survive adverse conditions.
Spore discharge	Spores may be discharged from parent mycelium by passive or active means. Passive mechanisms include using wind and water as dispersants; active mechanisms use explosive principles.
Air spora	Spores in the atmosphere can affect human, animal and plant health. They can cause allergies and spread plant disease.
Related topics	Fungal cell structure and growth (H1) Fungal nutrition (H2) Reproduction in fungi (H3) Detrimental effects of fungi in their environment (H7)

Fungal spores

There are two conflicting requirements fungi have for their spores. Spores must allow fungi to spread, but they must also allow them to survive adverse conditions. These requirements are met by different types of spores. Small, light spores are carried furthest from parent mycelium in air and these are the dispersal spores. They are usually the products of asexual sporulation, the **sporangiospores** and the **conidiospores,** and so spread genetically identical individuals as widely as possible. Genetic diversity is maintained by sexual reproduction, and the spore products are often large **resting spores** that withstand adverse conditions but remain close to their site of formation. Spores therefore vary greatly is size, shape and ornamentation, and this variation reflects specialization of purpose.

Spore discharge

Spores that have a dispersal function can be released from their parent mycelium by **active** or **passive** mechanisms (*Fig. 1*). As many spores are wind dispersed, they are produced in dry friable masses which are passively discharged by wind. Other spores are passively discharged by water droplets splashing spores away from parent mycelium.

Fungal spores can be actively discharged by **explosive** mechanisms. These mechanisms use a combination of an increasing turgor pressure within the spore-bearing hypha, combined with an in-built weak zone of the hyphal wall. This insures that when the hypha bursts the spore discharge is directed for maximum distance. Asci are usually dispersed in this way, and a few sporangia too. Basidiospores are also actively discharged, but by an as yet unknown mechanism.

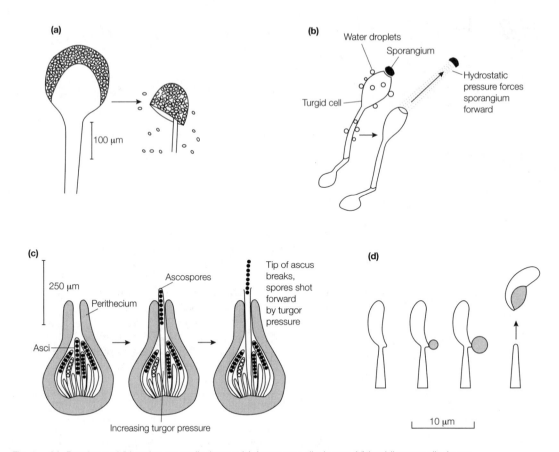

Fig. 1.   (a) Passive and (b) active spore discharge; (c) Ascospore discharge; (d) basidiospore discharge.

Air spora   Air-borne fungal spores can be carried great distances. Their presence in the air can have an impact on human health as they can cause **allergic rhinitus** (hay fever) and **asthma**. Many plant diseases that cause significant economic losses are air borne (see Topic H7). Spore clouds can be tracked across continents, and epidemic disease forecasts can be made, depending on weather conditions and air-spora counts.

H5 RELATED PHYLA: THE WATER AND SLIME MOLDS

Key Notes

Additional phyla	Convergent evolution has given rise to a number of groups of eukaryotic organisms which, while morphologically similar, are taxonomically distinct. All are unicellular for the greater part of their life cycles, and have an absorptive nutrition.
Oomycota and Hyphochytridio-mycota	These two phyla contain the water molds. Members of the Oomycota have cellulose-containing cell walls, tubular mitochondria and a diploid vegetative state. Members of the hyphochytrids also have cellulose-containing cell walls, but a haploid vegetative stage is dominant.
Plasmodio-phoromycota	Members of this group are intracellular parasites of plants. They form multinucleate unwalled protoplasts within the host cell, and cause hypertrophy and hyperplasia of surrounding host tissue.
Dictyosteliomycota and Acrasiomycota	These two phyla contain organisms that are termed cellular slime molds. They are soil dwelling, uninucleate amebae for most of their life cycle but form a plasmodium at sporulation.
Myxomycota	Myxomycota are acellular slime molds. They exist as haploid amebae for their vegetative stage, but fuse in pairs to form a diploid cell which undergoes repeated mitotic nuclear divisions without cell division, forming a plasmodium. The life cycle is then very similar to the cellular slime molds.
Related topics	Fungal cell structure and growth (H1) Reproduction in the algae (I3)
	Algal cell structure and growth (I1) Taxonomy of protozoa (J1)

Additional phyla　　There are a number of eukaryotic organisms that have a similar morphology to fungi, and have an absorptive nutrition, but have no close phylogenetic relationship with fungi. Their taxonomy is still uncertain, and in keeping with the desire to have monophyletic taxonomic groups, in this book they are placed within their own phyla, following the scheme adopted by Alexopoulous and Mimms. There are two major groups of phyla, the **water molds**, the Oomycota and the Hyphochytridiomycota, and the **slime molds.** Their morphological similarity has probably arisen because of convergent evolution, where their problems of being in the terrestrial environment, and being small and relatively immobile, lead to similar adaptations.

Oomycota and Hyphochytridio-mycota

Members of the Oomycota are common water molds, saprophytes or parasites of animals and plants. They share many morphological characters of the chytrids (see Topic H1), but differences include cellulose-containing cell walls, tubular mitochondrial cristae and a life cycle where the dominant somatic phase is diploid rather than haploid or dikaryotic. They also possess biflagellate zoospores instead of the single flagellate zoospores found in the chytrids. From DNA-sequence analysis their closest relatives appear to be the dinoflagellates (see Topics I1 and I3). These organisms are capable of causing devastating disease in both plants and fish.

The Hyphochytridiomycetes are also water molds, and these organisms have only recently been separated from the chytrids, Like the Oomycota, though they share morphological and nutritional similarities with the chytrid fungi, they have some cellulose in their cell walls, and DNA-sequence data also indicate a relationship with the dinoflagellate algae closer than that with the fungi.

Plasmodio-phoromycota

Members of this phyla of organisms are destructive **intracellular parasites** of plants and cause enlargement and multiplication of host cells, creating large and unsightly clubbed roots and often death of the plant. The best known disease is **Club root** of crucifers. The parasite forms multinucleate, unwalled protoplasts within the enlarged host cell, which have an absorptive nutrition. These organisms reproduce by cyst formation within the host cell, and these are released into the soil on breakdown of the plant root where they can persist for years. The presence of a suitable host plant root breaks dormancy, and the cyst germinates by producing anteriorly biflagellate zoospores that swim to the host root and actively penetrate it.

Dictyostelio-mycota and Acrasiomycota

These are the **cellular slime molds.** They exist for most of their life cycle as haploid, uninucleate amebae that feed within soil by engulfing bacteria. Although often grouped together, these two phyla have significant differences

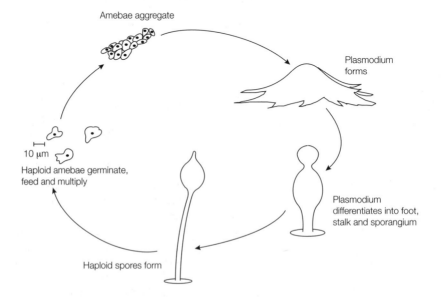

Fig. 1.   Life cycle of a cellular slime mold. Redrawn from Sleigh, M., Protozoa and Other Protists, *1991, with kind permission from Cambridge University Press.*

in their life cycles and structures. They are often included in the protozoa subphylum Sarcodina (see Topic J1).

Starvation of amebae leads to their aggregation into multicellular 'slug', or **plasmodium**, which eventually differentiate into foot, stalk and sporangium, where haploid spores are formed which are disseminated and give rise to haploid amebae (*Fig. 1*).

Myxomycota   Organisms in the myxomycota are termed the **acellular slime molds**. In the vegetative part of their life cycle they exist as haploid amebae, with a phago-cytic nutrition. However, under certain conditions, amebae fuse, or they produce flagellate cells which fuse, to form a diploid cell (*Fig. 2*). This diploid cell grows, feeds and undergoes repeated mitotic nuclear divisions without cell division, forming a large plasmodium. The fruit bodies are similar to those of the cellular slime molds with a foot, stalk and sporangium, but within the sporangium meiosis occurs and haploid spores are produced which germinate as amebae. Like the cellular slime molds, they are often included in the protozoan subphylum Sarcodina (see Topic J1).

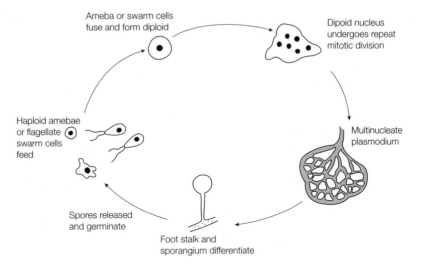

Fig. 2   Life cycle of an acellular slime mold. Redrawn from Sleigh, M., Protozoa and Other Protists, 1991, with kind permission from Cambridge University Press.

H6 BENEFICIAL EFFECTS OF FUNGI IN THEIR ENVIRONMENT

Key Notes

Bread and brewing	The products of yeast fermentation (CO_2 and alcohol) are exploited in bread making and alcohol brewing. Both processes enhance the value of the substrate but contribute little to its nutritional value.
Symbioses	Fungi can enter into specialized and intimate, mutually beneficial associations with higher plants, other microbes and animals. The associations can be external to the host cell, as in the ectomycorrhizae and lichens, or inside the cell as in the endomycorrhizae and endophytic fungi.
Decomposers	Fungi are the main agents of decay of plant wastes in the environment, decomposing substrates to CO_2, H_2O and fungal biomass, and releasing other nutrients back to the biosphere.
Biological control	Fungi can be used to control insect pests, weed plants and plant diseases by exploiting their natural antagonistic, competitive and pathological attributes.
Bioremediation	The degradative abilities of fungi can be exploited to decompose man-made pollutants such as hydrocarbons, pesticides and explosives. They may decompose substrates into CO_2 and H_2O by respirative pathways, or they may reduce toxicity by co-metabolic activity.
Industrially important natural products of fungi	Fungi naturally produce antibiotics, immunosuppressants, acids, enzymes and several other classes of useful natural products. They also can be used to produce large quantities of protein.

Related topics	Fungal cell structure and growth (H1)	Reproduction in fungi (H3)
	Fungal nutrition (H2)	Beneficial effects of fungi in their environment (H6)

Bread and brewing

The metabolic products of yeast metabolism have long been exploited by humans for **bread making** and **alcohol brewing**. Yeast metabolism of flour starch by respiration generates CO_2 which is trapped within the gluten-rich dough and forces bread to rise (see Topic B1). In the brewing of alcohol, yeasts are forced into fermentative metabolism in the sugar-rich, low-oxygen environments of beer wort or crushed grape juice. The fermentative metabolism is inefficient, and only partially metabolizes the available substrates, yielding CO_2

and **ethanol**. Both processes considerably enhance the value of the original substrate while contributing little to its nutritional status!

Symbioses

Fungi can enter into close associations with other microbes and with higher plants and animals. These beneficial associations are termed **symbioses**, and in most cases the symbiotic fungus gains carbohydrates from its associate, while the associating organism gains nutrients and possibly protection from predation and herbivory or plant pathogens (see Topic F1).

Fungal symbioses are common on plant roots, and the symbiotic roots are termed **mycorrhizae** (*Fig. 1*). Their presence enhances plant-root nutrient uptake and plant performance. The fungal association can be predominantly external to the root tissue. These associations are called **ecto**mycorrhizae and can be seen on beech and pine trees, for example. Other associations are predominantly within the plant root, and are termed **endo**mycorrhizae. These associations are seen on the roots of herbaceous species like grasses, but are also found in the roots of tropical species of tree and shrub.

Other associations between fungi and plants can occur within leaves or stems, and these are called **endophytic** fungi. They live almost all their life cycle within the host, grow very slowly and do not cause any signs of infection. They appear to protect their host from herbivory and fungal infection by the production of

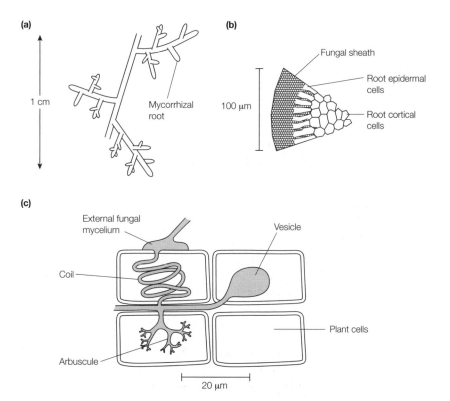

Fig. 1.   Structure of ecto- and endo-mycorrhizae. (a) Macromorphology of ectomycorrhiza; (b) micromorphology of ectomycorrhizae; (c) micromorphology of endomycorrhizae. Redrawn from Isaac, S., Fungal–Plant Interactions, 1991, with kind permission from Kluwer Academic Publishers.

metabolites. However, these products can have dramatic effects on herbivorous animals, causing symptoms of fungal toxicosis similar to St Anthony's Fire (see Topic H7).

Some fungi can form very intimate associations with algal species. These associations are termed **lichens**, and they have a form quite distinct from that of either component species, with **crustose**, **foliose** or **fruticose** thalli (*Fig. 2*). They are slow growing, and are adapted to occupy extreme or marginal environments, like bare rock faces, walls and house roofs. As they are under quite extreme stress they are highly sensitive to pollution, for instance, from acid rain or heavy metals, and their presence or absence in an environment has become a useful indicator of urban and industrial pollution.

Fungi can also associate with insect species in symbioses of varying intimacy. Some species of ants culture specific fungi on cut plant remains within their nests, and then browse on the fungal mycelium that develop. **Termites** have symbiotic fungi within their guts, which in association with a consortium of other microbes, including protozoa and bacteria, help the termite digest its woody gut contents.

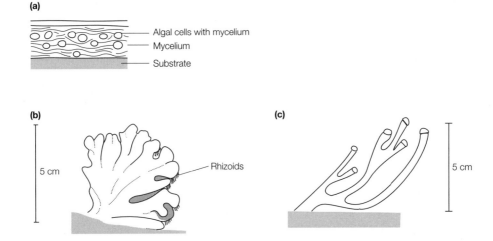

Fig. 2.   Lichen structure. (a) Crustose, tightly attached to substrate; (b) foliose, loosely attached to substrate; (c) fruticose, attached only at base.

Decomposers   The degradative processes that fungi perform with their extracellular enzymes are essential to the terrestrial biosphere (see Topics F1 and H2). They are the main agents of decay of cellulosic wastes produced by plants, which in the tropical rain forest can amount to $12\,000\,\text{kg hectare}^{-1}\,\text{year}^{-1}$. They decompose this material into CO_2, H_2O and fungal biomass, which in its turn is decomposed by other microbes, returning mineral nutrients like phosphorus, nitrogen and potassium to the biosphere.

Biological control   The natural attributes of fungi as disease-causing organisms can be exploited by humans to control weed plant populations and insect pests. They are even capable of parasitizing plant disease-causing fungi. This process is termed biological control and can be an alternative to the application of chemical pesticides. Applications of fungal propagules to targeted problem populations can cause epidemic disease or overwhelm and monopolize a niche.

Bioremediation

The degradative processes that fungal enzymes catalyze on their natural substrates can be used on other, man-made substrates to provide **biological cleanup** (see Topic F1). Hydrocarbons like oils can be degraded by fungi and other microbes to CO_2 and H_2O by aerobic respiration. These activities are termed bioremediation, and contaminated areas of land can be actively bioremediated by the addition of fungal propagules.

Pesticides, explosives and other recalcitrant molecules can be changed by **co-metabolic** activities of fungi, where enzymes normally used for one metabolic process within the fungus coincidentally catalyze another reaction. The products of co-metabolic reactions are not utilized any further by the fungus, but can sometimes be used by other microbes in the ecosystem. Reactions like these can lead to reduced toxicity of some contaminants, but in other cases can lead to **activation** of the pollutant, leading to an increase in the toxicity of the compound.

Industrially important natural products of fungi

Fungi are able to produce many different types of metabolite that are of commercial importance. These include **antibiotics** (e.g. Penicilliins and Cephalosporins, see Topic F7) and **immunosuppressants** (cyclosporins), important in medicine, **enzymes** that are used in the food industry (e.g. α-amylase, rennin), other enzymes (e.g. cellulases, catalase), **acids** (e.g. lactic, citric) and several other products.

Protein extracted from fungi is also an important commercial product. In the 1960s protein from yeasts (single-cell protein) was developed. Currently, fungal protein extracted from *Fusarium graminearum*, known as **Quorn**, is a useful meat alternative.

H7 DETRIMENTAL EFFECTS OF FUNGI IN THEIR ENVIRONMENT

Key Notes

Biodeterioration

Degradative activities of fungi in unwanted situations cause significant economic losses. Materials containing large quantities of cellulose, leather and hydrocarbons can be used as substrates by fungi, providing that there is an adequate water supply.

Plant disease

Fungi are capable of causing significant losses to crops both before and after harvest.

Animal and human disease

Fungi can cause both superficial and deep, life-threatening infections of both man and animals.

Fungal toxicosis

Ingestion of fungi or their secondary metabolic products, accidentally or deliberately, can cause intoxication and occasionally death of both humans and animals.

Related topics

Fungal cell structure and growth (H1)
Fungal nutrition (H2)

Reproduction in fungi (H3)
Beneficial effects of fungi in their environment (H6)

Biodeterioration

The same extracellular enzymes that are important in the degradation of leaf litter and the recycling of nutrients in the biosphere can cause massive economic losses when they occur in circumstances where they are not wanted. Fungi can attack and utilize as substrates paper, cloth, leather and hydrocarbons, but also can cause degradative change in other materials, for instance, glass and metal, because of their ability to produce acid as they grow. The supply of water is a key control point in these processes, and keeping substrates dry is an effective way of avoiding these changes.

Plant disease

Fungi are capable of attacking all plant species, causing serious damage and in some circumstances even death. In crop production over half of potential crop yield is lost to plant pathogens, most of it to **fungal disease**. In storage, up to one-third of the harvested product can be lost to post-harvest disease, again mostly as a result of the activities of fungi. Use of **fungicides** can reduce both pre- and post-harvest disease, and **plant-breeding** programs can introduce disease-resistant strains of crop plant. Post-harvest losses can be reduced by storage of products at low temperatures and low moisture levels.

Animal and human disease
Human and animal epidermis can be attacked by fungi, causing superficial damage and discomfort like **ringworm** infections, **athletes foot** and **thrush**. Other deeper, systemic fungal infections of the lung and central nervous and lymphatic systems cause much more serious diseases, for example, **aspergillosis, coccidiomycosis, blastomycosis, histoplasmosis** and **pneumocystis** pneumonia are all caused by fungi. Although most humans experience superficial fungal infections and survive, these deeper diseases are especially dangerous for the immunocompromised patient after transplantation, and the HIV-positive population.

Fungal toxicosis
Accidental or deliberate consumption of wild fungi or fungally contaminated food can lead to poisoning or **toxicosis** of the consumer because some fungi naturally contain toxic metabolites called **mycotoxins**. Deliberate toxicosis can arise from consumption of mushrooms and toadstools that are known to contain naturally hallucinatory drugs like **psilocibins**, which lead to euphoric states followed by extreme gastrointestinal distress. Accidental consumption of misidentified fungal fruit bodies can lead to fatal mushroom poisoning from fungal toxins, causing total liver failure between 8 and 10 h. Consumption of food accidentally contaminated by fungal metabolites also leads to human and animal death. For instance, rye flour contaminated by the ergots of the fungus *Claviceps pupurea* leads to the symptoms of **St Anthony's fire**, where peripheral nerve damage is caused by the presence of **ergometrine** in the fungal tissue. This can be followed by gangrene of the limbs and death. Detection of fungal mycotoxins such as **ochratoxin** in apple juice and **aflatoxin** in peanuts have also caused problems for food producers and consumers and forced improvements in product processing.

I1 ALGAL CELL STRUCTURE AND GROWTH

Key Notes

Algal structure

Most protistan algae are unicellular, photosynthetic organisms. A few have a filamentous or membranous morphology. They may have a cellulose cell wall or a proteinaceous pellicle. Many species are flagellate. Most contain chloroplasts, which vary in structure and pigment content.

Algal taxonomy

There are four major phyla in the protistan algae that have very different origins. The green, golden brown, euglenids and dinoflagellates are all thought to have arisen separately after a symbiotic association between a flagellate and a photosynthetic ancestor.

Algal growth

Growth in most unicellular algae is synonymous with longitudinal binary fission. Cenocytic, tubular or filamentous algae grow by tip growth like the fungi. Other filamentous or membranous algae grow by intersusception of new cells into the filament.

Kinetics of growth

The kinetics of growth are similar to those of fungi, but in addition to estimations of growth by mass measurement, cell counts and chlorophyll content can be assessed. Rapid cell division can lead to very high cell populations, only limited by nitrogen, phosphate or silicon availability.

Related topics

The microbial world (A1)	Related phyla: the water and slime
Bacterial cell structure (D2)	molds (H5)
Bacterial growth and cell cycle (D8)	Detrimental effects of algae (I5)
Cell division and ploidy (G3)	Cellular structure of protozoa (J2)
Fungal cell stucture and growth (H1)	

Algal structure

Most protistan algae are unicellular, photosynthetic organisms that multiply by binary fission. They may be motile or non-motile. In some algal species, daughter cells from division do not fragment, but instead aggregate to form **colonial**, **filamentous** or **membraneous thalli** (*Fig. 1*). However, protistan algae never form true tissues.

If present, most algal cell walls are formed from cellulose, sometimes impregnated with silica or calcium carbonate. Walls may be fibrillar, similar to those of the fungi, or they may be composed of distinctive **plates** secreted from the Golgi body. In one group there is no rigid cell wall, but the cell is protected by a flexible **proteinaceous pellicle** beneath the plasma membrane

Algal cells contain nuclei, mitochondria, ribosomes, Golgi and chloroplasts (Topic G2). The internal cell structure is supported by a network of microtubules

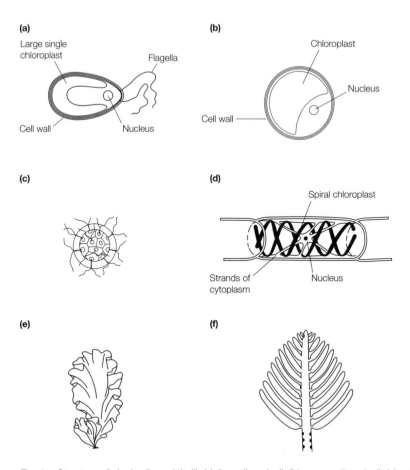

Fig. 1.   Structure of algal cells and thalli. (a) A motile unicell; (b) non-motile unicell; (c) colonial forms; (d) filmentous forms; (e) membranous (only two cells thick); and (f) tubular.
Reproduced from Clegg, C.J., Lower Plants, 1984, with kind permission from John Murray (Publishers) Ltd.

and endoplasmic reticulum. They often possess flagella that have a 9 + 2 micro-tubule arrangement within them (see Topic J2) and there may be one or two per cell. They may be inserted **apically**, **laterally** or **posteriorly** and trail or girdle the cell. The flagellum can be a single whiplash or it can have hairs and scales. The presence of **eyespots** near the flagellar insertion point allows the cell to swim towards the light. Movement may be by lateral strokes or by a spiral movement that can push or pull the cell through the water.

Chloroplasts are very variable structures in the algae and can be large and single, multiple, ribbon-like or stellate (*Fig.1*). Their shape and pigment content are useful distinguishing taxonomic features. The numbers of ER membranes that surround the chloroplast, and the numbers of thylakoid stacks within them, are also important indicators of phylogeny.

Algal taxonomy   Algae are a **polyphyletic** group that contain members that have a very different ancestry. The taxa Alga is used to denote photosynthetic protists, but many species within this kingdom have the ability to survive by photosynthetic, phagocytic or saprotrophic forms of nutrition. Molecular and ultrastructural

Table 1. Features of the main groups of Algae

Group	Pigments	Storage products	Flagellae	Cell wall	Chloroplast ER	Thylakoid stacks
Green algae (Chlorophyta)	Chl. a, b, caroteins, xanthophylls	Sugars, starch, fructosan	1, 2–8, equal apical/sub-apical	Cellulose, mannan, protein, $CaCO_3$	0	3–6
Golden brown algae (Chrysophyta incl. diatoms)	Chl. a, c1 and c2, carotene, fucoxanthin or phycobilins	Chrysolaminarin, oils	1–2, +/– equal or unequal, apical/sub-apical	Cellulose, silica, $CaCO_3$, chitin or none	2	3
Euglenids (Euglenophyta)	Chl. a,b, carotene, xanthophylls, carotene	Paramylon, oils and sugars	1–3, +/– apical	Proteinaceous pellicle	1	3
Dinoflagellates (Pyrrophyta)	Chl. a, c1 and c2, β-carotene, fucoxanthin, peridinin, dinoxanothin	Starch, glucan, oils	2, 1 trailing 1 girdling	Cellulose or absent	1	3

studies have revealed that some algae have arisen when colorless protists have aquired chloroplasts from different free-living photosynthetic species.

At present, the protistan algae include four major phyla. The **green algae** (Chlorophyta), the **golden brown algae** including the **diatoms** (Chrysophyta), the **euglenids** (Euglenophyta) and the **dinoflagellates** (Pyrrophyta). They have been divided into these phyla on the basis of their chloroplast structure and pigmentation, storage products, cell-wall structure and life history (*Table 1*).

Algal growth

Growth in unicellular algae is synonymous with binary fission. In most unicells, haploid or diploid nuclei undergo mitosis, and the cell then divides longitudinally to form two daughter cells. In some species there are two haploid divisions within the parent cell, followed by the formation of four, motile daughter cells (see Topic G3). Species that have shells or plates synthesize the missing sections on each daughter cell.

Some cenocytic filamentous algae grow from the tip of the filament in a way very similar to that of hyphal growth (see Topic H1). Others grow by division of vegetative cells within filaments or sheets.

Kinetics of growth

Accurate estimates of algal growth rates can be made by cell counting or by estimating chlorophyll content of a culture. Kinetics of growth of the algae are similar to those seen in the fungi (see Topic H1), but as algae are photosynthetic, depletion of nutrients other than carbon leads to culture limitations and the stationary and death phases. Nitrogen, phosphates or silicon are frequently limiting. Under optimum conditions, binary fission can be very rapid and can give rise to population explosions called **red tides** (see Topic I5), followed by population crashes.

12 ALGAL NUTRITION AND METABOLISM

Key Notes

Photosynthesis	• Algae are photosynthetic organisms and obtain their carbon and energy requirements by the fixation of CO_2, using photosynthesis. The reactions of photosynthesis can be divided into those that are light dependent, which take place in the thylakoid membrane, and those that are light independent, which take place in the cloroplast stroma. The light-dependent reactions of photosynthesis generate $NADPH + H^+$. The proton gradient generated by $NADPH + H^+$ formation is used to generate ATP by non-cyclic phosphorylation.
	• Light-independent reactions of photosynthesis use the energy generated from light-dependent reactions to synthesize carbohydrate from CO_2 and H_2O.
	• Many algae are not solely photosynthetic. They can also engulf their food (phagocytosis), or they can be saprophytic, with an absorptive nutrition.
Light and photosynthetic pigments	Light is an absolute requirement for photosynthesis. In a terrestrial habitat, light levels are usually adequate, but in the aquatic habitat light energy is rapidly absorbed in the top 0.5 m of the water column. Aquatic algae have evolved three chlorophylls, chl. *a,b* and *c*, and a large number of accessory pigments to allow them to extend the depth to which they can grow.
Oxygen and carbon dioxide	In the terrestrial environment algal CO_2 and O_2 requirements are almost always satisfied by atmospheric gases. In the aquatic environment at 0°C water saturates at 14 mg l^{-1} O_2. However, solubility of O_2 decreases with temperature and increasing dissolved CO_2 levels, and availability of O_2 therefore becomes limiting in warm waters.
Carbon metabolism	Algae use aerobic respiration via glycolysis and the citric acid cycle.
Nitrogen nutrition	Algae cannot fix nitrogen and they must, therefore, obtain it in a fixed, inorganic or organic form. Most can utilize nitrate or ammonia; some require organic compounds. Nitrogen levels may be limiting to growth in marine environments.
Macro-, micro-nutrients and growth factors	Most nutrients are available to excess in the aquatic environment, but phosphates and silicon are only poorly soluble in water and are often limiting to growth in fresh water. Some algae are autotrophic, but many require an external supply of amino acids, vitamins, nucleic acids and other growth factors.

| Water, pH and temperature optima | Algae do not survive severe desiccation and they are found in damp terrestrial habitats or in water. Phagocytic algae have an absolute requirement for liquid water. Most algae can tolerate a wide range of pH and temperature. Some algae are specialized and can inhabit extremely acidic, hot springs while others can complete their entire life cycle below 0°C. |

| Osmolarity | Fresh-water algae have a large difference between internal and osmotic pressure, and they must either have a rigid cell wall or a contractile vacuole to compensate for water uptake. Marine algae are roughly isotonic with sea water and therefore suffer less with this problem. |

Related topics

Heterotrophic pathways (B1)
Autotrophic metabolism (B3)
Beneficial effects of fungi in their
    environment (H6)
Algal cell structure and growth (I1)

Reproduction in the algae (I3)
Beneficial effects of algae (I4)
Taxonomy of protozoa (J1)
Cellular structure of protozoa (J2)

Photosynthesis

Algae are photosynthetic and therefore gain carbon and energy from the fixation of atmospheric or dissolved carbon dioxide using photosynthesis.

There are two sets of reactions in photosynthesis, those that are **light dependent** (light reaction) and those that are **light independent** (dark reaction). In the light-dependent reaction, energy in the form of ATP and NADPH + H$^+$ is generated from light. In the light-independent reactions this stored energy is used to synthesize carbohydrates from CO_2 and H_2O. The reaction can be summarized thus:

$$H_2O + CO_2 \rightarrow Carbohydrate + O_2$$

Light

The photosynthetic reactions take place in the chloroplasts (see Topic I1), the light reactions occurring in the chloroplast thylakoids, and the light-independent reactions occurring in the stroma (see Topic B3).

The chlorophyll pigments are membrane-bound within the thylakoids, and their properties include the ability to be excited by light. Different chlorophylls accept light energy of different wavelengths, depending on their structure. Accessory pigments, the **carotenoids, phycobilins** and **xanthophylls**, also absorb light energy of different wavelengths and pass their excitation to chlorophyll, maximizing the breadth of wavelength over which light energy can be absorbed. In the algae the role of accessory pigments is particularly important as they allow algae to live at different depths in water.

Photosynthesis occurs within **antenna complexes** and **reaction centers** in the thylakoid membrane. Antenna complexes are formed from several hundred chlorophyll molecules plus accessory pigments. Light excitation of the chlorophyll molecule results in an electron in a chlorophyll molecule being excited to a higher orbit, and this energy is transferred between chlorophylls until it is channeled into the chlorophyll molecules of the reaction center.

The reaction center contains two photosystems, called **photosystem I (PS I)** and **photosystem II (PS II)**, with different light-energy absorption maxima.

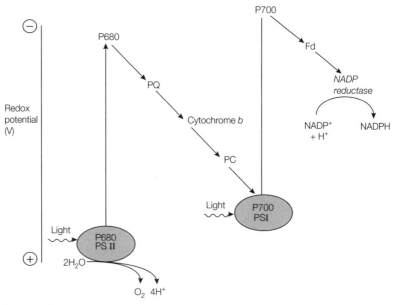

Fig. 1.　The Z scheme for noncyclic photophosphorylation. From Hames, B.D. et al., Instant Notes in Biochemistry, © BIOS Scientific Publishers Limited, 1997.

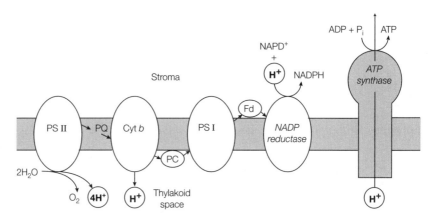

Fig. 2.　Formation of the proton gradient and ATP synthesis. From Hames, B.D. et al., Instant Notes in Biochemistry, © BIOS Scientific Publishers Limited, 1997.

PS I absorbs at 700 nm, and PS II at 680 nm. The reaction centers are linked by other electron carriers, and if the components are arranged by their redox potentials they assume a Z shape, so the scheme is called the **Z scheme** (*Fig. 1*).

The reactions of the Z scheme generate NADPH from NADP. ATP is generated by non-cyclic phosphorylation reactions because of the creation of a **proton gradient** between the thylakoid space and the stroma by the reactions of PS I and PS II. An ATP synthase is present in the thylakoid membrane, and H$^+$ is pumped from the stroma into the thylakoid space, generating ATP (*Fig. 2*).

PS I may operate without PS II in some circumstances, and in this reaction no O_2 is produced; only ATP is produced via the proton gradient.

The light-independent reactions of photosynthesis use the NADPH and ATP generated from the light reactions to synthesize carbohydrates from CO_2 and water. This is called the **Calvin cycle**. A key enzyme in this cycle is **rubisco**, ribulose bisphosphate carboxylase, a large, multi-component enzyme. This enzyme incorporates CO_2 into ribulose 1,5-bisphosphate to form first a six-carbon compound, which then splits to form two three-carbon molecules, 3-phosphoglycerate. Subsequent reactions regenerate ribulose 1,5-bisphosphate from one of the 3-phosphoglycerate molecules, to continue the cycle (*Fig. 3*). The other molecule of 3-phosphoglycerate is transported to the cytosol and used in respiration and to produce storage sugars.

Many algae are not wholly phototrophic. They can be phagocytic, engulfing particulate food through a cytostome (see Topic J2), or they can be saprobic, absorbing food through their cell walls as fungi do.

Light and photosynthetic pigments

Light is an absolute requirement for photosynthetic organisms. Chlorophylls *a,b* and *c* are the photosynthetic pigments in the algae, with chlorophyll *a* being the primary photosynthetic pigment. The absorption **maxima** of chlorophyll *a* are at 430 and 663 nm (*Fig.4a*).

Other chlorophylls are accessory pigments and have slightly different absorption maxima: chlorophyll *b* at 435 and 645 nm, chlorophyll *c*1 at 440, 583 and 634 nm, and *c*2 at 452, 586 and 635 nm. Phycobilins absorb light energy at

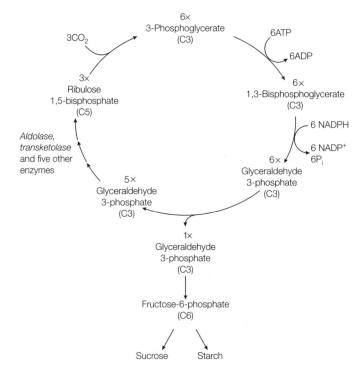

Fig. 3.   The Calvin cycle. From Hames, B.D. et al., Instant Notes in Biochemistry, © BIOS Scientific Publishers Limited, 1997.

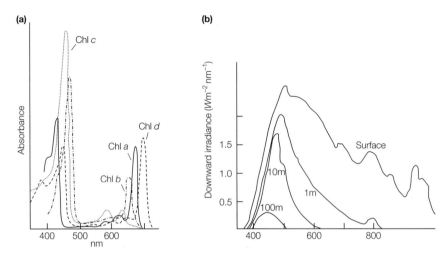

Fig. 4.   (a) Absorption spectra of chlorophylls, and (b) downwards penetration of different light wavelengths.

565 nm. There are also other accessory pigments found in the different algal groups, the carotenes and xanthophylls. Depending on the different accessory pigments present in individual species, light of different wavelengths can be absorbed and utilized in photosynthesis. Light of different wavelengths penetrates water to different degrees (*Fig. 4b*). Red light is absorbed rapidly by water, blue light least.

Differences in pigmentation therefore allow species of algae to occupy different parts of the **photic zone**, the shallow layer of water where sufficient light penetrates to support photosynthesis. The actual depth to which light will penetrate varies with turbidity and dissolved organic matter content (see Topic I4).

Oxygen and carbon dioxide

In an aquatic environment temperature influences levels of dissolved oxygen. Water is saturated with oxygen at 14 mg O_2 l^{-1} at 0°C, but only 9 mg l^{-1} at 20°C. Oxygen utilization by algae and other organisms increases with a rise in temperature; thus, availability of oxygen is likely to limit growth in warmer waters.

Carbon dioxide, the levels of which in water vary inversely with dissolved oxygen, provides carbon to autotrophic, aquatic algae for photosynthesis. Dissolved carbon (as **carbonic acid**, HCO_3^-) is usually present at between 2.2 and 2.5 $nmol^{-1}$, while CO_2 is present at only 10 μmol l^{-1}. Most algae utilize carbonic acid. Anaerobic photosynthesis occurs in a few specialist algae, using hydrogen sulfide or carbon dioxide as terminal acceptors.

The absolute requirement for light in photosynthetic algae means that algae must have adaptations that allow them to remain in the photic zone of their environment. On land this is not problematic, but in the aquatic ecosystem there is a natural tendency for cells to sink, and thus there are cell modifications that counter this. Many algal cells, zoospores and gametes are flagellate and can swim towards the light (see Topic I1) . Other species of algae have structural modifications like spines that increase the resistance of the cell to sinking, and catch water uplift. Some have **gas vacuoles** that increase bouyancy.

Carbon metabolism

Like the fungi, within the cell carbon and energy metabolism is by glycolysis and the citric acid cycle (see Topic B1). Almost all algae are aerobic and use mitochondrial respiration with oxygen as the terminal electron acceptor. A few algae that live in anerobic lake sediments can use alternative electron acceptors such as nitrate.

Nitrogen nutrition

Eukaryotic algae are heterotrophic for nitrogen and must obtain it in a fixed form such as nitrate, ammonia or amino acids. Amino acid requirement is common amongst the algae (see growth factors). Poor availability of nitrogen is a common limiting factor to marine algal growth.

Macro-, micro-nutrients and growth factors

Carbon is rarely limiting for algal growth, but nitrogen and phosphorus often are. Fresh-water productivity is often limited by phosphate availability. Silicon is often limiting to diatom growth in nutrient poor, **oligotrophic** lakes. Most other nutrients are present to excess in the aquatic environment.

Although some algae are completely **autotrophic**, many require an external supply of **vitamins**, frequently thiamin, biotin, B_{12} and riboflavin, purines, pyrimidines and other classes of growth factors. This requirement is termed **auxotrophy**, and it reflects the abundance of dissolved organic matter to be found in their environment, which allows the selection of populations of algae that do not synthesize all their own metabolic requirements (see Topic D7).

Water, pH and temperature optima

Almost all algae are limited to damp environments, and phagocytic algae need to be in water to feed. A few algae that are in lichen symbiosis (see Topic H6) are protected from desiccation and can survive extremely dry conditions.

Most algae are tolerant of a wide range of pH; some are specialized and can inhabit highly acidic environments like those of hot sulfur springs. In highly illuminated, oligotrophic lakes, large changes in pH can be caused by variations in dissolved CO_2 concentrations and these changes can have detrimental effects on algal populations.

Some dormant stages of algae are capable of withstanding 100°C for several hours. Other species can grow and divide at –2°C in sea water, and specialized snow algae have growth optima between 1° and 5°C. Most algae have a temperature optimum between 5° and 50°C.

Osmolarity

Seawater and fresh water have very different **osmolarities**. Fresh water algae have an internal osmotic pressure of 50–150 mOs ml^{-1} while fresh water osmolarity is <10 mOs ml^{-1}. Those species that do not have a rigid cell wall to prevent excessive water uptake have instead contractile vacuoles that collect and expel excess water from the cell (see Topic J2). Marine organisms have cytoplasm that is roughly isotonic with seawater and therefore usually do not require contractile vacuoles.

I3 REPRODUCTION IN THE ALGAE

Key Notes

Life cycles	Asexual reproduction is associated with growth in unicells, with mitotic nuclear divisions occurring after a period of cell growth. Sexual reproduction is associated with the formation of gametes, followed by zygote formation. In organisms with a haploid vegetative phase, meiosis immediately follows zygote formation. Diploid organisms do not undergo meiosis until gamete formation.
Reproduction in Chlorophyta (green algae)	Green algae can be unicellular, filamentous, sheet-like or tubular in their cellular organization. All have a haploid vegetative phase, and motile or non-motile gametes. Gametes are formed by the differentiation of a vegetative cell. Zygote formation is followed by meiosis, producing haploid progeny cells.
Reproduction in Chrysophyta (golden brown algae)	Golden brown algae are almost always unicellular; a few are colonial or filamentous. They have a diploid vegetative phase, and meiosis and gamete formation precede sexual reproduction.
Reproduction in Pyrrophyta (dinoflagellates)	Dinoflagellates have a haploid, unicellular, motile vegetative phase, and sexual reproduction is preceded by the production of gametes which fuse to form a zygote. Meiosis occurs and can be followed either by the formation of a resting cyst, or by a new haploid vegetative phase.
The Euglenophyta (euglenids)	The euglenids are unicellular, motile algae. They appear to have many chromosomes, and polyploidy appears common. Mitotic division is followed by longitudinal cell division, but sexual processes have not been observed.
Related topics	Algal cell structure and growth (I1) Taxonomy of protozoa (J1) Algal nutrition and metabolism (I2) Cellular structure of protozoa (J2)

Life cycles

The dominant vegetative phase of protistan algae can be haploid or diploid. Vegetative growth in the protistan algae is associated with mitotic cell division. In algae with a haploid life history, meiosis occurs at zygote germination, and the cells remain haploid for the whole of their vegetative life. Diploid formation occurs only in the zygote. Algae with a diploid life history are only haploid at gamete formation and after zygote formation continue their life cycles as diploids.

Reproduction in Chlorophyta (green algae)

The green algae range in complexity from unicellular motile or non-motile organisms to **sheets**, **filaments** and **cenocytes**. They are found in fresh and salt water and soil and on and in plants and animals. They have cellulose-containing cell walls, chloroplasts with chlorophylls *a* and *b* and carotenoids and they store **starch**. They have a vegetative phase that is haploid, and sexual reproduction occurs when cells are stimulated to produce gametes instead of normal vegetative cells at binary fission.

Motile compatible gametes first entangle flagellae, cells then **conjugate** and nuclei fuse to form a zygote. The zygote may remain motile, or it may form a thick-walled resting cyst. Meiosis occurs within the zygote, and haploid, flagellate cells are released (*Fig. 1*). Similar events are seen in the colonial forms of the green algae; all cells of a colony may develop into free-swimming gametes after breakdown of the colony structure.

Filamentous green algae may reproduce sexually by conjugation. In this process, two vegetative cells form a **conjugation tube** between them and fuse. The cellular content from one cell then moves into the other, where nuclear fusion and zygote formation occur. The zygote encysts and meiosis occurs before the emergence of a haploid new filament (*Fig. 2*).

Other filamentous species produce motile gametes of two mating types from different vegetative cells. Often one gamete is considerably larger (**macrogamete**) than the other (**microgamete**). In some species, only one motile microgamete is produced and it fuses with a non-motile gamete cell called an **oogonium** to form the zygote.

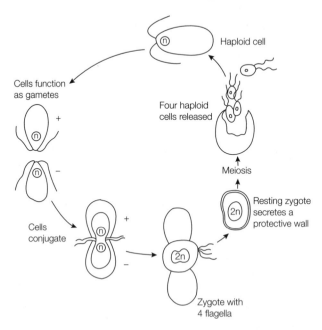

Fig. 1. Sexual reproduction in the unicellular algae. Reproduced from Gross, T., Faull, J.L., Kettridge, S. and Springham, D., Introductory Microbiology, *1995, with kind permission from Kluwer Academic Publishers.*

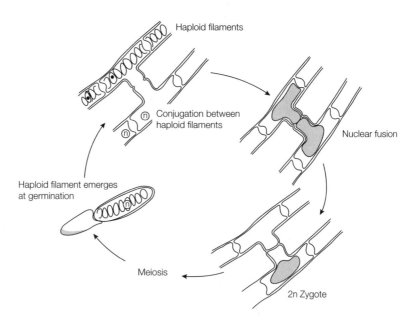

Fig. 2. *Conjugation of two filamentous green algal cells. Reproduced from Gross, T., Faull, J.L., Kettridge, S. and Springham, D.,* Introductory Microbiology, *1995, with kind permission from Kluwer Academic Publishers.*

Reproduction in Chrysophyta (golden brown algae)

The golden brown algae contain several groups of algae that contain chlorophylls *a* and *c* and various accessory pigments. All have arisen after the symbiotic association between a protozoan and **red** or **brown** unicellular algae some time ago in evolutionary history. The pigment content, the numbers of ER layers around the chloroplast, and the numbers of thylakoid lamellae confirm this hypothesis (see Topic I1). They are often described and classified as pigmented protozoa (see Topic J1). This group includes the **cryptophytes**, **chrysophytes**, **xanthophytes**, **bacillariophytes** (diatoms) and **primnesiophytes** (coccolithophorids). Those species that contain pigments that have come from red algae, the cryptophytes, can occupy the very deepest layers of the photic zone (see Topic I2).

Most species in this group are unicellular, but some are colonial or filamentous. Life cycles are similar to those of green algae, but the dominant vegetative stage is diploid. In the chrysophytes spore formation is termed **intrinsic** and it is independent of external conditions. Around 10% of the population will encyst as a zygotic spore in any one generation, allowing populations in optimal growth conditions to maintain genetic diversity and reduce intraspecific competition.

In other members of the group sexual encystment is **extrinsic**, associated with physiological stress. The zygotic spore serves to allow the population to survive stressful environmental conditions. The life cycle of diatoms serves as a general example of sexual reproduction in this group of algae (*Fig. 3*).

Diatoms differ from other members of the golden brown algae because their vegetative stage often lacks flagellae, but they have a have a **gliding motility** on solid surfaces. They have a silica-containing cell wall composed of a pair of 'nested' shells called **frustules** with a **girdle band** around them. The large half of the shell is termed the **epitheca**, the smaller the **hypotheca**. Vegetative cells are diploid, and repeated mitotic divisions leads to a reduction in cell volume as

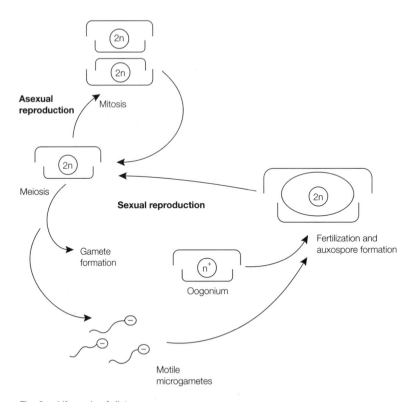

Fig. 3.   Life cycle of diatoms.

daughter cells synthesize new frustules that fit within the inherited parent frustule. Once a 30% reduction in volume has been reached diatoms either produce
a resting spore (or **auxospore**) to regain cell size or they reproduce sexually.

Sexual reproduction begins by the formation of gametes after meiosis. In
round, centric diatoms a single or multiple macrogamate forms within an **oogonium**. Motile microgametes are formed within the other mating-type diatom.
These microgametes are released and fertilization of the oogonium leads to the
formation of a zygote within an auxospore. The auxospore enlarges and secretes
a new pair of full-size frustules. The diploid, vegetative life cycle then
continues.The long, thin diatoms, termed pennate, do not form motile gametes,
but after a meiotic division fuse somatically to form a zygote.

**Reproduction in
Pyrrophyta
(dinoflagellates)**

Dinoflagellates are predominantly unicellular, marine, free-living, motile algae
that contain chlorophylls *a*, *c*1 and *c*2 and the carotenoid **fucoxanthin**. They
store **chrysolaminarin**, a β1–3 linked glucose. Some are capable of phagotrophy;
others live within marine invertebrates and are termed **zooxanthallae** (see Topic
I4). They have a unique cell surface where the cell plasma membrane is under-
lain by a layer of vesicles called **alveoli**, which contain cellulose plates or scales
generated from the Golgi. These vesicles fuse with the plasma and provide the
cell with an effective cell wall. The plates can be impregnated with silica or
calcium carbonate.

Although unicellular, the dinoflagellates are structurally an extremely diverse
group of unicellular organisms. Most have two flagellae inserted into the cell

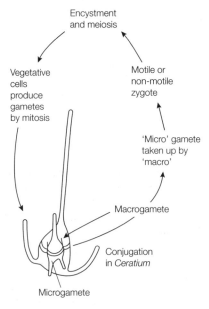

Fig.4.   Reproduction in the dinoflagellates. Reproduced from Gross, T., Faull, J.L., Kettridge, S. and Springham, D., Introductory Microbiology, 1995, with kind permission from Kluwer Academic Publishers.

at right angles to each other, around the midline of the cell. One is wrapped around the waist of the cell in a groove, the other extends from the posterior of the cell.

Dinoflagellates are haploid and have unusual chromosomes which are condensed throughout their life cycle. The chromosomes contain very little histone. Sexual reproduction commences when motile cells differentiate to become macro- or micro-gametes *(Fig. 4)*. The gametes fuse to from a zygote, and meiosis occurs followed by the degeneration of three of the four nuclear products. A haploid, motile cell then emerges from the zygote.

The Euglenophyta (euglenids)

Euglenids are unicellular, motile algae that contain chlorophylls *a* and *b* and store **paramylon** starch. They probably arose when a flagellate protozoan ingested a green alga, and even now Euglena cells can be 'cured' of their chloroplasts, leaving them with a saprophytic and phagocytic nutrition. They inhabit nutrient rich (**eutrophicated**) fresh and brackish water. A few species are parasitic. They have a proteinaceous flexible pellicle that lies underneath the cell membrane very like that of the closely related protozoan species (see Topic J2). The flexibility of this pellicle allows euglenids a characteristic flexibility termed **euglenoid movement** when cells are on solid substrates. Many species are capable of phagotrophy, and some are wholly saprophytic and devoid of chloroplasts and classed as protozoa (see Topic J1). The nuclei appear to contain many chromosomes, and polyploidy appears common. Binary fission is by longitudinal division, beginning at the flagellar insertion, but sexual processes have not yet been confirmed.

14 BENEFICIAL EFFECTS OF ALGAE

Key Notes

Primary productivity	Primary productivity in the oceans due to algae is estimated to be $\sim 50 \times 10^9$ tonnes per annum. Carbon can be released from the algae either as dissolved organic matter (80–90%) or as particulates (10–20%). The algal cells, associated saprophytic bacteria and dissolved organic matter together form the basis of the aquatic food chain.
Symbiosis	Symbiotic associations can occur between algae, fungi and animals. In these associations, carbohydrates from algal photosynthesis are exchanged for nutrients from the fungus or animal partner.
Diatomaceous earth	Diatomaceous earth is formed from the silica-containing shells of diatoms. It has several commercial uses that take advantage of its chemically inert but physically abrasive qualities.
Bioluminescence	Bioluminescence is found in several of the algal phyla and is associated with luciferin–luciferase reactions that occur within scintillons in the algal cell. The function of this reaction is unknown.

Related topics

Beneficial effects of fungi in their environment (H6)	Reproduction in the algae (I3)
Algal cell structure and growth (I1)	Detrimental effects of algae (I5)
Algal nutrition and metabolism (I2)	Beneficial effects of protozoa: symbiotic relationships (J6)

Primary productivity

In the ocean the total plant biomass (as **phytoplankton**) is estimated to be 4×10^9 tonnes (dry weight). Annual net **primary production** is around 50×10^9 tonnes, diatoms accounting for 31%, dinoflagellates 28%, flagellates and coccoids 25% and blue–green bacteria 16% of productivity. Productivity ranges from 25 g carbon m^{-3} in oligotrophic waters to 350 g carbon m^{-3} in eutrophic waters. This is released from living algae as **dissolved organic matter** (**DOM**), which accounts for 80–90% of organic matter present in the sea. The DOM is used as a nutrient source by heterotrophic bacteria; populations between 10^4 and 10^6 ml^{-1} are commonly found in nutrient-poor (**oligotrophic**) waters, and much higher populations are found in nutrient-rich (**eutrophic**) waters (see Topic I5). The algal cells (both alive and as dead particulates), DOM and bacterial populations form the base of the aquatic food chain.

Symbiosis

Both dinoflagellates and green algae can form symbiotic relationships with fungi as lichens (see Topic H6) and with many animals. The zooxanthellae are symbiotic dinoflagellates that are found as coccoid cells within animal cells. They are enclosed in intracellular double membrane-bound vacuoles which remain undigested. They are found in protozoa, ciliates, radiolaria, hydroids, sea anemonies, corals and clams where they provide glycerol, glucose and organic acids for the animal, and the symbiotic alga gains CO_2, inorganic nitrogen, phosphates and some vitamins from the animal.

Zoochlorellae, green algal symbionts, are found in other ciliates, amebae, hydrae, celenterates, flatworms, bivalves and foraminifera. Within their host they provide a supply of maltose, alanine and glycollic acid from photosynthesis, and gain CO_2, inorganic nitrogen and some other macronutrients from the animal partner.

Reef-building corals are only able to build reefs if they have their symbiotic algal partner, and many of the animal hosts are at least partly dependent on the algal partner for carbohydrates. **Radiolaria**, responsible for massive primary productivity in the oceans, are wholly dependent on their algal symbiont for carbohydrate.

Diatomaceous earth

At death, diatom cells fall through the water column to the sea bed. The inert nature of the frustule silica, silicon dioxide (SiO_2), means that it does not decompose but accumulates, eventually forming a layer of diatomaceous earth. This material has many commercial uses to humans, including filtration, insulation and fire-proofing and as an active ingredient in abrasive polishes and reflective paints. Recently, it has been used as an insecticide, where the abrasive qualities of diatomaceous earth are used to disrupt insect cuticle waxes, causing desiccation and death.

Bioluminescence

Many of the marine dinoflagellates are capable of **bioluminescence**, where chemical energy is used to generate light. The light is in the blue–green range, 474 nm, and can be emitted as high intensity short flashes (0.1 sec) either spontaneously, after stimulation, or continuously as a soft glow.

Bioluminescence is created by the reaction between a tetrapyrrole (**luciferin**), and an oxygenase enzyme (**luciferase**). The entire reaction is held within membrane-bound vesicles called **scintillons**, where the luciferin is sequestered by **luciferin-binding protein** (**LBP**) and held at pH 8. Release of light is stimulated by either a cyclical or mechanical stimulation of the scintillon, which leads to pH changes within it. Luciferin is released from LBP, and luciferase is able to activate it. Activated luciferin exists for a very brief time before it returns to its inactivated form, releasing a photon of light energy (*Fig. 1*).

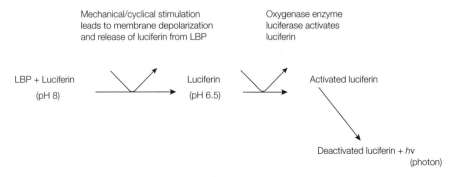

Mechanical/cyclical stimulation
leads to membrane depolarization
and release of luciferin from LBP

Oxygenase enzyme
luciferase activates
luciferin

LBP + Luciferin
(pH 8)

Luciferin
(pH 6.5)

Activated luciferin

Deactivated luciferin + *h*v
(photon)

Fig. 1. Luciferin–luciferase reaction within scintillon.

The distribution of scintillons within the algal cell varies over 24 h. During the night they are distributed throughout the cytoplasm, but in daylight they are tightly packed around the nucleus.

The function of bioluminescence for the algae is uncertain. It appears to be useful in defense against predators. It has become important in medicine and science because the coupled system of luciferin/luciferase can be used to mark cells. Once tagged with the bioluminescent marker, marked cells can be mechanically sorted from other cells, visualized by microscopy or targeted for therapy.

15 DETRIMENTAL EFFECTS OF ALGAE

Key Notes

Toxic blooms	Exponential growth of toxin-containing algae can lead to the accumulation of their toxins in filter-feeding molluscs and fish. The toxins may directly affect these animals or, when they are consumed by higher animals, they can have severe, even fatal effects on the consumer.
Eutrophication	Eutrophication is the over-enrichment of water by the presence of excess of carbon and nutrients. Oxygen consumption by high numbers of algae and bacteria growing in eutrophic conditions leads to oxygen depletion and anaerobic conditions.
Protothecosis	A few algae are capable of infecting humans and animals. Prototheca is a colorless saprophytic, soil-dwelling alga which can cause skin and occasionally deeper lesions when it enters wounds.
Related topics	Algal cell structure and growth (I1) Unusual metabolic pathways (J4)

Toxic blooms

When environmental conditions for growth are optimal, protistan algae can grow exponentially, leading to high local populations (see Topic I1). This can occur in several species of marine dinoflagellates that contain poisons which are toxic to fish, invertebrates or mammals, depending on the species of alga and the class of compound they produce. The toxins are accumulated in the digestive glands of shellfish and when consumed by man cause **paralytic shellfish poisoning**. Many of these compounds are neurotoxic. **Saxitoxin**, accumulated in shellfish and accidentally consumed, causes numbness of the mouth, lips and face which reverses after a few hours.

Toxins can also accumulate in higher animals like fish. For example **ciguatoxin** from *Gambierdiscus toxicus* accumulates in muscle tissue of grouper and snapper and when eaten cause gastric problems, central nervous sytem damage and respiratory failure.

Toxins can also be formed by other groups of algae. Some species of diatoms can produce **domoic acid**, which can accumulate in mussels and, when consumed by humans, causes **amnesic shellfish poisoning**, a short-term loss of memory but which can occasionally cause death. Some of the golden brown algae that give rise to these types of poisoning are highly pigmented members of the Chrysophycae, and high cell densities can be seen as so-called **red tides** in the sea. Other toxins from species of the Prymnesiophycae can affect gill function of fish and molluscs.

Eutrophication   The presence of high populations of algae, their DOM products and the high
                       numbers of bacteria that can be supported by large amounts of organic matter
                       lead to a condition known as **eutrophication**. Water is in a nutrient-rich state,
                       biomass is high and oxygen demand exceeds supply. Anoxic conditions rapidly
                       develop, leading to the death of aerobic organisms.

                       The decomposition of dead material by bacteria leads to further demands
                       for oxygen, and the entire environment becomes anaerobic, allowing for the
                       growth of anaerobic bacteria, and the production of methane, hydrogen sulfide,
                       hydrogen and many other products of anaerobic metabolism (see Topic J4).
                       Eutrophication is commonly seen around untreated sewage outfalls and dairy
                       farm waste run-off, and in fresh water streams polluted by nitrate-rich agri-
                       cultural field run-off.

Protothecosis   Prototheca is a common soil-dwelling alga that has lost its chlorophyll and lives
                       saprophytically. It is an opportunist pathogen that can enter wounds on the
                       feet, where it can cause a subcutaneous infection. The initially small lesion can
                       spread through the skin, producing a crusty, warty lesion. It may spread to
                       lymph nodes and the disease can become debilitating. In rare cases it can grow
                       rapidly in the bloodstream, causing rapid death in ailing or immunosuppressed
                       animals and humans.

J1 TAXONOMY OF PROTOZOA

Key Notes

Protozoa	The protozoa are a subkingdom comprising a large group of unicellular organisms, which exhibit a wide range of size and morphology. The subkingdom can be divided into six phyla.
Phylum Sarcomastigophora	This phylum is characterized by flagellate or ameboid organisms. Members have one type of nucleus and some may contain photosynthetic organelles. This phylum is subdivided into three subphyla, Mastigophora, Opilinata and the Sarcodina, which include several important parasites.
Phylum Labyrinthomorpha	Members of this phylum exhibit ameboid motility and possess a unique organelle, the sagenetosome. Sexual reproduction is typical, with some species producing motile (flagellate) spores.
Phylum Apicomplexa	The Apicomplexa are mostly parasites, characterized by the presence of specialized organelles known as the apical complex. *Plasmodium*, the malaria-causing parasite, is the best known of this phylum.
Phylum Microspora	The organisms in this group are intracellular parasites of animals. Spores are formed during development. Reproduction is via binary fission
Phylum Myxozoa	This phylum contains parasites of cold-blooded vertebrates and annelids, which are similar to microspora
Phylum Ciliophora	Members of this phylum are ciliate during at least part of their life cycle. The phylum is clearly distinguished from other protozoans by the presence of a macro- and micro-nucleus. The outer membrane and associated cilia in this phylum is known as the pellicle. The phylum is divided into three subphyla.

Related topics	Heterotrophic pathways (B1)	Cellular structure of protozoa (J2)
	Autotrophic metabolism (B3)	Reproduction (J5)
	Bacterial cell structure (D2)	Beneficial effects of protozoa: symbiotic relationships (J6)

Protozoa

The **subkingdom Protozoa** consists of a large group of organisms that exhibit a wide diversity of size and morphology. However, all protozoa can be described as unicellular colonial organisms that possess a eukaryotic-like cellular structure. In some groups such as the ciliates there are species that contain bacterial symbionts (symbiosis is a mutually advantageous relationship between two species). The subkingdom can be divided into six phyla.

**Phylum
Sarcomastigo-
phora**

The Sarcomastigophora contain both flagellates and ameboid organisms. These organisms contain only one form of nucleus, but may be multinucleate. This phylum is divided into three subphyla.

Subphylum Mastigophora
These organisms are characterized by the presence of a flagellum (see Topic J2). Some of the species in this group lack flagella during some phases of their life cycles. Reproduction is typically by binary fission. There are two major classes within the subphylum: the **phytoflagellates** (Phytomastigophora) contain pigmented or colorless organelles termed **plastids** and this group exhibits both **autotrophic** (see Topics D3 and I2) and/or **heterotrophic** (see Topic B1) metabolism. The phytoflagellates are usually considered to be algae and are dealt with in further detail in Topic I3. The **zooflagellates** (Zoomastigophora) are obligately heterotrophic. Some species within this group are capable of ameboid locomotion and some produce colonial forms; others possess a siliceous outer skeleton. The range of morphologies exhibited by this group is shown in *Fig. 1*.

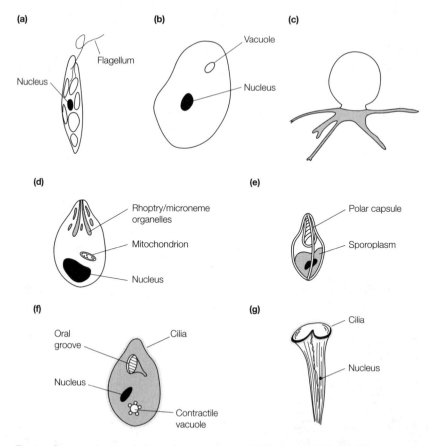

Fig. 1. Comparative morphology of some protozoa: (a) Euglena gracillis, *25–30 μm;
(b)* Entamoeba histolytica, *15–20 μm; (c)* Cryptodifflugia ouiformis, *with a smooth calcareous shell, 30–50 μm; (d) merozoite stage of* Plasmodium talciparum, *1–3 μm; (e) spore of the myxosporan* Thelohanellus, *4–6 μm; (f)* Tetrahymena patula, *80–100 μm; (g)* Stentor roeseli, *600–700 μm.*

Subphylum Opilinata

This is a well-defined group containing organisms bearing numerous closely spaced flagella covering the surface.

Subphylum Sarcodina

This is a large group, all producing pseudopodia (finger-like cytoplasmic extensions). Flagella may be present in some developmental stages. Reproduction in this group is by fission; however, some species such as the slime molds (**Mycetozoea**) have a sexual element in the life cycle. This group also aggregate, producing multicellular forms. These organisms are often considered to be close to the fungi and are frequently grouped with them. They are discussed in further detail in Topic H5. The subphylum also contains the *Foraminifera*; these are sarcodina enclosed within an organic or mineralized **test** (outer shell). Also, the *Radiolaria* which have a siliceous skeleton. Both of these are major components of marine zooplankton. This subphylum also includes several important parasites such as *Entamoeba histolytica*.

Phylum Labyrinthomorpha   Members of this phylum exhibit ameboid motility and possess a unique cell surface organelle (**sagenetosome**) associated with the cytoplasmic network. Sexual reproduction is typical within this group, with flagellate spores formed.

Phylum Apicomplexa   The apicomplexa are parasites characterized by the presence of a group of specialized organelles know as the **apical complex**, located at one end of the organism. *Plasmodium* (the malarial parasite of humans and other animals) is the best known member of this phylum (*Fig. 1*). These organisms exhibit both sexual and asexual reproduction.

Phylum Microspora   This group are intracellular parasites of animals. During development within the host cell, spores are formed and externalized. The thick-walled spore contains the infective sporoplasm and a coiled hollow tube (polar filament) through which the sporoplasm is injected into its host (*Fig. 1*). Reproduction is by binary fission.

Phylum Myxozoa   This group contains parasites of cold-blooded vertebrates and annelids and are similar to microspora in that spores are produced during development.

Phylum Ciliophora   Members of this phylum are ciliate during at least part of their life cycle. The arrangement of hair-like organelles (cilia) on these organisms is used to differentiate organisms within the phylum. The cilia range in size from 5 to 20 μm. Food ingestion can occur by phagotrophy where food particles are taken up through specialized structures such as the **cytosome**. The membranous system and associated cilia surrounding the cell is known as the **pellicle**. Cilia are arranged into motile units, a **kinetid**. Cilia in groups of two are termed **dikinetid** and single cilia **monokinetid**. The morphology of some common ciliates is shown in *Fig. 1*. These organisms also possess a macro- and micronucleus; these have distinct functions. This phylum is divided into three subphyla.

Subphylum Postciliodesmatophora

Ciliates in this group are dikinetid. They possess rows of cilia within the oral cavity.

Subphylum Rhabdophora

Members of this subphylum are monokinetid except around the oral structure where they are dikinetid. The surface pellicular structure is not as well developed as that observed in the other subphyla.

Subphylum Cyrtophora

These are identified as having various forms of ciliary arrangement, monokinetid, dikinetid and polykinetid. The oral cavity bears rows of cilia.

J2 CELLULAR STRUCTURE OF PROTOZOA

Key Notes

Protozoal structure	Protozoa exhibit many eukaryotic characteristics, but possess many unique structures.
Outer surface	A plasma membrane forms the outer boundary of most protozoans. Some also have organic material or minerals such as silica deposited on their outer surface.
Nuclei	The DNA in protozoa is arranged in chromosomes, contained within nuclei. Some protozoa have two or more identical nuclei. Others have distinct types of nuclei; a macronucleus that is associated with cellular function and a micronucleus that controls reproduction.
Flagellum	The flagellum is a filamentous structure that moves in an undulating manner and propels the cell. Its structure resembles a membrane-bound cylinder made up of pairs of microtubules with two central and nine surrounding. Microtubules are made from a protein, tubulin, which has two subunits α and β. The protein dyenin forms arms between outer microtubules and is involved in producing flagellar movement from ATP hydrolysis. The kinetosome is the portion of the flagellum anchored within the cytoplasm.
Cilia	Cilia are smaller than flagella (5–20 μm) in length and propel the cell by beating. The fine structure of cilia is very similar to that of the flagellum.
Contractile vacuoles	Contractile vacuoles are found in free-living protozoa, particularly fresh-water species. They function to regulate osmotic pressure within the cell by expelling water.
Mitochondria	Protozoan mitochondria show some variation from that of typical eukaryotes. Kinetoplasts have branched mitochondria, and mitochondria in other organisms frequently do not produce ATP.
Glycosomes	Glycosomes are unique to the Kinetoplastida and contain the enzymes of glycolysis to the level of phosphoglycerate kinase. This organelle functions closely with the mitochondria.
Hydrogenosome	A small number of anaerobic protozoa possess hydrogenosomes. This organelle contains the enzymes involved in the latter stages of carbohydrate fermentation.

Apical complex	This complex of organelles and structures is uniquely found in the Apicomplexa. The complex consists of an apical ring and rhoptry and microneme organelles and is thought to aid cell invasion.
Other membrane-bound organelles	Protozoans exhibit a variety of membrane-bound structures such as vacuoles that aid buoyancy and extrusomes, which are involved in the secretion of mucous and thus aid motility.
Related topics	The microbial world (A1) Eukaryotic cell structure (G2)
	Heterotrophic pathways (B1) Taxonomy of protozoa (J1)

Protozoal structure

Protozoa exhibit many eukaryotic characteristics (see Topics A1 and G2). However, there are many unique structures found in the protozoa. Some protozoa do not possess mitochondria; these are termed **amitochondrial**.

Outer surface

The plasma membrane forms the outer boundary of protozoa, and these membranes exhibit a typical bilayer structure. However, many have a number of bilayer boundary membranes. Some have a **thick layer** of organic material or minerals such as silica deposited on their outer surface (see Topics I3 and J1). These deposits may occur throughout the life cycle of some species or may be limited to one life cycle form such as the cyst.

Nuclei

The DNA in protozoa is arranged in **chromosomes**. The number of chromosomes varies between species. Chromosomes are contained within nuclei. The structure of the nucleus is similar to that in other eukaryotes (see Topic G2). However, some protozoa have two or more identical nuclei and others have two distinct types of nuclei. The **macronucleus** is the larger of the two and is associated with cellular function. The **micronucleus** functions in controlling reproduction.

Flagellum

The flagellum is a long (50–200 μm) filamentous structure that moves in an undulating manner. The flagellum resembles a membrane-bound cylinder and located in the center of this is the **axoneme** which is made up of two central protein fibers (the fibers are tube like and termed **microtubules**) surrounded by nine pairs of outer microtubules (9 + 2 arrangement). Each outer pair has arms projecting towards a neighboring doublet (*Fig. 1*) and a spoke extending to the inner pair of microtubules. Microtubules are formed from a protein, **tubulin.** Tubulin is composed of two subunits α and β arranged in a helical fashion (*Fig. 1*). The projecting arms between outer subunits are made up of the protein, **dyenin**. This protein is involved in converting the energy released from ATP hydrolysis into mechanical energy for flagellar movement. Movement is produced by the interaction of the dyenin arms with one of the microtubules of adjacent doublets. A basal body (**kinetosome**) anchors the flagellum within the cytoplasm. Variation in the structure of flagella can be observed in the protozoa with some species exhibiting a 9 + 1 arrangement of microtubules. Others possess strengthening fibers that run parallel to the axoneme, and some flagella possess bristle-like projections.

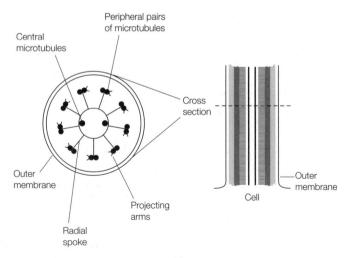

Fig. 1.   General structure of a flagellum or cilium.

Cilia

Cilia are typically 5–20 μm in length and propel the cell by beating. The beating of cilia in these organisms is usually co-ordinated and can be seen as waves over the surface of the organism. The structure of cilia is very similar to that of the flagellum

Contractile vacuoles

Contractile vacuoles are found in free-living protozoa, particularly freshwater species. Their function is to regulate osmotic pressure within the cell by expelling water. These structures consist of a central contractile vacuole, reinforced by bands of microtubules and surrounded by collecting canals that collect fluid from the cytoplasm (*Fig. 2*). Water is expelled from the vacuole through a pore in the outer surface.

Mitochondria

Protozoan mitochondria show variation from that of typical eukaryotes. In the phylum Kinetoplastida, each organism contains only one large tubular, branched mitochondrium. Mitochondria in other groups have few or no

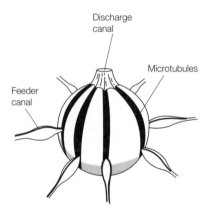

Fig. 2.   Diagram of a contractile vacuole.

cristae (internal membranous projections); these mitochondria frequently do not perform ATP synthesis, but are required for production of metabolic intermediates for synthetic reactions.

Glycosomes

Glycosomes are unique to the Kinetoplastida. This organelle possesses a single unit membrane and contains the **enzymes of glycolysis** (see Topic B1) to the level of phosphoglycerate kinase. The organelle functions closely with the mitochondria–kinetoplast complex.

Hydrogenosome

The hydrogenosome is a unique membrane-bound organelle found in a small number of anaerobic protozoans that lack mitochondria. This organelle contains enzymes involved in the latter stages of **carbohydrate fermentation** (the incomplete breakdown of carbohydrates) (see Topic B1) and the formation of ATP. Molecular hydrogen is a characteristic product of metabolism by this organelle.

Apical complex

This complex of organelles and structures is uniquely found in the Apicomplexa (see Topic J1). The complex comprises an **apical ring** and **rhoptry** and **microneme** organelles. These structures function to aid cell invasion by these parasitic organisms.

Other membrane-bound organelles

Protozoans exhibit a variety of unusual membrane-bound structures within the cell. Examples of these include vacuoles, thought to aid buoyancy. Other groups of protozoa have specialized granules observed at the periphery of structures involved in ameboid or gliding motility. These are termed **extrusomes** and release a mucous secretion to aid motility.

J3 PHYSIOLOGY OF PROTOZOA

Key Notes

Sensation and coordination	Protozoa respond to a range of environmental stimuli. These responses are mostly linked to motility. Membrane-bound proteins are involved in the translation of the stimulus into an intracellular event. There is evidence that Ca^{2+} movement is a major effector in environmental response.
Flagellar motion	Flagella exhibit a wave-like motion that propels water either away or towards the cell. Movement away from the cell causes movement of the cell in a direction opposite to the flow of water.
Ciliary motion	Cilia exhibit an oar-like motion. During the power stroke, the stiffened cilium propels water lateral to its surface and parallel to the cell surface. For structures as small as cilia, water is highly viscous and thus cilia exert a considerable propelling force on the surrounding water.
Ameboid locomotion	Ameboid movement is thought to involve the streaming of endoplasm from one end of a cell to the other. This is driven by unequal pressure applied by a viscous and filamentous ectoplasm (a viscous region of cytoplasm close to the outer cell surface). This type of movement requires that the cell is in contact with a solid surface.
Feeding and nutrition	Protozoa have a wide range of feeding mechanisms. Ingestion of particulate material requires phagocytosis (engulfment of particles by the cell) and digestion. Most protozoa are heterotrophic (utilize organic sources for energy), but some can also function as phototrophs (obtain energy from light). Anabolic (synthetic) metabolism in the protozoa is diverse, but some organisms that are involved in parasitic or symbiotic relationships have a reduced synthetic capacity as synthetic precursors can be obtained from either the host or from the other partner in the symbiotic relationship.
Respiration	Aerobic protozoa must obtain oxygen by simple diffusion. Heterotrophs will utilize oxygen, if available, to maximize ATP generation. In anaerobic environments heterotrophs will exhibit a fermentative metabolism.
Osmoregulation	Most freshwater protozoa are hyperosmotic (intracellular solutes and ions present at high concentrations relative to their environment) and hence accumulate water by osmosis. Water must be removed to prevent lysis of the cell. The contractile vacuole is the main organelle for expulsion of water and waste.

Related topics	Heterotrophic pathways (B1)	Autotrophic metabolism (B3)
	Electron transport and oxidative phosphorylation (B2)	Cellular structure of protozoa (J2)

Sensation and coordination

Protozoa, like other eukaryotic cells, respond to environmental stimuli such as chemicals (**chemotaxis**), gravity (**geotaxis**) and light (**phototaxis**). These sensory responses are mostly linked to motility. Translation of the stimulus to an intra-cellular event involves specific proteins in the outer membrane of the organism, and the influx (entry) and efflux (exit) of ions such as Ca^{2+} into or out of the cell. There is evidence that movement of Ca^{2+} coupled with the activity of **calmodulins** (small Ca^{2+}-binding proteins) activates the ATPase (see Topic B2) activity of dyenin in flagella and cilia (See Topic J2).

Flagellar motion

Flagella typically exhibit a sinusoidal motion in propelling water parallel to their axis (*Fig. 1a*). The undulating action of the flagellum either propels water away from the surface of the cell body (**pulsellum**) or draws water towards and over the cell body (**tractellum**). A pulsellum produces cellular motion very much like that of a fish tail. The propelled water leaving the flagellum produces an equal and opposite force on the cell, causing it to move in a direction oppo-site to the flow of water.

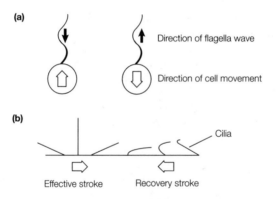

Fig. 1.   Motility in (a) flagellates and (b) ciliates.

Ciliary motion

Cilia exhibit an oar-like motion. During the **power stroke**, the stiffened cilium propels water lateral to its surface and parallel to the cell surface. For struc-tures as small as cilia, water is highly viscous and the cilium exerts considerable propelling force on the surrounding water as it undergoes an oar-like motion (*Fig. 1b*). However, during the recovery phase when the cilium is returning to a stiffened configuration, it gradually unfolds towards the tip, thus avoiding the heavy resistance of the surrounding water. Owing to the viscous proper-ties of the surrounding water, the action of the individual cilia become coordinated. The water molecules provide a liquid matrix that constrains the motion of the cilia. Thus, eventually, the individual activities of the cilia become coupled, increasing efficiency and conserving energy.

Ameboid locomotion

Ameboid movement results from cytoplasmic streaming within the cell and formation of finger-like extensions (**pseudopodia**). This type of movement requires that the cell is in contact with a solid surface. The cell is partitioned into two regions, a viscous **ectoplasm** and a less viscous **endoplasm**. Within

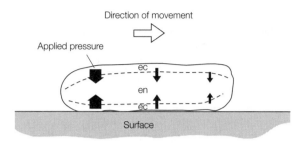

Fig. 2. A proposed method for ameboid movement. Symbols: ec = ectoplasm; en = endoplasm.

the ectoplasm are myosin and actin filaments. In the endoplasm only unpolymerized actin is present. To produce ameboid movement it has been suggested that the microfilaments in the ectoplasm produce a pressure on the endoplasm (*Fig. 2*). If this pressure is unequal across the cell, endoplasm will move to one end of the cell, producing extension at one end and contraction at the other. The extension and contraction allows the outer membrane to move with the pseudopodia without changing surface area. This process is influenced by Ca^{2+} concentration and pH.

Feeding and nutrition

Protozoa have a wide range of feeding mechanisms. Some species living in organic-rich environments assimilate dissolved nutrients by absorption through the plasma membrane. In bacteria-enriched environments, specialized modes of particle ingestion are used. When larger prey are present, there are additional strategies for snaring and engulfing these. Ingestion of particulate material requires **phagocytosis** (engulfment of particles by the cell) and formation of digestive vacuoles within the cell. Enzymic and chemical degradation releases soluble materials which can then be transported across the vacuole membrane for utilization within the cell. Protozoa typically exhibit heterotropic (obtain energy from organic material) nutrition (see Topic B1); however, those that possess photosynthetic capacity can function as phototrophs (utilize light for the fixation of CO_2) (see Topics I2 and B3). Anabolic (synthetic) metabolism in the protozoa is diverse, with free-living species exhibiting a wide range of synthetic capabilities. Those protozoa living within an organism such as parasites or those involved in symbiotic relationships (beneficial association between two organisms) will have a limited synthetic capacity as synthetic precursors can be obtained from either the host or from the other partner in the symbiotic relationship.

Respiration

Aerobic protozoa must obtain oxygen by simple diffusion. The utilization of oxygen by respiration drives oxygen into the cell. However, in large protozoa such as the ciliates it is unlikely that oxygen penetrates fully into the cell. Heterotrophs will use oxygen, when available, to generate ATP by oxidative phosphorylation. In anaerobic environments heterotrophs will exhibit a fermentative metabolism (incomplete oxidation of substrates, Topic B1)

Osmoregulation

Osmoregulation is a vital function for those organisms that lack a rigid outer wall. Most freshwater protozoa are **hyperosmotic** (solutes and ions present at high concentrations relative to their environment) and hence accumulate water by **osmosis**. This water must be removed to prevent lysis of the cell. The contractile vacuole is the main organelle for expulsion of water and waste. The expansion and contraction of the vacuole is energy requiring, and energy-requiring active transport is also used to adjust the ionic composition of the fluid expelled. The frequency of vacuolar contraction is inversely related to external ionic strength; as the osmotic strength of the external medium increases, less water is taken up into the cell and the frequency of vacuolar contraction decreases. Most marine organisms lack a contractile vacuole as the cells will be **hypoosmotic** (solutes and ions present at low concentration relative to their environment).

J4 UNUSUAL METABOLIC PATHWAYS

Key Notes

Protozoan metabolism	Protozoa exhibit a wide range of metabolic diversity, presumably in response to the range of habitats that these organisms exploit.
Glycolysis	Most aerobic protozoa possess a typical glycolytic pathway. However, some anaerobic organisms such as *Entamoeba* spp. have a modified pathway that utilizes PPi (inorganic pyrophosphate) rather that ATP for some reactions. One particular group, the *Typanosomatids*, have most of the glycolytic pathway sequestered in an organelle, the glycosome.
Mitochondrial metabolism	Protozoal mitochondria typically have modified electron-transport pathways. These pathways are usually branched. This is thought to allow the mitochondria to utilize a wide range of substrates and to function at very low levels of oxygen. These mitochondria, however, are less efficient at producing ATP.
Hydrogenosomal metabolism	The hydrogenosome is found in a few anaerobic protozoa. Pyruvate and malate are metabolized in the hydrogenosome to a range of incompletely oxidized compounds such as acetate and molecular hydrogen, with the formation of ATP.
Nitrogen metabolism	Protozoa can utilize a wide range of inorganic and organic sources of nitrogen when available. Amino acids are accumulated within the cell and used as a nitrogen reserve. Amino acids are frequently end products of metabolism in most protozoa and may be excreted; storage and excretion of amino acids are thought to play a role in osmoregulation in protozoans lacking contractile vacuoles.
Related topics	Heterotrophic pathways (B1) Autrotrophic metabolism (B3) Electron transport and oxidative Biosynthetic pathways (B4) phosphorylation (B2) Cellular structure of protozoa (J2)

Protozoan metabolism

Protozoa inhabit a wide range of habitats and can utilize a large number of substrates. This has produced significant diversity in their metabolism. Most of the **catabolic** (pathways involved in degradation) and **anabolic** (synthetic) pathways are similar to those found in other eukaryotes and bacteria (see Section B). Protozoa do, however, possess a number of unique or modified pathways.

Glycolysis

Most aerobic protozoa possess a typical glycolytic pathway (see Topic B1). However, some anaerobic organisms (e.g. *Entamoeba* spp.) utilize **inorganic**

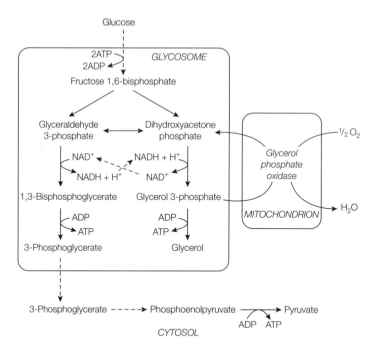

Fig. 1.   Aerobic glycolysis in bloodstream forms of trypanosomes, showing compartmentalization of much of the pathway in the glycosome. Dotted lines indicate several reaction steps.

pyrophosphate (PPi) instead of ATP for some of the reactions. For example **pyrophosphate-dependent pyruvate phosphate dikinase** catalyzes the conversion of:

$$\text{Phosphoenolpyruvate} + \text{AMP} + \text{PPi} \leftrightarrow \text{Pyruvate} + \text{ATP} + \text{Pi}$$

This enzymes replaces the glycolytic enzyme **pyruvate kinase**. Because the PPi-dependent enzymes are reversible, glucose can be formed from other compounds such as amino acids via this modified pathway. The most unusual modification is seen in the *Trypanosomatids*. Here, most of the glycolytic pathway is sequestered in an organelle, the **glycosome** (see Topic J2). Glycosome metabolic pathways and the link between the glycosome and the mitochondria are shown in *Fig. 1*. It is thought that this adaptation allows glycolysis to function optimally in the presence of oxygen and provides glycerol-3-phosphate, a major substrate for the trypanosome mitochondria.

Mitochondrial metabolism

Many protozoa that contain mitochondria have modified electron-transport pathways which are typically branched like those of the bacteria (see Topic B2). Electron transport is thus divided into '**mammalian like**' respiration sensitive to cyanide and having a cytochrome aa_3 as terminal oxidase (see Topic B2) and an '**alternative**' cyanide-insensitive pathway with an alternative terminal oxidase. The alternative pathways allow mitochondria to utilize other substrates and to function at very low levels of oxygen where the mammalian-type pathways are less active. However, it is still not known if any of the alternative pathways are involved in the production of ATP.

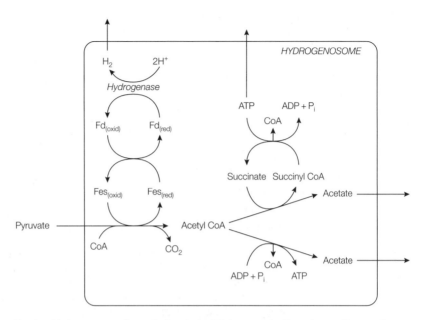

Fig. 2.   Hydrogenosomal metabolism in the Trichomonads. Fes = Iron–sulfur proteins; Fd = ferridoxins; oxid = oxidized state; red = reduced state.

Hydrogenosomal metabolism

The hydrogenosome (see Topic J2) is found in a few anaerobic protozoa. Although hydrogenosomal metabolic pathways vary between species, some aspects of metabolism are common. Typically, pyruvate and malate, formed by cytosolic glycolytic pathways, enter the hydrogenosome where they are converted by fermentation reactions into a range of incompletely oxidized compounds such as acetate, with the formation of ATP (*Fig. 2*). During the formation of acetate, molecular hydrogen is formed by the action of an anaerobic electron-transport pathway (see Topic B2).

Nitrogen metabolism

Free-living protozoa can utilize inorganic as well as organic sources of nitrogen when available. Of the inorganic sources, ammonium is preferentially used; however, many species have **nitrate reductase** (converts nitrate into nitrite) and **nitrite reductase** (catalyzes the reduction of nitrite to ammonium), suggesting that these organisms will utilize many available sources of nitrogen. Most parasitic protozoa and those involved in symbiotic relationships will exclusively use amino acids as sources of nitrogen. Parasites secrete **protease enzymes** that degrade protein, releasing small peptides and amino acids. Amino acids are accumulated within the cell in many free-living organisms such as *Euglena* spp. These function as a nitrogen reserve. Amino acids may also be end products of metabolism and may be excreted; storage and excretion of amino acids is thought to play a role in osmoregulation in protozoans lacking contractile vacuoles.

J5 REPRODUCTION

Key Notes

Asexual reproduction	Protozoa exhibit either fission or budding as mechanisms of asexual reproduction. Fission produces two daughter cells from a parent cell. This requires mitotic division (division producing two identical sets of chromosomes). Budding is a specialized type of fission with daughter nuclei produced within the parent cell, which migrate into cytoplasmic buds that are released by fission.
Sexual reproduction	Protozoa produce gametes by meiosis (reduction division). When two gametes unite external to the parent organism this is called syngamy. If the gamete nuclei unite within the parent protozoa, this is called autogamy. When there is an exchange of gametes between protozoa of complementary mating types (equivalent to male and female), this is called conjugation.
Related topic	Cell division and ploidy (G3)

Asexual reproduction

Asexual reproduction occurs in most protozoa by either fission or budding. Fission usually produces **daughter cells** from a **parent cell** by **mitotic** division (division producing two identical sets of chromosomes) (see Topic G3). Several kinds of fission are observed. **Symmetrogenic fission**, typical of flagellates and most non-ciliate protozoa, produces division along the long axis of the cell, forming daughter cells which are mirror images. **Homothetogenic fission** occurs in ciliates, and division occurs across the narrow part of the cell. **Binary fission** occurs in ameba, and the plane of division occurs through the midline of the cell. This will produce two similar, but not identical, daughter cells. **Multiple fission** is also common; this involves the parent cell undergoing repeated mitotic division, producing many nuclei; these give rise to multiple progeny by repeated fission. **Budding** can be thought of as a specialized type of fission. Mitotic division produces daughter nuclei, and each migrates into a cytoplasmic bud that is ultimately released by fission.

Sexual reproduction

In protozoa that exhibit sexual reproduction, the haploid (only contain half the number of chromosomes) gamete nuclei, **pronuclei**, are borne in specialized gamete cells. These are formed by **meiosis** (reduction division) (see Topic G3). If the gametes are identical morphologically and physiologically, they are called **isogametes**; if they are dissimilar, they are called **anisogametes**. Usually, the larger anisogamete is the **macrogamete** and the smaller, the **microgamete**. **Syngamy** results when the two haploid gametes unite externally to the parent organism. If the haploid gamete nuclei unite within the parent protozoa, this is called **autogamy**. Exchange and fusion of gametes between protozoa of complementary mating types (equivalent to male and female) is called conjugation.

Conjugation is most common among ciliate protozoa. Typically, two ciliates unite, fusing their pellicles (outer surface) at the contact point. The macronucleus in each is degraded. The individual micronuclei divide twice by meiosis to form four haploid pronuclei, three of which disintegrate. The remaining pronucleus divides again mitotically to form two gametic nuclei, a stationary one and a migratory one. The nuclei pass into the respective conjugates. The ciliates then separate, the pronuclei fuse, and the resulting diploid nucleus undergoes three rounds of mitosis. The eight resulting nuclei have different fates: one nucleus is retained as a micronucleus, three others are destroyed, and the four remaining nuclei develop into macronuclei. Each separated conjugate now undergoes cell division. Eventually, progeny (offspring) with one macronucleus and one micronucleus are formed.

J6 BENEFICIAL EFFECTS OF PROTOZOA: SYMBIOTIC RELATIONSHIPS

Key Notes

Endosymbionts	Symbiosis is the close association between two different types of organisms. Endosymbionts (symbionts within the cytoplasm of the protozoan) are common and include a range of bacteria and eukaryotes . These associations provide the endosymbiont with a favorable and protected environment. In turn, the symbiont provides benefits to the protozoan. Autotrophic endosymbionts share some of their photosynthetic products with the host.
Ectosymbionts	Protozoa can be ectosymbionts of higher animals. The animal provides a safe habitat and nutrients for the protozoa, and the protozoa provide useful products for the higher animal. Typically, ectosymbionts are found in the gut of the animal.
Related topics	Autotrophic metabolism (B3) Taxonomy of protozoa (J1)

Endosymbionts

Symbiosis, the close association between two or more organisms of different genetic composition, is widespread among protozoa and assumes highly diverse forms. Bacteria are **endosymbionts** (symbionts within the cytoplasm of the protozoan) of divergent groups of protozoa. Algae are by far one of the most common endosymbionts, dwelling within the cytoplasm of sarcodina (see Topic J1). In all of these associations, the endosymbiont is within a favorable environment, protected from predators, and often assimilates host-produced substances (sometimes metabolic wastes) that promote growth. In turn, the symbiont provides benefits to the host. **Autotrophic endosymbionts** (see Topic B3) share some of their photosynthetic products with the host, and thus contribute to a stable association. In some cases, the presence of the symbionts contributes to maintenance of host metabolic processes. The host must, however, regulate the density of the symbionts to prevent an overpopulation of the cell. The presence of *Chlorella* within *Paramecium bursaria* is perhaps one of the most intensively studied endosymbiotic associations. Endosymbionts are enclosed by vacuolar membranes that separate the host cytoplasm from the foreign cells. This allows careful control of exchange between the diverse organisms, resulting in a benefit to both species. The pathways showing exchange of material between a ciliate and an alga are shown in *Fig. 1*. It is clear that in many cases the host–symbiont association is highly specific, and only one symbiont species

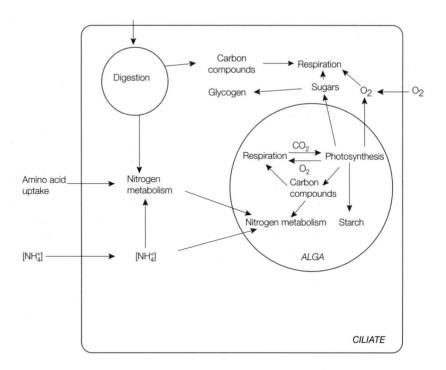

Fig. 1.   Typical interaction between an alga and a freshwater ciliate. The algal symbiont is enclosed in a perialgal vacuole.

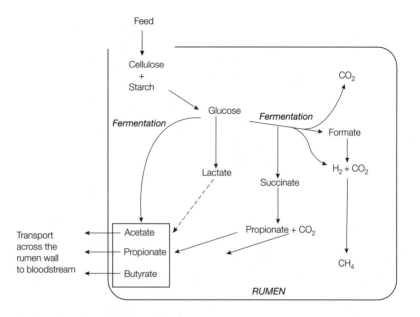

Fig. 2.   Microbial metabolism in the rumen.

is adopted. This indicates that the protozoan has the ability to associate selectively with a given symbiont.

Ectosymbionts   Protozoa can also be **ectosymbionts** of higher animals. The animal provides a safe habitat and nutrients for the protozoa, and the protozoa provide useful products for the higher animal. Typically, the ectosymbionts are found in the gut of the animal. Well studied systems include the gut of termites and the rumen of herbivorous mammals. The **rumen** (a large chamber in the stomach of ruminants) contains a complex microflora consisting of bacteria, fungi and ciliate protozoa. The function of the protozoa in this complex system is to control the microbial population by feeding, and perform some aspects of rumen metabolism. The bacteria and fungi degrade cellulose and other polymers from the plant material to sugars, which are then fermented by the bacteria and protozoa to produce acetate, propionate and butyrate (*Fig. 2*). These are used by the ruminant. Methane production by the protozoa in the rumen involves the action of endosymbiotic **methanogenic bacteria** (synthesize methane from H_2 and CO_2).

J7 DETRIMENTAL EFFECTS OF PROTOZOA: PARASITIC RELATIONSHIPS

Key Notes

Parasitism	Parasitism is an interaction between two organisms where the parasite derives nutrients and a stable environment from the 'host' organism. Some life cycles require more than one host for completion. Hosts are identified as either intermediate or final. Intermediate hosts frequently aid the transmission of the parasite.
Diseases caused by parasitic protozoans	Parasites that infect the intestine typically produce a diarrheal disease; however, infections caused by *Entamoeba histolytica* and *Sarcocystis* sp. spread from the initial site of infection to other organs and tissues. Transmission of these organisms is usually via fecal-contaminated food and water, and many are zoonotic (organisms can be transmitted from animals to humans). Parasites that infect blood and tissues include *Plasmodium*, the causative agent of malaria. Four species of *Plasmodium* infect man: *P. falciparum* causes the most severe infections; *P. malariae* is transmitted by the *Anopheles* mosquito and produces the mildest of the malaria infections, which may last for up to 30 years if untreated. Other blood and tissue infections include African and American trypanosomiasis, caused by *Trypanosoma brucei* and *T. cruzi*, respectively. Both cause a generalized infection that is fatal if untreated. The tsetse fly is the vector for *T. brucei* and the triatomine bug for *T. cruzi*. Leishmaniasis is caused by *Leishmania* spp. found in the Americas, Africa, Asia, the Middle East and southern Europe. Infections occur at various sites on the body and are self-limiting (self-healing). Visceral leishmaniasis is a more disseminating infection that is often fatal. Other infectious protozoans include *Trichomonas vaginalis*, a common sexually transmitted organism, and *Toxoplasma gondii* which causes a systemic infection in humans; infections during pregnancy are a problem as the parasite can be transmitted to the fetus.
Physiology of parasitic protozoa	Parasitic protozoa exhibit a wide range of motility, with some parasites showing more than one form of motility. Parasitic protozoa exhibit heterotrophic nutrition. Many parasites have mitochondria (some lack mitochondria), but energy is usually produced by fermentation (incomplete oxidation of substrates). Most protozoan parasites have limited biosynthetic ability, and metabolic precursors are scavenged from their environment. Secretion in pathogenic protozoa is important for infection and survival. Reproduction within this group can be asexual or sexual; some exhibit both types in their life cycles.

Habitats	Parasites inhabit a wide range of habitats within host organisms. Some parasites will inhabit only one site throughout their life cycle, but many move to various sites to evade host immune attack. The changes in habitat elicit major changes in the physiology of the parasite.
Life cycles	Life cycles of parasitic protozoa can be simple, involving one host and two parasite forms, or more complex life cycles requiring several hosts and multiple life-cycle forms.
Strategies of immune evasion	Once a parasite enters the host it is recognized as foreign and subjected to a wide range of host responses. There are three basic strategies used by parasitic protozoans to overcome host responses: (1) non-induction of host response by invading sites not protected by the host immune response; (2) control of host response by producing immune modulators; (3) immunological evasion by shedding of surface and degradation of antigen–antibody complexes or by over-production of protective enzymes in the parasite.
Opportunistic protozoan parasites	Some protozoans commonly found in the environment can produce infection in humans; these are called opportunistic parasites. For these organisms to cause infection, unusual conditions must exist such as the host being immune suppressed or the organisms being present in large numbers.
Control	Control of protozoan infections can be achieved in several ways. The eradication of vectors or intermediate hosts in the environment is effective in controlling some disease; however, resistance of vectors to insecticides poses a problem. Drugs represent the best approach for infection control or prophylaxis (prevent infection). Drug resistance in protozoans is a serious worry. There are no vaccines currently available for the control of human infection.

Related topics	Bacterial disease – an overview (F3) Physiology of protozoa (J3)
	Human host defense mechanisms (F4) Unusual metabolic pathways (J4)
	Control of bacterial infection (F7) Reproduction (J5)
	Cellular structure of protozoa (J2)

Parasitism
Parasitism is a specific type of interaction between two organisms. However, unlike other types of interaction, one organism, the parasite, obtains nutrients and shelter from the other '**host**' organism; the parasite derives all the benefit from the association. Protozoans parasitize all higher animals and some plant species. Some life cycles require more that one host for completion. In these cycles there is an **intermediate host** as well as a **final host**. The intermediate host may also act as a vector for transmission.

Diseases caused by parasitic protozoans
Intestinal infections
Of the organisms that cause intestinal infection, **amebic dysentery**, *Entamoeba histolytica* is probably the most well known. This organism invades the intestinal wall, causing a severe diarrheal disease. The infection spreads from the initial

site of infection to the liver and other organs. Transmission of this organism is usually via fecal contaminated food and water. Other intestinal protozoans transmitted by fecal contamination include *Giardia intestinalis* and *Cryptosporidium parvum*. Both are **zoonotic** (organisms can be transmitted from animals to humans) and cause **self-limiting** (infections that self-cure) diarrheal disease. Another member of this group *Sarcocystis* sp. initially invades the intestine; however, parasites migrate and develop in other organs such as the liver and in tissues such as skeletal or cardiac muscle.

Blood and tissue infections

Malaria, the most well known of this group, is caused by four species of *Plasmodium*. **Falciparum** malaria, caused by *P. falciparum*, is the most severe of these infections, causing a variety of major clinical problems such as cerebral infection and low blood pressure. **Vivax** (**tertian**) and **ovale** malaria, caused by *P. vivax* and *P ovale*, can also produce serious illness; however, fewer clinical complications occur. Infections can last for up to 3 months if untreated, with episodes of severe illness occurring every 3–6 weeks. **Quartan** malaria is caused by *P. malariae* and is the mildest of the malaria infections, which may last for up to 30 years if untreated. Malaria is transmitted by some 60 species of *Anopheles* mosquito. Trypanosomiasis is another major infection. African trypanosomiasis (**sleeping sickness**) is a zoonotic infection caused by either *Trypanosoma brucei rhodesiense* (east and south-east Africa) or by *T.b. gambiense* (west and central Africa). The infection is transmitted by tsetse flies and soon, after infection, parasites invade all tissues except initially those of the central nervous system (CNS). These are invaded after some time (weeks–months), producing a generalized meningoencephalitis. Both forms of sleeping sickness are fatal if left untreated. American trypanosomiasis (**Chagas' disease**) is caused by *T. cruzi*. Parasites are transmitted by the **triatomine** bug, which deposits infected feces on the skin and this contaminates vector bite wounds. Movement of the parasites to tissues in the body induces a variety of responses including enlargement of organs. The major problems associated with infection are damage to heart muscle and meningoencephalitis. Infections are fatal if untreated. A wide range of Leishmania species infect humans. These zoonotic diseases are divided on the basis of geography and infection type. Infections can be cutaneous, mucocutaneous or diffuse. New world (the Americas) infections are caused by *Leishmania mexicana* and *L. brasiliensis*. Old world (north and east Africa, Asia, the Middle East and southern Europe) infections are caused by *L. tropica*, *L. major* and *L. donovani*. Visceral (**Kala Azar**) leishmaniasis is associated with old world infections. Parasites are transmitted by a variety of fly vectors. The invasion of tissues by parasites typically causes the formation of lesions which self-cure after prolonged periods of up to 2 years. Visceral leishmaniasis occurs when *L. donovani* infects cells of the bone marrow, liver and spleen, producing a slowly progressing disease that is often fatal.

Other protozoan diseases

Trichomonas vaginalis is a common sexually transmitted protozoan that causes acute vaginitis (vaginal infection) and infects the urethra in men and women. Toxoplasmosis is a systemic protozoal infection in humans caused by *Toxoplasma gondii*. Infections in adults manifest as a mild general infection with flu-like symptoms. In pregnant women, however, the parasite can be transmitted to the fetus. Infection of the fetus causes damage to tissues such as the brain and eyes.

Physiology of   **Motility.**   Like their free-living counterparts, parasitic protozoa utilize flagella,
parasitic protozoa cilia or ameboid movement for motility. The complexity of some parasite life
cycles means that some parasites may exhibit, at different times, more than one
form of motility. Chemotaxis plays a significant role in directing this process.

Nutrition.   All pathogenic protozoa are **heterotrophs**, using carbohydrates or
amino acids as their major source of carbon and energy.

Respiration.   Some parasitic protozoa can utilize oxygen to increase energy
production (see Topic B2), but many of these organisms lack mitochondria or
have mitochondria that do not function like typical mammalian mitochondria.
These parasites survive in environments where oxygen is present occasionally
or at low levels, and therefore exhibit a fermentative metabolism. Since oxygen
is toxic to such organisms, they may apparently reduce oxygen in order to
remove it and thus maintain an anaerobic metabolism.

Excretion.   Excretion in parasitic protozoa occurs by a variety of mechanisms
such as diffusion or osmotic pressure (movement of compounds driven by the
difference in osmotic pressure between the inside of the cell and the environment).

Secretion.   Pathogenic protozoa secrete a range of hemolysins (molecules that
lyse red blood cells), cytolysins (which cause cell and tissue lysis), proteolytic
enzymes and toxins. Antigenic and immunomodulatory (reduce host immune
response) molecules are also produced.

Reproduction.   The survival of parasites is partly due to their high rate of
reproduction. This may be either sexual or asexual (see Topic J5). Some species
such as *Plasmodium* exhibit both forms of reproduction in their life cycle.

Metabolism.   The metabolism of parasites is highly adapted with many
synthetic pathways found in other eukaryotes absent. This is because most metabolic intermediates or precursors such as lipids, amino acids or nucleotides are
actively scavenged from the parasite's environment (see Topic J4). This mini-
mizes energy expenditure, which is finely balanced in parasitic protozoa.

Habitats   Parasites inhabit a wide range of habitats within their hosts. Some parasites
will inhabit only one site throughout their life cycle, but many move to various
sites within the body. Such movement may require the formation of motile
forms and will produce a significant change in the physiology and morphology

Table 1. Oxygen, carbon dioxide and pH values for various sites within mammals

Habitat	O_2 tension (mmHg)	CO_2 tension (mmHg)	pH
Skin	50–100	20–30	6.8–7.4
Subcutaneous tissue	20–43	20–45	6.8–7.4
Arterial blood	70–100	15–25	6.8–7.4
Venous blood	40–70	34–42	6.8–7.4
Stomach	0–70	20–400	1–3
Small intestine	0–65	20–600	5–7.8
Large intestine	0–5	200–600	6–7.5

of the parasite as a result of environmental change. For example, parasites moving from the gut to tissues or organs such as the liver will encounter higher levels of oxygen, changes in pH and significant exposure to the host immune response. The physiological parameters of various sites in humans are shown in *Table 1*. The reasons for movement between sites in the body include evasion of the host immune attack, which can be achieved by the parasites moving to areas not affected by the immune response, such as the CNS, or by entering host cells.

Life cycles

Life cycles of parasitic protozoa can be simple, involving only one host and two morphologically distinct forms, typically a **cyst** and **trophozoite**. The life cycle of *Giardia intestinalis* is shown in *Fig. 1*. This type of simple life cycle is typical for intestinal and luminal protozoa. More complex life cycles requiring several hosts and multiple life-cycle forms are typical of blood and tissue parasites. These parasites undergo development in each host, but both hosts are required for completion of the life cycle. The life cycle of *Plasmodium*, the human malarial parasite, is shown in *Fig. 2*. It should be remembered that different species of *Plasmodium* have modified life cycles. These changes include the presence of dormant states, and the timing of some aspects of the cycle. The life cycle of another tissue parasite *T. cruzi* involves an arthropod intermediate host but this cycle, unlike that of *Plasmodium*, does not have a sexual component.

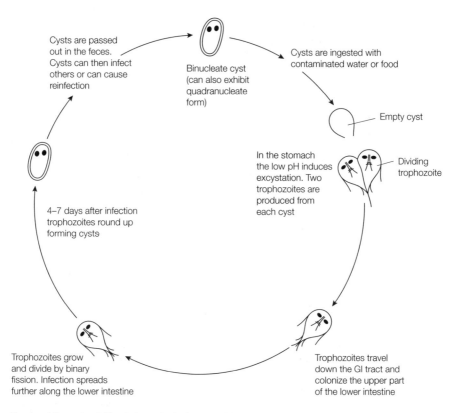

Cysts are passed out in the feces. Cysts can then infect others or can cause reinfection

Binucleate cyst (can also exhibit quadranucleate form)

Cysts are ingested with contaminated water or food

Empty cyst

In the stomach the low pH induces excystation. Two trophozoites are produced from each cyst

Dividing trophozoite

4–7 days after infection trophozoites round up forming cysts

Trophozoites grow and divide by binary fission. Infection spreads further along the lower intestine

Trophozoites travel down the GI tract and colonize the upper part of the lower intestine

Fig. 1. Life cycle of Giardia intestinalis *(syn. lamblia, duodenalis). Life cycle requires only one host.*

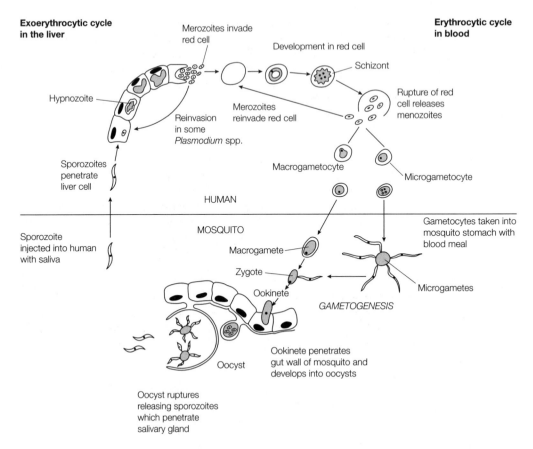

Fig. 2.   Life cycle of Plasmodium *spp.*

Strategies of immune evasion

Once a parasite enters the host, it is recognized as foreign and subjected to a wide range of host responses. There are three basic strategies used by parasitic protozoans to overcome host immune response (see also Topic F5):

(i)   Non-induction of host response.
(ii) Control of host response.
(iii) Other methods of immunological evasion.

Non-induction of host response
A parasite can avoid producing an immune response by becoming intracellular. Examples of this include *Leishmania* spp. and *Trypanosoma* spp. Both these parasites invade host cells where a significant portion of their life cycle is spent. Other parasites such as *Naegleria* spp. invade the CNS and brain. These sites are not protected by the host immune response and are thus termed **immune privileged sites**.

Control of host response
Control of the host immune response can be achieved by several methods. A good example is that of *Leishmania* spp. This parasite reduces the production

of γ-**interferon** by the host, which is required to activate **macrophages**. Activation of these cells is required for the killing of parasites.

Immunological evasion

Many parasites can produce a wide range of surface antigen types, which are expressed during infection. These have the effect of making the host immune response ineffective as antibodies produced by the host in response to one antigenic type will not necessarily bind to a different surface antigen. This is termed **antigenic variation** and involves the random activation or rearrangement of genes that code for surface antigen. The best example of this is found in African Trypanosomes. Other mechanisms in this group include the **shedding of surface antigen** (*E. histolytica*). Shedding of antigen and antigen–antibody complex directs the immune response away from the parasite. The formation of antigen–antibody complex can also be affected by the action of parasite **proteases** (enzymes involved in the breakdown of proteins) which degrade the complex or the antibody before it binds to the parasite surface.

Opportunistic protozoan parasites

Some protozoans commonly found in the environment such as *Acanthamoeba castellani* and *Naegleria fowleri* can produce infections in humans. However, for infection to occur, a variety of factors must be fulfilled. These may include long-term exposure of the host to large numbers of the parasite or appropriate environmental conditions such as defined levels of humidity and temperature. More recently the use of the term opportunistic has been used to describe parasitic infections of immune compromised hosts. Because of an impaired immune response, these hosts are more susceptible to parasite infections.

Control

Control of parasitic diseases can be achieved in a variety of ways. Diseases that have a complex life cycle can be controlled by eradicating the intermediate hosts or vector. The spraying of chemical insecticides such as DDT and benzene hexachloride has been useful in controlling the *Triatomine* vector of Chagas' disease and the *Anopheles* mosquito; however, resistance to these agents is a major

Table 2. *The spectrum of activity and mode(s) of action of the major antiprotozoal drugs*

Drug	Mode of action	Spectrum of activity
Sulfonamides	Inhibition of folic acid synthesis	*Plasmodium* *Toxoplasma gondii*
Proguanil	Inhibition of folic acid synthesis	*Plasmodium*
Chloroquine	Inhibition of DNA synthesis	*Plasmodium*
Mefloquine	Inhibition of DNA synthesis	*Plasmodium*
Metronidazole	Inhibition of DNA synthesis	*Giardia intestinalis* *Entamoeba histolytica*
Pentamidine	Inhibition of DNA synthesis	*Trypanosoma* *Leishmania*
Amphotericin B	Disrupts membrane function	*Leishmania*
Eflornithine	Inhibition of protein synthesis	*Trypanosoma*
Tetracycline	Inhibition of protein synthesis	*Plasmodium*
Albendazole	Inhibition of microtubule assembly	*G. intestinalis*
Buparvaquone	Inhibition of energy production	*Plasmodium*
Meglumine antimonate	Inhibition of energy production	*Leishmania*
Metronidazole	Inhibition of energy production	*G. intestinalis* *E. histolytica*
Primaquine	Inhibition of energy production	*Plasmodium*

problem. Biological control of vectors using natural parasites and predators has proved to be effective. Vaccination is another approach to control; however, there are no vaccines available for the control of human infection. Our inability to develop useful vaccines is due to the complexity of many life cycles and the ability of parasites to overcome the host's immune defense system. Drugs represent the best approach for disease control today; however, many of the drugs currently available are of limited use owing to significant human toxicity or resistance. The drugs used to treat some protozoan infections and their modes of action are shown in *Table 2*. Most of these drugs are taken for the treatment of infection but some may be taken for **prophylaxis** (to prevent infection). Resistance to drugs is a major problem in the treatment of protozoan infections.

K1 VIRUS STRUCTURE

Key Notes

Definitions	Viruses are obligate intracellular parasites and vary from 20–200 nm in size. They have varied shape and chemical composition, but contain only RNA or DNA. The intact particle is termed a 'virion' which consists of a capsid that may be enveloped further by a glycoprotein/lipid membrane. Viruses are resistant to antibiotics.
Methods of study	Virus morphology has been determined by electron microscopy (EM) (using negative staining), thin-section EM (using negative staining), immunoelectron microscopy (using negative staining), electron cryo-microscopy and X-ray crystallography.
Virus symmetry	Virus capsids have helical or icosahedral symmetry. In many cases the capsid is engulfed by a membrane structure (the virus envelope). Helical symmetry is seen as protein sub-units arranged around the virus nucleic acid in an ordered helical fashion. The icosahedron is a regular shaped cuboid which consists of repetitions of many protein sub-units assembled so as to resemble a sphere.
Virus envelopes	Virus envelopes are acquired by the capsid as it buds through nuclear or plasma membranes of the infected cell. Envelopes may contain a few glycoproteins, for example, human immunodeficiency virus (HIV), or many glycoproteins, for example, Herpes simplex virus (HSV). The virus envelope contains the receptors, which allow the particle to attach to and infect the host cell.
Related topic	Bacteriophage (E7)

Definitions

Viruses are obligate intracellular parasites which can only be viewed with the aid of an electron microscope. They vary in size from approximately **20–200 nm**. In order to persist in the environment they must be capable of being passed from host to host and of infecting and replicating in susceptible host cells. A virus particle has thus been defined as a structure which has evolved to transfer nucleic acid from one cell to another. The nucleic acid found in the particle is either **DNA** or **RNA**, is single- or double-stranded and linear or segmented. In some cases the nucleic acid may be circular. The simplest of virus particles consists of a protein coat (sometimes made up of only one type of protein which is repeated hundreds of times) which surrounds a strand of nucleic acid. More complicated viruses have their nucleic acid surrounded by a protein coat which is further engulfed in a membrane structure, an envelope consisting of virally coded glycoproteins derived from one of several regions within the infected

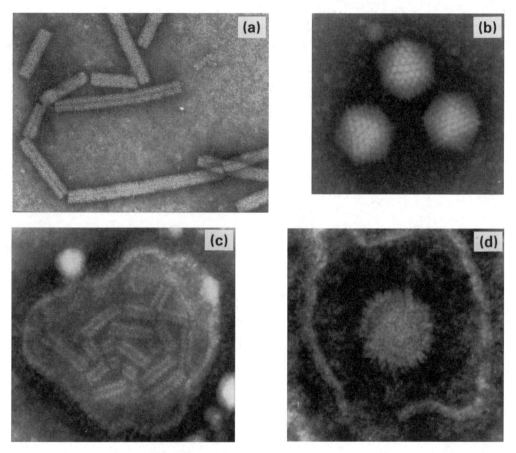

Fig. 1.   Examples of viruses from main groups according to 'standard' morphology: (a) unenveloped/helical (tobacco mosaic virus); (b) unenveloped/icosahedral (adenovirus); (c) enveloped/helical (paramyxovirus); (d) enveloped/ icosahedral (herpesvirus). From Harper, D., Molecular Virology, 2nd edn, © BIOS Scientific Publishers Limited, 1998.

cell during the maturation of the virus particle. The genetic material of these complex viruses encodes for many dozens of virus specific proteins.

The complete fully assembled virus is termed the **virion**. It may have a glyco-protein envelope which has **peplomers** (projections) which form a 'fringe' around the particle. The protein coat surrounding the nucleic acid is referred to as the **capsid** (see *Figs 1* and *2*). The capsid is composed of morphological units or capsomers. The type of capsomer depends on the overall shape of the capsid, but in the case of **icosahedral** capsids the capsomers are either **pentamers** or **hexamers**. Capsomers themselves consist of assembly units that comprise a set of structure units or **protomers**. **Structure units** are a collection of one or more non-identical protein subunits that together form the building block of a larger assembly complex (e.g. virus proteins VP1, VP2, VP3 and VP4 of picornaviruses). The combined nucleic acid–protein complex which comprises the genome is termed the **nucleocapsid**, which is often enclosed in a core within the virion.

Methods of study   While the **electron microscope** (EM) had been known for many years, the inven-tion of the **negative-staining technique** in 1959 revolutionized studies on virus structure. In negative-contrast EM, virus particles are mixed with a heavy metal

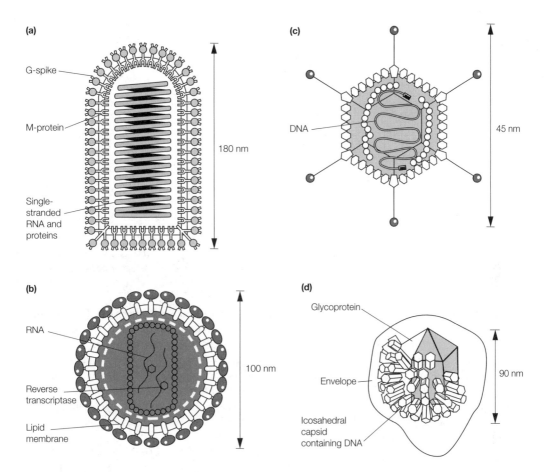

(a)

G-spike

M-protein

Single-
stranded
RNA and
proteins

180 nm

(b)

RNA

Reverse
transcriptase

Lipid
membrane

100 nm

(c)

DNA

45 nm

(d)

Glycoprotein

Envelope

Icosahedral
capsid
containing DNA

90 nm

Fig. 2. Diagrammatic representation of the structure of four virus particles. Rabies virus (a) and HIV (b) have tight-fitting envelopes, herpes simplex (d) has a loose-fitting envelope, whereas adenovirus (c) is non-enveloped. The diagrams are based on electron microscopic observation and molecular configuration exercises. Redrawn from Phillips and Murray, Biology of Disease, 1995, with permission from Blackwell Science Ltd.

solution (e.g. sodium phosphotungstate) and dried onto a support film. The stain provides an electron-opaque background against which the virus can be visualized. Observation under the electron microscope has thus allowed the definition of **virus morphology** at the 50–77Å resolution level. In addition, negative staining of thin sections of infected cells has allowed definition of structures which appear during virus maturation and their interactions with cellular proteins. **Immunoelectron microscopy** is used to study those viruses which may be present in low concentrations or grow poorly in tissue culture (e.g. Norwalk virus). These clumps of virus are more readily observed. **Electron cryomicroscopy** reduces the risk of seeing artifacts as may be unavoidable by negative staining. High concentrations of virus are rapidly frozen in liquid ethane while on carbon grids. Electron micrographs can be digitized and three-dimensional reconstruction performed. Resolutions of 9Å have been achieved using this method. **X-ray diffraction** of virus crystals is the ultimate in determining the ultrastructure of virion morphology. At present, only simple viruses can be crystallized. More complex viruses are analyzed by attempting to form

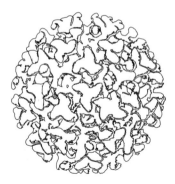

Fig. 3. Three-dimensional reconstruction of an icosahedrally symmetric virus particle.

crystals of sub-particular molecules. The X-ray diffraction pattern of the virion particle allows mathematical processing, which can predict the molecular configuration of the virus particle (*Fig. 3*).

Virus symmetry The capsids of virions tend to have one of two symmetries – **helical** or **cuboid.** Helical symmetry can be loosely described as having a 'spiral staircase' structure. The structure has an obvious axis down the center of the helix. The subunits are placed between the turns of the nucleic acid. A diagram of such a structure (tobacco mosaic virus) is shown in *Fig. 4*, and an electron micrograph in *Fig. 1a*. Animal viruses with a similar capsid structure include measles, rabies and influenza. Most animal viruses have spherical or cuboid symmetry. Obtaining a true sphere is not possible for such structures and hence subunits come together to produce a cuboid structure which is very close to being spherical. The 'closed shell' capsid is usually based on the structure referred to as an icosahedron. A regular icosahedron, formed from assembly of identical subunits, consists of **20 equilateral triangular faces, 30 edges and 12 vertices and exhibits 2-, 3- and 5-fold symmetry** (see *Fig. 4*). The minimum number of capsomers required to construct an icosahedron is 12, each composed of five identical subunits. Many viruses have more than 12. A model of such a

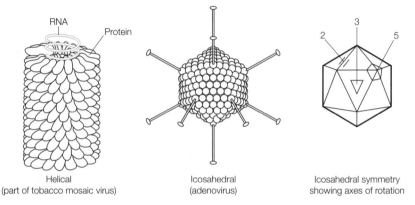

Fig. 4. Diagrammatical representation of helical and icosahedral symmetry. From Harper, D. Molecular Virology, 2nd edn, © BIOS Scientific Publishers Limited, 1998.

structure is shown in *Fig. 4*, although in adenoviruses projecting fibers are also present, which distinguishes this capsid from that of other viruses. The maturation and assembly of these structures is very complex; indeed, much of how it happens is unknown. Icosahedral structures are, however, usually formed via a complex but structured array of molecular-assembly procedures which eventually give rise to the mature capsid. These may be self-assembly processes or may involve virus non-structural proteins acting as **scaffolding** proteins which do not finish up in the mature capsid.

Virus envelopes   Many viruses in addition to having a capsid also contain a virus-encoded **envelope**. Most enveloped viruses bud from a cellular membrane (plasma membrane, e.g. influenza virus, or nuclear membrane, e.g. herpes simplex virus). Within this virus lipid/protein bilayer are a number of inserted virus-encoded **glycoproteins**. The envelopes can be amorphous (e.g. the herpes virion, *Figs 1* and *2*) or tightly bound to the capsid (e.g. HIV). Thus, the lipid of the envelope is derived from the cell, the glycoprotein being encoded by the virus. Quite how the process of budding occurs is largely unknown.

K2 VIRUS TAXONOMY

Key Notes

Virus taxonomy	The classification of viruses involves the use of a wide range of characteristics (morphology, genome, physicochemical and physical properties, proteins, antigenic and biological properties) to place viruses in orders, families, genera and species.
Virus orders	These are groupings of families of viruses that share common characteristics.
Virus families	These are groupings of genera of viruses that share common characteristics and are distinct from the members of other families. They are designated by the suffix -viridae.
Virus genera	These are groupings of species of viruses which share common characteristics. These are designated by the suffix -virus.
Virus species	These represent a polythetic class of viruses that constitutes a replicating lineage and occupies a particular ecological niche.

Related topics	Bacterial taxonomy (D1)	Reproduction in the algae (I3)
	Taxonomy (G1)	Taxonomy of protozoa (J1)
	Reproduction in fungi (H3)	

Virus taxonomy

Initial attempts to classify viruses were based on their pathogenic properties, and the only common feature to many of the viruses placed together in such groupings was that of **organ tropism** (e.g. viruses causing hepatitis or respiratory disease). There were more important aspects (e.g. **virus structure** and **composition**) which led virologists to believe that these initial attempts of classification were far from adequate. In the late 1950s and early 1960s, hundreds of new viruses were beginning to be isolated and the need for a different classification method became essential. In 1966 the **International Committee on Nomenclature of Viruses (ICNV)** was established at the International Congress of Microbiology held in Moscow. As the present classification scheme has evolved, an acceptance of the characteristics to be considered and their respective weighting has become universal.

In 1995 the sixth report of the **ICTV (International Committee on Taxonomy of Viruses)** recorded a universal taxonomy scheme consisting of one order, 71 families, 11 subfamilies and 164 genera (with more than 4000 members). The system still contains hundreds of unassigned viruses, largely because of lack of data. Also, new viruses are still being discovered.

Table 1.   Some properties of viruses used in taxonomy

Virion properties
 Morphology
   Virion size
   Virion shape
   Presence or absence and nature of peplomers
   Presence or absence of an envelope
   Capsid symmetry and structure
 Physicochemical and physical properties
   Virion molecular mass (M_r)
   Virion buoyant density (in CsCl, sucrose, etc.)
   Virion sedimentation coefficient
   pH stability
   Thermal stability
   Cation stability (Mg^{2+}, Mn^{2+})
   Solvent stability
   Detergent stability
   Irradiation stability
 Genome
   Type of nucleic acid (DNA or RNA)
   Size of genome in kb/kbp
   Strandedness: ss or ds
   Linear or circular
   Sense (positive-sense, negative-sense, ambisense)
   Number and size of segments
   Nucleotide sequence
   Presence of repetitive sequence elements
   Presence of isomerization
   G + C content ratio
   Presence or absence and type of 5′ terminal cap
   Presence or absence of 5′ terminal covalently linked protein
   Presence or absence of 3′ terminal poly (A) tract
 Proteins
   Number, size and functional activities of structural proteins
   Number, size and functional activities of nonstructural proteins
   Details of special functional activities of proteins, especially transcriptase, reverse
    transcriptase, hemagglutinin, neuraminidase and fusion activities
   Amino acid sequence or partial sequence
   Glycosylation, phosphorylation, myristylation property of proteins
   Epitope mapping
 Lipids
   Content, character, etc.
 Carbohydrates
   Content, character, etc.
 Genome organization and replication
   Genome organization
   Strategy of replication
   Number and position of open reading frames
   Transcriptional characteristics
   Translational characteristics
   Site of accumulation of virion proteins
   Site of virion assembly
   Site and nature of virion maturation and release
Antigenic properties
 Serologic relationships, especially as obtained in reference centers
Biologic properties
 Natural host range
 Mode of transmission in nature
 Vector relationships
 Geographic distribution
 Pathogenicity, association with disease
 Tissue tropisms, pathology, histopathology

Table 2. *Taxonomic chart of selected virus families*

Family	Characteristics	Typical members	Diseases caused
Poxviridae	dsDNA, 'brick'-shaped particles; largest virus	Vaccinia Variola	Laboratory virus Smallpox (now eradicated)
Herpesviridae	dsDNA, icosahedron capsid enclosed in an envelope, latency in host common	Herpes simplex Varicella-zoster Cytomegalovirus	'Cold' sores, genital infections Chicken pox, shingles Febrile illness or disseminated disease in immunosuppression
		Epstein–Barr virus	Glandular fever. Virus is also associated with certain malignancies, e.g. Burkitt's lymphoma
Adenoviridae	dsDNA, icosahedron with fiber structures, non-enveloped	Adenoviruses (many types)	Respiratory and eye infections, tumors in experimental animals
Papovaviridae	ds circular DNA, 72 capsomeres in capsid, non-enveloped	Human papilloma viruses	Warts, association with some cancers (e.g. cervical cancer)
Hepadnaviridae	One complete DNA minus strand with 5' terminal protein, DNA circularized by an incomplete plus strand, 42 nm enveloped particle	Hepatitis-B virus	Serum hepatitis, association with hepatocellular carcinoma
Paramyxoviridae	ssRNA, enveloped particles with 'spikes'	Parainfluenza virus Measles virus Respiratory syncytial virus	Respiratory tract infection ('croup') Measles Bronchiolitis

Orthomyxoviridae	Eight segments of ssRNA, enveloped particles with 'spikes', helical nucleocapsid	Influenza virus	Influenza
Reoviridae	10–12 segments of dsRNA, icosahedron, non-enveloped	Rotavirus	Infantile diarrhea
Picornaviridae	ssRNA, 22–30 nm particle of cubic symmetry, non-enveloped	Poliovirus Coxsackie virus Rhinovirus Heptatitis-A virus	Poliomyelitis Myocarditis Common cold Infectious hepatitis
Togaviridae	ssRNA, enveloped particles, icosahedron nucleocapsid	Rubella virus Arbovirus	German measles Yellow fever
Rhabdoviridae	ssRNA, bullet-shaped, enveloped particle	Rabies virus	Rabies
Retroviridae	ssRNA, enveloped particles with icosahedral nucleocapsid, employ reverse transcriptase enzyme to make DNA copy of genome on infection	Human T lymphotropic virus-1 Human immunodeficiency virus (HIV)	Adult T cell leukemia and lymphoma Acquired immune deficiency syndrome (AIDS)

As techniques in molecular biology, serology and EM have advanced, taxonomy has been based on more and more characteristics. Thus, factors now taken into consideration (see *Table 1*) include virion morphology, physico-chemical and physical properties, the genome, virus proteins, lipids, carbo-hydrates, and antigenic and biological properties. It has, however, been esti-mated that to define a virus appropriately some 500–700 characters must be determined. Over the next few years the ICTV plan is to create a readily acces-sible database which will allow cataloging of viruses down to strains. *Table 2* is a brief summary of some well known animal viruses and the diseases they cause.

Virus orders

These represent groupings of families of viruses that share common character-istics which make them distinct from other orders and families. Orders are designated by the suffix **-virales**. One order has been approved by the ICTV: **Mononegavirales,** which consists of the families Paramyxoviridae, Rhabdo-viridae and Filoviridae (these are single-stranded negative-sense non-segmented enveloped RNA viruses).

Virus families

These represent groupings of genera of viruses that share common character-istics and are distinct from the members of other families. Families are designated by the suffix **-viridae**. Families as groupings have proved to be excel-lent models for classification. Most of the families have **distinct virion morphology**, **genome structure** and **strategies of replication**. Examples include Picornaviridae, Togaviridae, Poxviridae, Herpesviridae and Paramyxoviridae. In some families (e.g. Herpesviridae) the complex relationships between indi-vidual members has led to the formation of sub-families, which are designated with the suffix **-virinae**. Thus, the Herpesviridae are further classified into the Alphaherpesvirinae (e.g. herpes simplex virus) the Betaherpesvirinae (e.g. cytomegalovirus) and the Gammaherpesvirinae (e.g. Epstein–Barr virus).

Virus genera

Virus genera are groupings of species of viruses which share common charac-teristics and are distinct from the members of other genera. They are designated by the suffix **-virus** (e.g. genus Simplexvirus and genus Varicellovirus of the Alphaherpesvirinae). The criteria for designating genera vary from family to family, but include genetic, structural and other differences.

Virus species

A virus species is defined as 'a polythetic class of viruses that constitutes a replicating linage and occupies a particular ecological niche'. Members of a polythetic class are defined by more than one property. At present the ICTV is examining carefully the properties which can be included in determining species. The division between species and strains is a difficult one.

K3 VIRUS PROTEINS

Key Notes

Overview	Viral proteins, coded by the viral genome, are either structural (capsid, envelope) or non-structural (e.g. enzymes, oncogenes, inhibitors of cell macromolecular synthesis and interaction with MHC presentation). They may be essential or non-essential in tissue culture replication.
Methodology	Structural proteins are studied, following virus purification, using a range of techniques including sodium dodecyl sulfate (SDS)–polyacrylamide gel electrophoresis, Western blotting and immunoprecipitation. Non-structural proteins are examined by, for example, pulse–chase experiments, use of protease and glycosylation inhibitors.
Protein synthesis and complexity	Viral proteins are synthesized by the translation of viral mRNAs on cellular ribosomes. Proteins are often processed following synthesis (e.g. proteolytic cleavage, glycosylation, myristylation, acylation and palmitoylation). In many viruses, protein synthesis is controlled at the levels of transcription and translation, which makes virus replication quite an efficient process. This control is usually directed by the virus genome.
Structural proteins	Structural proteins are either nucleocapsid, matrix or envelope proteins. They have a role in protecting the viral genome and in delivering the genome from one host to another via receptors on host cells. They also have a major role in the assembly of the virion. In most viruses, structural proteins are produced in abundance late in the replicative cycle.
Non-structural proteins	These may be carried in the virion (but are not part of the virion architecture) where they have enzymic activity which is necessary for initiating infection. Others are not destined for the virion but have roles in the infected cell. These roles include switching off host-cell nucleic acid and protein synthesis, polymerase, protease and kinase activities, DNA binding activity and gene regulation.
Related topics	Virus replication (K7) Virus vaccines (K10) Viruses and the immune system (K9) Antiviral chemotherapy (K11)

Overview

The **coding capacity** of viral genomes varies from < 5 to >100 genes. This in turn is reflected in the different complexities of virus particles and their respective replicative cycles. Viral proteins are either **structural** (part of the virion architecture), **non-structural** but in the virion (usually enzymes) or **non-structural** and present only in infected cells, never the virion (these have a range of functions). The limited number of virus proteins synthesized means that many of them are multifunctional.

Virus proteins are often referred to as being **essential** or **non-essential**. The former are an absolute requirement for the virus in order for it to complete a replicative cycle and to produce infectious virions. The latter can be deleted from the virus genome without seriously affecting virus growth in tissue culture. However, this may not reflect the *in vivo* situation.

Methodology

The study of virus structural proteins as assembled virions has required the **purification of viruses** from infected cells or infected-cell supernatants, ensuring that they are free from surrounding contaminating proteins. This is usually achieved by a series of **differential centrifugation** steps, followed by **sucrose gradient-density centrifugation**. Centrifugation through a sucrose gradient usually results in a sharp band of virus at a specific location on the gradient. This is harvested for further studies. It is normal to study **radiolabeled virions** by a variety of techniques including **SDS–polyacylamide gel electrophoresis** (separation based on size), **Western blotting** (reaction with antibodies) and **immunoprecipitation** (precipitation of proteins with antibodies).

The location of virus proteins within cells can be determined by **differential staining** and **immunofluorescence** techniques, with monoclonal antibodies to specific virus protein epitopes being the key reagents. Use of, for example, **pulse–chase** experiments, **protease** or **glycosylation inhibitors** have allowed studies on protein processing.

Gene sequencing and **amino acid prediction** of proteins has allowed a number of predictions to be made concerning virus protein structure and function.

Protein synthesis and complexity

As was outlined earlier, all proteins are synthesized from a mRNA template which is translated on the host-cell ribosomes. The mRNA may be (a) transcribed from the viral genome DNA (e.g. HSV), (b) complementary to the viral genome RNA (**negative-strand RNA viruses,** e.g. influenza) or (c) **viral** genomic RNA (**positive-strand RNA viruses,** e.g. polio)

During the growth cycle, in more complex viruses, virus specific proteins are synthesized at various times post-infection, for example, structural proteins may be produced late on in infection. Simple viruses (e.g. polio) do not have such levels of **temporal control.**

The amino acid sequence of the protein contains a number of **motifs** which determine its **post-translational modification**, its location in the cell (by **signaling**) and its **secondary** and **tertiary** structure.

Proteins may be the result of **proteolytic cleavage** of a larger **precursor molecule** (e.g. polio polyprotein, HIV gag-polymerase complex). They may be **phosphorylated**, the degree of phosphorylation often determining the functional activity of the protein (e.g. N, NS proteins of rhabdoviruses). The attachment of sugar residues to proteins (**glycosylation**) is either N- or O-linked and for many virus proteins (usually envelope glycoproteins) the sugar residues may account for 75% of the protein weight. **Myristylation**, **acylation** and **palmitoylation** are other post-translational modifications.

Structural proteins

Whatever the complexity, size or shape of a virus particle it is the role of the structural protein to provide **protection** for the viral genome and to allow delivery of virus particles from one host to another. Structural proteins are in either the **nucleocapsid**, **matrix** or **envelope** of the particle (*Fig. 1*).

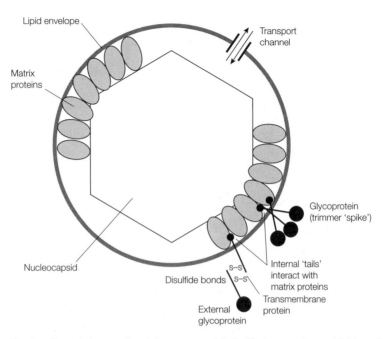

Fig. 1.   Several classes of proteins are associated with virus envelopes. Matrix proteins link the envelope to the core of the particle. Virus-encoded glycoproteins inserted into the envelope serve several functions. External glycoproteins are responsible for receptor recognition and binding, while transmembrane proteins act as transport channels across the envelope. Host-cell derived proteins are also sometimes found to be associated with the envelope, usually in small amounts. Redrawn from Cann, A., Principles of Molecular Virology, 1993, with permission from Academic Press.

Nucleocapsid proteins either self-assemble (e.g. TMV, polio) or assemble with the help of scaffolding proteins (e.g. HSV) to form the helical and icosahedral structures described earlier (Topic K1). The polio capsid is relatively simple with four proteins, VP1, VP2, VP3 and VP4. These assemble via a precursor structure (the procapsid) which contains VP0 (a precursor protein) and VP1 and VP3. VP0 cleaves to give VP2 and VP4 as the capsid acquires its nucleic acid. The reovirus capsid is much more complex and is composed of a double-shelled icosahedron structure (*Fig. 2*). Capsid proteins, in addition to protecting the genome and 'assembling' to form structural virions, must also interact with the nucleic acid genome during the assembly process. **Packaging** of the large piece of nucleic acid into a confined area requires complex folding of the genome and intimate association through chemical bonding with selected capsid proteins. Examples are rhabdovirus N protein and influenza virus NP protein whose positively charged amino acids interact with the negatively charged nucleic acid to encourage packaging. Nucleic acids also contain **packaging sequences** which result not only from nucleotide sequence but also from the **secondary** and **tertiary structures** of the genome. Packaging becomes even more complex when one considers a multigenome virus (e.g. influenza) where eight different segments of RNA have to be packaged into the virion nucleocapsid.

The **envelope** which surrounds many virus particles is acquired by nuclear membrane, plasma membrane or endoplasmic reticulum budding. Within the envelope are glycolipid proteins.

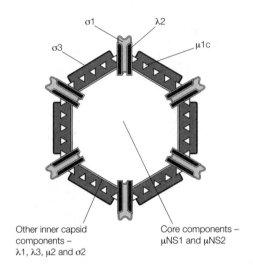

Fig. 2.   Reovirus particles consist of an icosahedral, double-shell arrangment of proteins
surrounding the core. Redrawn from Cann, A., Principles of Molecular Virology, 1993, with
permission from Academic Press.

Matrix proteins are internal proteins whose function, when present, is to link the internal nucleocapsid proteins to the envelope. They are usually not glycosylated and may contain **transmembrane anchor domains** or are associated with the membrane by hydrophopic patches on their surface or by protein–protein interactions with envelope glycoproteins. In HSV the space between the envelope and the capsid is referred to as the **tegument**.

External glycoproteins are anchored in the envelope by **transmembrane domains**. Most of the structure of the protein is on the outside of the membrane, with a relatively short internal tail. Many of these glycoproteins are **monomers** which group together to form the spikes visible in the electron microscope. They are the major **antigens** of enveloped viruses and include the G spikes of rabies virus, the gp120 spike of HIV and the HA spike of influenza.

The detailed structure of many of these molecules has been determined by **X-ray crystallography** and **cryoelectron microscopy**. One of the first molecules to be analyzed in this detail was the **hemagglutinin** of influenza. This excellent work resulted in the detailed molecule shown in *Fig. 3*.

Certain structural proteins in either naked capsids (e.g. polio) or enveloped viruses (e.g. HSV) initiate viral infection by attaching to **receptor sites** on host cells. Immune responses to these virus proteins often protect against viral infection and our understanding of this interaction forms the basis of modern vaccine design (see Topic K10). *Table 1* lists examples of specific receptor sites available on host cells for attachment by viruses.

Non-structural proteins

Non-structural proteins may be carried within the virus particle (but not as part of the 'architecture') or appear only in infected cells. The number and function of these proteins varies greatly from one virus family to another, depending on the complexity of the viral genome and the replicative cycle. Many have enzymic activity (e.g. the **reverse transcriptase protease** and **integrase** of retroviruses, the **thymidine kinase** and **DNA polymerase** of HSV). These enzymes are key targets for antiviral drugs (see Topic K11). Others have a role in virion assembly, acting as scaffolding proteins upon which capsids are assembled.

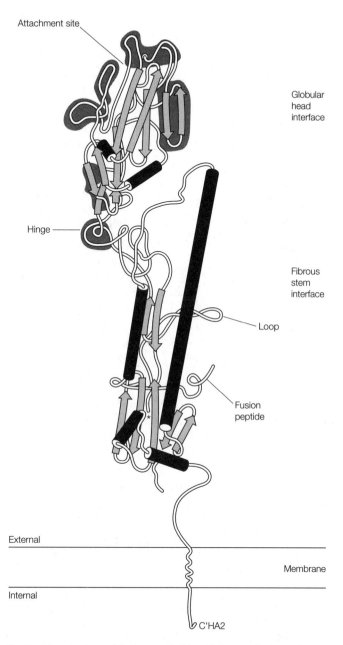

Fig. 3.   The structure of the hemagglutinin of influenza. Redrawn from Wiley et al., Nature, vol. 289, pp. 373–378, 1981, with permission from Macmillan Magazines Limited.

Table 1.   *Examples of receptors for viruses that infect humans*

Family	Virus	Cellular Receptor
Adenoviridae	Adenovirus type 2	Integrins $\alpha_v\beta_3$ and $\alpha_v\beta_5$
Coronaviridae	Human coronavirus 229E	Aminopeptidase N
Coronaviridae	Human coronavirus OC43	*N*-Acetyl-9-*O*-acetylneuraminic acid
Hepadnaviridae	Hepatitis B virus	IgA receptor
Herpesviridae	Herpes simplex virus	Heparan sulfate proteoglycan plus mannose-6-phosphate receptor
Herpesviridae	Varicella zoster virus	Heparan sulfate proteoglycan?
Herpesviridae	Cytomegalovirus	Heparan sulfate proteoglycan plus second receptor
Herpesviridae	Epstein–Barr virus	CD21 (CR2) complement receptor
Herpesviridae	Human herpesvirus 7	CD4 (T4) T-cell marker glycoprotein
Orthomyxoviridae	Influenza A virus	Neu-5-Ac (neuraminic acid) on glycosyl group
Orthomyxoviridae	Influenza B virus	Neu-5-Ac (neuraminic acid) on glycosyl group
Orthomyxoviridae	Influenza C virus	*N*-Acetyl-9-*O*-acetylneuraminic acid
Paramyxoviridae	Measles virus	CD46 (MCP) complement regulator
Picornaviridae	Echovirus 1	Integrin VLA-2 ($\alpha_2\beta_1$)
Picornaviridae	Poliovirus	IgG superfamily protein
Picornaviridae	Rhinoviruses	ICAM-1 adhesion molecule
Poxviridae	Vaccinia	Epidermal growth factor receptor
Reoviridae	Reovirus serotype 3	β-Adrenergic receptor
Retroviridae	Human immunodeficiency virus	CD4 (T4) T-cell marker glycoprotein
Rhabdoviridae	Rabies	Acetylcholine receptor

Table 2.   *Proteins of the human immunodeficiency virus (HIV) type 1*

gag	Polyprotein, cleaved by the viral aspartyl proteinase, produces the core proteins
pol	Polymerase/reverse transcriptase, aspartyl proteinase and integrase
env	Polyprotein (gp160), cleaved by cellular enzymes to produce gp41 and gp120

Auxiliary genes

vif	Required for production of infectious virus
vpr	Function unknown
tat	Up-regulates viral mRNA synthesis by binding to the TAR (*trans*-activator response element) in the transcripts from the long terminal repeat. Made from a spliced mRNA
rev	Required by mRNA transport. Made from a spliced mRNA
vpu	Virion release, receptor degradation
nef	Down-regulator of viral metabolism

Many (e.g. HIV **Tat** HSV **tegument proteins**) have regulatory roles in transcription and others form part of nucleic acid synthesis complexes (e.g. **DNA helicase**, DNA **binding proteins**). The shut-off of host cell macromolecular synthesis is usually the function of viral non-structural proteins. Viruses which induce tumors in their host by **transforming** normal cells do so by producing **viral oncogenes** (non-structural virus proteins) or by activating cellular oncogenes. Other proteins associated with cell transformation include, for example, the large **T antigen** of SV40 and the **EBNA** protein of Epstein–Barr virus. More recently, non-structural proteins have been associated with **anti-apoptosis**, **anti-cytokine** activity and interference with MHC antigen presentation (i.e. immune evasion – see Topic K9): *Table 2* lists the functions of HIV non-structural proteins as an example of selected activities.

K4 VIRUS NUCLEIC ACIDS

Key Notes

Types of viral genome	Viral genomes are diverse in size, structure and nucleotide make up. They are linear, circular, dsDNA, ssDNA, dsRNA, ssRNA, segmented or non-segmented.
Techniques of study	Many viral genomes have been sequenced and the repertoire of their coding potential determined. Genomes are studied by a range of techniques including restriction enzyme analysis, buoyant density, thermal denaturation, nuclease sensitivity and EM.
Large DNA viruses	Herpesviruses and adenoviruses have large DNA genomes which vary in their structures. DNA viruses often have terminal inverted repeats as part of their genome structure. They may have coding capacity for up to 80 virus proteins.
Small DNA viruses	Parvovirus (ssDNA) and polyoma viruses (dsDNA) are small DNA genomes. The coding capacity may be increased by use of overlapping genes and both strands of DNA.
RNA viruses (non-segmented)	These are either positive sense (e.g. picornaviruses, coronaviruses, flaviviruses and togaviruses) or negative sense (e.g. orthomyxoviruses, paramyxoviruses and rhabdoviruses). Positive-sense genomes act as mRNAs, negative sense genomes need a cRNA to act as mRNA.
RNA viruses (segmented)	Segmented viral genomes (e.g. orthomyxoviruses) are those which are divided into two or more physically distinct molecules of nucleic acid packaged into a single virion.
Related topics	Structure and organization of DNA (C1) DNA replication (C2)

Types of viral genome

The viral genome carries the nucleic acid sequences which are responsible for the **genetic code** of the virus. In infected cells the genome is **transcribed** and **translated** into those amino acid sequences which make up the viral proteins, be they structural or non-structural products. The genome, or in some cases a transcript of it, forms the basis of the template upon which new genomes are synthesized before assembly of progeny virions. Viral genomes are either **linear** or **circular** and composed of **RNA** or **DNA**. They vary in size from 3500 nucleotides (e.g. small phage) to 280 kb pairs, that is, 560 000 nucleotides (e.g. some herpesviruses). These sequences must be capable of being decoded by the host cell and thus the control signals must be recognized by host factors usually in association with viral proteins. Because of their small size, viral genomes have evolved to make **maximal** use of their nucleotide coding potential. Thus, **overlapping** genes and **spliced mRNAs** are common.

Characterization of the viral genome is based on a number of parameters: (1) **composition of the nucleic acid (i.e. DNA or RNA)**; (2) **size and number of strands**; (3) **terminal structures**; (4) **nucleotide sequence**; (5) **coding capacity**; (6) **regulatory signal elements, transcriptional enhancers, promoters and terminators.**

Some points of note are:

● dsDNA genomes (e.g. Poxviridae, Herpesviridae and Adenoviridae) are usually the largest genomes, ssDNA genomes (e.g. Parvoviridae) being smaller.
● dsRNA genomes (e.g. Reoviridae) are all segmented.
● ssRNA genomes are classified as being positive (+) sense or negative (–) sense. Viral genomes that are + sense can act as mRNAs and are usually infectious without the nucleocapsid proteins (e.g. Picornaviridae, Caliciviridae, Coronaviridae, Flaviviridae and Togaviridae). Negative-sense RNA genomes are usually not infectious unless accompanied by nucleocapsid proteins which have enzymic activity (transcriptases). These enzymes transcribe the negative-sense RNA into a complementary strand (cRNA) which acts as mRNA (e.g. Orthomyxoviridae, Paramyxoviridae, Rhabdoviridae and Filoviridae).
● Most ssRNA genomes are single molecules with the exception of, for example, Orthomyxoviridae (influenza) which have segmented genomes. Retroviridae have a ssRNA viral genome which replicates via a dsDNA intermediate, this being facilitated by a reverse transcriptase enzyme carried within the virion.
● The DNA genome of Hepadnaviridae is synthesized within hepatocytes via an RNA intermediate. These different modes of viral replication are summarized in *Fig. 1*.

Techniques of study

The advances in molecular-biological techniques over the last decade has made analysis of viral nucleic acids simpler, more efficient and quick. Viral genomes from most of the families have been totally **sequenced** and the open reading frames, and in many cases the gene products, characterized. This allows comparison with known sequences of genes stored on computer databases and has revealed fascinating similarities between viral and various eukaryotic genes which have been conserved through the process of evolution. Genes can be cloned into a variety of vectors and analyzed by a number of techniques (e.g. site-directed mutagenesis) to study the role of individual amino acids in determining the structural and functional integrity of the coded protein (see *Instant Notes in Biochemistry* in this series for an excellent review of these techniques).

DNA viruses are often shown diagramatically as linear molecules with **restriction enzyme** sites scattered throughout the genome. There are dozens of restriction enzymes, which are used to digest DNA into small segments at very specific nucleotide sequences. These segments are separated by size on agar gels following electrophoresis. Each DNA genome has a restriction enzyme **map** which characterizes it. This was not possible for RNA genomes until a reverse transcriptase enzyme was used to make a copy (c) DNA molecule from the RNA template, the cDNA being 'cut' with restriction enzymes.

Viral nucleic acids can be characterized by their melting temperature (T_m), their buoyant density in cesium chloride gradients, their 'S' value in sucrose gradients, their infectivity (or lack of), their nuclease sensitivity and their electron microscopic appearance.

(a) dsDNA genome: pox, herpes, adeno, papova, irido

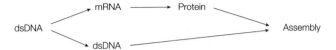

(b) ssDNA genome: parvo

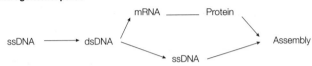

(c) ss/dsDNA genome using ssRNA intermediate: hepadna

(d) dsRNA genome: reo, birna

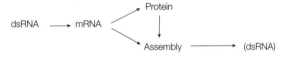

(e) +sense ssRNA genome: picorna, calici, corona, toro, toga

(f) –sense ssRNA genome: orthomyxo, paramyxo, rhabdo, filo (bunya, arena)

(g) ssRNA genome using ds DNA intermediate: retro

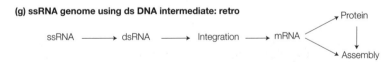

Fig. 1. General methods of viral replication. From Harper, D., Molecular Virology, *2nd edn,* ©
BIOS Scientific Publishers Limited, 1998.

Large DNA viruses

Herpesviridae vary in size from herpes simplex virus and varicella zoster virus (120–180 kbp) to cytomegalovirus and HHV-6 (180–230 kbp). The DNA codes for > 30 virion proteins and > 40 non-structural proteins (found only in infected cells). The structure of the DNA, while showing minor variations between members of the group is unique in that several **isomers** of the same molecule can exist. The genome of herpes simplex virus consists of two covalently-joined sections, a **unique long** (U_L) and a **unique short** (U_S) region, each bounded by **inverted repeats**. These repeats allow structural rearrangements of the unique regions, thus creating four isomers all of which are functionally equivalent (*Fig. 2*). Herpesvirus genomes also contain **multiple repeated sequences** which vary greatly between viruses, thus giving them a larger or smaller molecular weight than average.

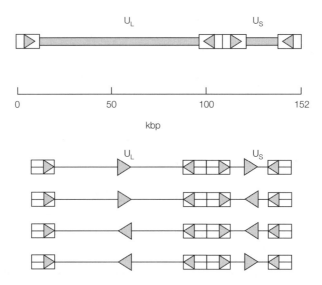

Fig. 2.   (a) Some herpes virus genomes (e.g. herpes simplex virus) consist of two covalently joined sections, U_L and U_S, each bounded by inverted repeats. (b) This organization permits the formation of four different forms of the genome. Reproduced from Cann, A., Principles of Molecular Virology, 1993, with permission from Academic Press.

The genome of adenoviruses is smaller, with a size of 30–38 kbp. Each virus contains 30–40 genes. The terminal sequence of each strand is a 100–140 bp **inverted repeat** and thus the denatured single strands can combine to form a **pan-handle** structure, which has an important role in replication. The genome has a 55k terminal protein at the 5′ end which acts as a primer for the synthesis of new DNA strands (*see Fig. 3*).

Whereas in herpesviruses each gene has its own promoter the adenovirus genome has **clusters** of genes which are served by a common promoter.

Small DNA viruses

Parvovirus genomes are linear non-segmented ssDNA of about 5 kb. Most of the strands that appear in virions are of negative (–) sense. These very small genomes contain only two genes, **rep**, which encodes proteins involved

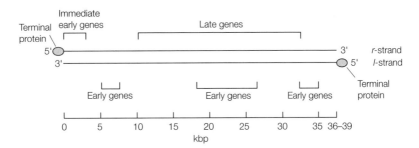

Fig. 3. Organization of the adenovirus genome. Reproduced from Cann, A., Principles of Molecular Virology, 1993, with permission from Academic Press.

in transcription, and **cap,** which encodes the coat proteins. The ends of the genomes have **palindromic sequences** of about 115 nucleotides which form hairpin structures essential for the initiation of genome replication.

Polyoma viruses have dsDNA, 5 kbp in size and circular. The DNA in the virion is **super-coiled** and is associated with four cellular histones (H2A, H2B, H3 and H4). The virus has six genes, this having been achieved by using **both strands** of the DNA for coding and making use of **overlapping genes** (see *Fig. 4*).

RNA viruses (non-segmented)

RNA viruses tend to have smaller genomes than most DNA viruses. They vary in size from 30 kb (for example, coronaviruses) to 3.5 kb (small phage, for example, Qβ). They are either **positive sense** or **negative sense** as described earlier. Most (not all) positive-sense ssRNA genomes have a **cap** that protects

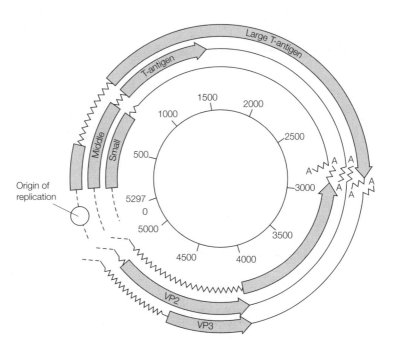

Fig. 4. The complex organization of the polyomavirus genome results in the compression of much genetic information into a relatively short sequence. Reproduced from Cann, A., Principles of Molecular Virology, 1993, with permission from Academic Press.

the 5′ terminus of the RNA from attack by phosphatases and other nucleases and promotes mRNA function at the level of translation. In the 5′ cap the terminal 7-methylguanine and the penultimate nucleotide are joined by their 5′-hydroxyl groups through a triphosphate bridge. This 5′–5′ linkage is inverted relative to the normal 3′–5′ phosphodiester bonds in the remainder of the polynucleotide chain. In most picornaviruses this **cap** is replaced by a small protein, VPg. The 3′ end of most positive-sense viral genomes, like most eukaryotic mRNAs, is **polyadenylated.**

Viruses with negative-sense RNA genomes are usually larger and encode more genetic information than positive genomes. The genomic organization of selected RNA viruses is shown in *Fig. 5*.

RNA viruses (segmented)

Segmented virus genomes are those which are divided into two or more **physically distinct molecules** of nucleic acid, all of which are packaged into a single virion. They have the advantage of carrying genetic information in smaller

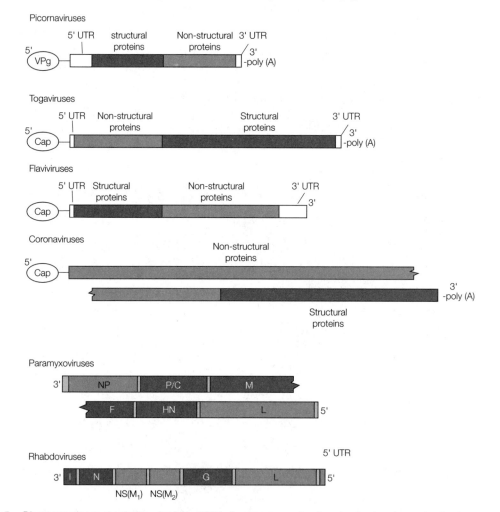

Fig. 5.   Diagrammatic representation of selected RNA virus genomes showing structural and non-structural genes. 5′ UTR are regions of the genes that are not translated into proteins. Reproduced from Cann, A., Principles of Molecular Virology, 1993, with permission from Academic Press.

Orthomyxoviruses

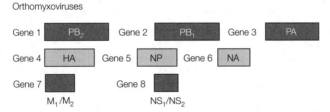

Fig. 6. Relative size and coding of orthomyxovirus RNA segments. Reproduced from Cann, A., Principles of Molecular Virology, 1993, with permission from Academic Press.

Table 1. Segments of the influenza virus genome

Segment	Size (nt)	Polypeptide(s)	Function
1	2341	PB_2	Polymerase – cap binding
2	2341	PB_1	Polymerase – endonuclease (?)
3	2233	PA	Polymerase – elongation
4	1778	HA	Hemagglutinin
5	1565	NP	Nucleoprotein
6	1413	NA	Neuraminidase
7	1027	M_1	Matrix protein (non-structural)
		M_2	Non-structural – function unknown
8	800	NS_1	Non-structural – function unknown
		NS_2	Non-structural – function unknown

strands, which makes them less likely to break due to shearing. This breakage is a hazard for large RNA molecules. On the downside, of course, is the fact that these viruses must have an elaborate packaging mechanism which allows a representative of each strand to be packaged into the virion. Influenza virus, a negative ssRNA virus, has eight segments each coding for one or two viral proteins (*Fig. 6* and *Table 1*).

K5 CELL CULTURE AND VIRUS GROWTH

Key Notes

Historical perspective

The major thrust in virology research was only possible after the derivation of defined synthetic growth medium and the *in vitro* culture of cells. This allowed viruses to be grown, purified and studied outside of the host, leading to the development of the first virus vaccines.

Method of cell culture

Single-cell suspensions are seeded into tissue culture quality glass or plastic vessels where they attach to the surface and divide to form a monolayer of cells. Cells can be grown in a full range of vessels but must be carefully nurtured and repassaged, ensuring they are free from contamination. Most cells grow at 37°C.

Media and buffer

Defined growth media contain a balanced salts solution supplemented with amino acids, vitamins, glucose, serum, antibiotics, a buffer and pH indicator. Such media are used routinely to support the growth of tissue-culture cells and can be either manufactured from the individual constituents or bought ready-made commercially.

Cell culture types

Cells are either primary (very limited cell passage), diploid cell strains (up to 50–60 cell passages) or continuous cell lines (unlimited cell passage).

Virus growth in culture

Small aliquots of virus are added to cell monolayers at low or high multiplicity of infection and, following replication, progeny virus is harvested and assayed. The virus is either cell-associated or related into the culture medium. Following titration, the virus is stored at either –20°C or –70°C until required.

Use of embryonated eggs

For some viruses, growth in embryonated hens eggs is the preferred culture method. Influenza virus is grown in the allantoic cavity of the embryonated egg.

Related topics Virus assay (K6) Virus replication (K7)

Historical perspective

In addition to helping solve the mysteries of virus diseases, knowledge of virus structure, replication and host interactions, virus research has extended our understanding of the fundamentals of eukaryotic biology (e.g. protein:protein interactions, nucleic acid transcription and translation, evolution). Few of these studies would have been possible without the means of growing viruses outside their normal hosts. For most viruses this means growing them in **cell culture**. Cell culture is the art of growing cells *in vitro*. In addition to deriving cells for culture, this technique relied heavily on the development of **media** and **buffers** that supported cell growth *in vitro*. Most of these techniques did not become available until the early 1950s, which consequently saw an upsurge in the study of viruses.

There are now over 3200 characterized cell lines derived from over 75 species, which are kept in, for example, the American Type Culture Collection and the European Collection of Animal Cell Cultures.

Method of cell culture

Cells are propagated on **glass** or **plastic flasks** of various sizes (as required) or in vast vessels or vats. The technique relies on good **aseptic technique**, thus keeping the cultures free of fungal and bacterial contamination.

Single-cell suspensions of cells of known concentration are **seeded** into a sterile flask along with appropriate **growth medium**. The flask, usually a plastic or glass bottle, is incubated at the appropriate temperature (usually 37°C) in the flattened position. Cells adhere to the surface and begin to replicate, forming a **monolayer** of cells which adhere to each other, in addition to being anchored to the surface of the flask. After a few days the metabolic activity of the cells means that the growth medium will be 'spent' and the cells, unless **reseeded**, will deteriorate and die. Thus, the cell monolayer is treated with trypsin and/or versene solution to create single cells again. These cells are used for **seeding** into fresh flasks. The continued seeding of cells is referred to as cell passage. Cell monolayers are used for virus growth and assay and to examine many aspects of virus–host interactions (see *Fig. 1*).

In addition to growing as a monolayer, some cell types may also be able to grow in **suspension**, where they do not anchor themselves to the surface of the flask or adhere to each other (e.g. hybridoma cells which secrete monoclonal antibodies).

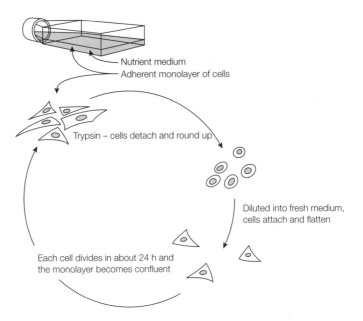

Nutrient medium
Adherent monolayer of cells

Trypsin – cells detach and round up

Diluted into fresh medium, cells attach and flatten

Each cell divides in about 24 h and the monolayer becomes confluent

Fig. 1. Cell culture. Redrawn from Dimmock and Primrose, 1994, Introduction to Modern Virology, *4th edn, with permission from Blackwell Science Ltd.*

Media and buffer

Most growth media in use today are chemically defined, but are usually supplemented with 5–20% serum (which contains the stimulants necessary for cell division). Serum-free medium with added stimulants is used for some purposes. Media contain an isotonic **balanced salts solution**, supplemented with amino acids, vitamins and glucose, for example, Eagle's minimal essential medium (MEM), formulated by Eagle in the 1950s. In addition to serum, MEM is also supplemented with antibiotics, usually penicillin and streptomycin, to help prevent bacterial contamination. Generally, cells grow well at pH 7.0–7.4 and hence phenol red is added to act as a pH indicator. It is red at pH 7.4, orange at pH 7.0 and yellow at pH 6.5. In the alkaline direction it becomes bluish at pH 7.6 and purple at pH 7.8.

Culture media also require buffering under two sets of conditions: (a) use of open flasks, in which exposure to oxygen causes the pH to rise; and (b) high cell concentrations when CO_2 and lactic acid are produced, causing the pH to fall. A buffer is thus incorporated into the medium, and in 'open flask' conditions the incubator is purged with an exogenous source of CO_2. The buffer used by most laboratories relies on the bicarbonate – CO_2 system, and hence bicarbonate solution is added to the growth medium.

Reagents for use in media preparation and cell culture must be sterilized. This is achieved by autoclaving (moist heat), hot-air ovens (dry heat), membrane filtration or, in the case of plastic ware, by irradiation.

Cell culture types

Primary cells are freshly isolated cells that are derived directly from the tissue of origin. The tissue source for the majority of primary cell cultures is either laboratory animals (e.g. monkey kidney cells) or human pathology specimens (e.g. human amnion cells). Tissue samples are incubated with a proteolytic enzyme (usually trypsin) overnight with a number of washes, etc., in order to produce a single-cell suspension. The harvested cells are then seeded into appropriate flasks with growth medium. Cell cultures are usually either epithelial (cuboid shaped) or fibroblastic (spindle shaped), but it is usual for primary cultures to contain both morphological types. Primary cultures are sensitive to a wide range of viruses and are routinely used in diagnostic laboratories for growth of fresh virus isolates (from patients). Unfortunately, primary cells usually die after only a few passages.

Many **cell lines** will continue to grow for more than four to five passages but subsequently die after 50–60 passages. These cell lines, usually referred to as **diploid cell strains**, are often derived from fetal lung tissue. They have the normal chromosome number and are usually fibroblastic (e.g. MRC-5 cells).

Continuous cell lines are capable of continued passage in tissue culture. They may be epithelial (e.g. Hela cells) or fibroblastic (e.g. baby hamster kidney cells, BHKs) and are derived from tumorous tissue (e.g. Hela cells) or by the sudden **transformation** of a primary cell (e.g. BHK cells). They are **heteroploid** (i.e. have aberrant chromosome numbers).

Virus growth in culture

Most experiments in virology have required virus growth in culture although, nowadays, more experiments rely entirely on cloned genes and expressed proteins outside of cell cultures. Historically, however, it has been those viruses which grow well in cell culture that have been studied in most detail. Lack of *in vitro* growth has seriously curtailed progress in research, vaccine production and development of antiviral drugs for hepatitis B and C viruses.

Viruses are grown in cultures to create **virus stocks**. The passaged virus is stored at –70°C and referred to as a **master-stock, sub-master stock**, etc., depending on its **passage number**. It is important in virology to record the passage history in tissue culture of the virus being used.

Virus stocks are grown by infecting cells at a low **multiplicity of infection (m.o.i.)** that is, approximately 0.1–0.01 infectious units per cell. Virus attaches to cells and goes through several replicative cycles in the cell culture. After a few days, virus is harvested from the extracellular medium surrounding the cultured cells or from the cells themselves, which are lysed by freezing and thawing or by using an ultrasonic bath. The virus so derived is quantitated by an infectivity assay.

When large numbers of virus particles are required (e.g., for virus purification), cell cultures are infected at a high m.o.i. (e.g. 10 infectious units per cell). This ensures that all cells are synchronously infected and that only one round of replicative cycle will ensue. Virus is harvested as above at the end of the cycle. Infected cells yield various numbers of new (progeny) virus particles ranging from 10–10 000 particles per cell.

Use of embryonated eggs

For some viruses (e.g. influenza virus) cell culture is not the chosen procedure for virus growth and instead a fertilized chick embryo is used. The fertilized embryo has a complex array of membranes and cavities which will support the growth of viruses (see *Fig. 2*). Small aliquots of influenza virus are inoculated into the allantoic cavity of the egg. The virus then attaches to and replicates in the epithelial cells lining the cavity. Virus is released into the allantoic fluid and harvested after two days growth at 37°C. Influenza vaccines are propagated in this way.

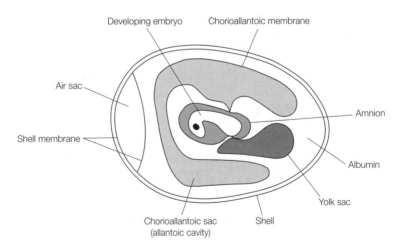

Fig. 2.   The embryonated egg.

K6 VIRUS ASSAY

Key Notes

Virus infectivity	The ability of a virus particle to attach, penetrate and undergo an infectious cycle in a susceptible host cell, usually resulting in cell damage.
Virus dilution	Virus is diluted using factors of two, five or ten in an appropriate solution (e.g. buffer or growth medium) prior to an infectivity assay.
Plaque assay	This is a focal assay used to detect zones of cytopathic effect (CPE) in a monolayer of healthy cells (plaques). One infectious virus produces one plaque.
$TCID_{50}$	The tissue culture infective dose$_{50}$ ($TCID_{50}$) is defined as that dilution of virus which will cause CPE in 50% of a given batch of cell cultures.
Particle counting	The total number of virus particles (infectious and non-infectious) is determined by counting with the aid of an electron microscope. This allows determination of the particle/infectivity ratio.
Hemagglutination	The agglutination of red blood cells (RBCs) by some viruses is referred to as hemagglutination (HA). This agglutination forms the basis of an assay which measures the number of HA units per unit volume in a given suspension.

Related topics   Cell culture and virus growth (K5)   Virus replication (K7)

Virus infectivity   In order to 'reproduce', viruses need to be capable of replicating in a susceptible host cell. This **replicative cycle** is accompanied by a number of biochemical and morphological changes within the cell which usually results in the death of the cell. The accompanying morphological changes (e.g. cell rounding or fusion) are referred to as the cytopathic effect (CPE). A particular type of CPE is often a characteristic of specific virus growth and can be used when attempting to identify an unknown virus. The appearance and detection of CPE regularly forms the basis of **infectivity assays**, designed to determine the number of **infectious units of virus per unit volume**, is the infectivity **titer** (e.g. **plaque forming units** (pfus) per milliliter). An infectious unit is thought of as being the smallest amount of virus that will produce a detectable biological effect in the assay (e.g. a pfu). Infectivity assays are either **quantal,** an 'all or none' approach (e.g. tissue culture infective dose 50 ($TCID_{50}$)) or **focal,** detection of a focus of infection (e.g. a plaque assay).

Virus dilution

Virus titers are determined by making accurate serial dilutions of virus suspensions. Such dilutions are usually done using factors of two, five or ten. For routine use, **10-fold dilutions** are usually carried out. It is important to use a new sterile pipette for the transfer of volumes between each dilution and to mix thoroughly the dilution before further transfer. Once diluted, virus should be assayed as soon as possible as most viruses rapidly lose infectivity at room temperature.

Plaque assay

The plaque assay quantifies the number of **infectious units** in a given suspension of virus. **Plaques** are localized discrete foci of infection denoted by zones of cell lysis or cytopathic effect (CPE) within a monolayer of otherwise healthy tissue culture cells. Each plaque originates from a **single infectious virion,** thus allowing a very precise calculation of the virus titer. The most common plaque assay is the **monolayer** assay. Here, a small volume of virus diluent (0.1 ml) is added to a previously seeded confluent tissue culture cell monolayer. Following adsorption of virus to the cells, an **overlay medium** is added to prevent the formation of **secondary** plaques. Following incubation, the cell sheets are 'fixed' in **formol saline** and stained, and the plaques counted. For statistical reasons, 20–100 plaques per monolayer are ideal to count, although the actual number that can be easily counted is often dependent on the size of the plaque and the size of the vessel used for the assay. Typical plaques are shown in *Fig. 1*, and the assay procedure is summarized in *Fig 2*. The **infectivity titer** is expressed as the number of plaque forming units per ml (**pfu ml^{-1}**) and is obtained in the following way:

$$\text{pfu ml}^{-1} = \frac{\text{plaque number}}{\text{dilution} \times \text{volume (ml)}}$$

For example, if there is a mean number of 100 plaques from monolayers infected with 0.1 ml of a 10^{-6} dilution then the calculation is:

$$\frac{100}{10^{-6} \times 0.1} = 1 \times 10^9 \text{ pfu ml}^{-1}$$

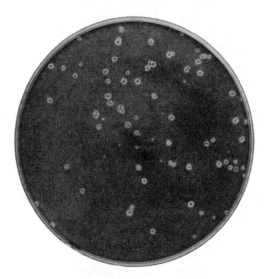

Fig. 1. Herpes virus plaques on a tissue culture monolayer.

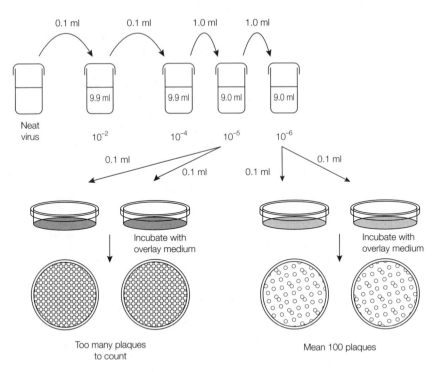

Fig. 2.   Diagrammatic representation of virus plaque assay (see text for calculation).

TCID$_{50}$

The TCID$_{50}$ is defined as **that dilution of virus required to infect 50% of a given batch of inoculated cell cultures**. The assay relies on the presence and detection of CPE. Host cells are grown in confluent healthy monolayers, usually in tubes, to which aliquots of virus dilutions are made. It is usual to use either five or ten repetitions per dilution. During incubation the virus replicates and releases progeny virus particles into the supernatant which in turn infect other healthy cells in the monolayer. The CPE is allowed to develop over a period of days, at which time the cell monolayers are observed microscopically. Tubes (the 'test units') are scored for the presence or absence of CPE. In this quantal assay the data are used to calculate the TCID$_{50}$ i.e. the dilution of virus which will give CPE in 50% of the cells inoculated. *Table 1* shows some typical data. By using the data in *Table 1* the following calculation can be made:

$$\text{TCID}_{50} = \log_{10} \text{ of highest dilution giving 100\% CPE} + \frac{1}{2} -$$

$$\frac{\text{total number of test units showing CPE}}{\text{number of test units per dilution}}$$

$$= -6 + \frac{1}{2} - \frac{4}{5} = -7.3 \text{ TCID}_{50}$$

or $10^{-7.3}$ TCID$_{50}$ unit vol.$^{-1}$

The titer is therefore $10^{7.3}$ TCID$_{50}$ per unit vol.$^{-1}$

Particle counting

Not all virus particles are infectious. Indeed in many cases for every one infectious particle up to 100 or more **non-infectious particles** may be produced from

Table 1. Data used to calculate TCID$_{50}$ (See text for calculation)

Log$_{10}$ of virus dilution	Infected test units (e.g. infected tubes)
–6	$^5/_5$
–7	$^3/_5$
–8	$^1/_5$
–9	$^0/_5$

an infected cell. The total number of particles can only be determined by counting them with the aid of an **electron microscope**. The counting procedure relies on the use of **reference particles** which are usually latex beads of uniform diameter. The principle is that if viruses can be mixed with reference particles of known concentration (i.e. a number per unit volume), a simple determination of the ratio of virus to reference particles will yield the virus count. Latex and virus particles are distinguished after **negative staining** with phosphotungstate. The ratio of total particles to infectious particles is termed the **particle/infectivity ratio,** which is important to know when, for example, monitoring virus purification, or determining the state or age of a virus suspension.

Hemagglutination Many viruses have the ability to agglutinate RBCs, this being referred to as **hemagglutination**. In order for the reaction to occur, the virus should be in sufficient concentration to form cross-bridges between RBCs, causing their agglutination. Non-agglutinated RBCs will form a **pellet** in a hemispherical well, whereas agglutinated RBCs form a **lattice-work** structure which coats the sides of the well. This phenomenon forms the basis of an assay which determines the number of **hemagglutinating particles** in a given suspension of virus. It is not a measure of infectivity, but is one of the most commonly used **indirect methods** for the determination of virus titer. The assay is done by **end-point titration**. Serial two-fold dilutions of virus are mixed with an equal volume of RBCs and the wells are observed for agglutination. The end point of the titration is the **last dilution showing complete agglutination,** which by definition is said to contain one **HA unit**. The HA titer of a virus suspension is therefore defined as being the reciprocal of the highest dilution which causes complete agglutination and is expressed as the number of HA units per unit volume. An example upon which a calculation of the HA titer can be made is shown in *Fig. 3*. The end point in this figure is $^1/_{512}$. If 0.2 ml virus dilution was added per well the HA titer would be 512 HA units per 0.2 ml or 2560 HA units ml^{-1}.

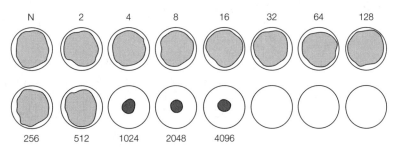

Fig. 3. Diagram of a sample hemagglutination assay. Serial doubling dilutions of virus shows agglutination end-point at 1:512.

K7 VIRUS REPLICATION

Key Notes

Replicative cycle	As obligate intracellular parasites, viruses must enter and replicate in living cells in order to 'reproduce' themselves. This 'growth cycle' involves specific attachment of virus, penetration and uncoating, nucleic acid transcription, protein synthesis, maturation and assembly of the virions and their subsequent release from the cell by budding or lysis.
Attachment, penetration and uncoating	Attachment is a very specific interaction between the virus capsid or envelope and a receptor on the plasma membrane of the cell. Virions are either engulfed into vacuoles by 'endocytosis' or the virus envelope fuses with the plasma membrane to facilitate entry. Uncoating is usually achieved by cellular proteases 'opening up' the capsid.
Transcription and translation	Using cellular and virus-encoded enzymes and 'helper' proteins, nucleic acid is usually transcribed in a controlled fashion. Control is also exercised at the level of mRNA concentration (apart from some simple viruses, e.g. polio). Nucleic acid is synthesized by virus-encoded enzymes. Translated proteins may undergo post-transitional modification (e.g. cleavage, glycosylation, phosphorylation).
Maturation, assembly and release	Subunits of capsids assemble via 'sub-assembly' structures, with or without the help of scaffolding proteins. Envelopes, when present, are acquired by capsids budding through the nuclear or plasma membrane.
Related topics	Cell culture and virus growth (K5) Antiviral chemotherapy (K11) Virus assay (K6)

Replicative cycle The complexity and range of virus types is echoed in the various strategies they adopt in their replicative cycles. Viruses, as **obligate intracellular parasites, must attach** to or **enter** host cells in order to undergo a 'reproductive' cycle. This cycle is highly dependent on the metabolic machinery of the cell, which in most cases the virus takes over and orchestrates towards its own replication, usually **inhibiting host-cell protein and nucleic acid synthesis**. The outcome is the production of hundreds of progeny virions which leave the infected cell (by **lysis** or **budding**), killing the cell and spreading to infect more host cells and tissues. This replicative or **growth cycle** can be analyzed in tissue culture cells and is often referred to as the **one-step** growth cycle. The cycle has a number of stages – attachment and penetration, nucleic acid synthesis and transcription, protein synthesis, maturation, assembly and release. A typical pattern for a growth curve is shown in *Fig. 1*. Following attachment and penetration

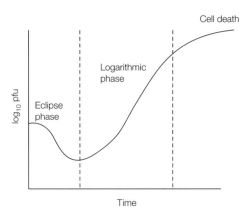

Fig. 1. A typical 'one-step' growth or replicative cycle of a virus.

by virus, cells are lysed and titrated for infectious virus particles (pfus) at various times post-infection. Plotting $\log_{10}$ pfu versus time gives the characteristic curve which has an **eclipse period** (where no new virions have been formed) followed by a **logarithmic expansion phase** until **peak** virus titers are reached when the cell usually dies and virions are released. The shape of this curve varies greatly between viruses – for most bacteriophage it takes less than 60 min and for many animal viruses it can exceed 24 h before maximum titers are reached. The replication cycles of three animal viruses have been selected below for further study and to compare their strategies – herpes simplex (a DNA virus), poliovirus (RNA virus) and HIV (a single-stranded RNA virus which replicates via a DNA intermediate). Their strategies are typical of many viruses, although divergent strategies also exist!

Attachment, penetration and uncoating

Attachment is mediated by a **specific interaction** between the virus and a **receptor** on the plasma membrane of the cell (*Fig. 2*). Indeed it is the presence of such a receptor that determines the **cell tropism** and **species tropism** of the virus. These receptors have cellular functions other than providing a binding site for viruses, but have been paramount in affecting virus evolution. Thus, the herpes simplex virus (HSV) binds to a heparin sulfate proteoglycan plus mannose-6-phosphate receptor via at least two virus-coded envelope glycoproteins. One of the four proteins in the poliovirus capsid attaches to a receptor of the **Ig protein superfamily** (only on primate cells). HIV, via its major envelope glycoprotein (**gp120**), attaches to the **CD4 receptor** found predominantly on human T4 lymphocytes. Recently, two further ligand receptors were shown to have a role in HIV attachment.

For HSV and HIV, penetration of the virus across the plasma membrane is achieved by **fusion** of the viral envelope with the membrane, releasing the nucleocapsid into the cytoplasm of the cell (*Fig. 2*). The naked capsid of poliovirus, however, is taken up by the process of **endocytosis**, the membrane invaginating to engulf the capsid, resulting in the formation of a **vacuole** which transports the capsid into the cytoplasm. The virion is later released from this vacuole (*Fig. 2*).

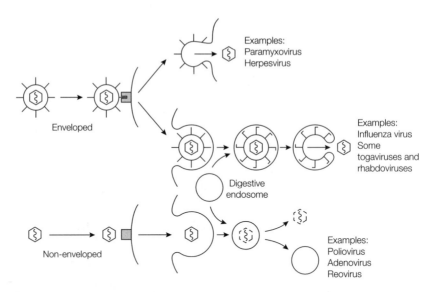

Fig. 2.   Methods of virus entry. From Harper, D., Molecular Virology, *2nd edn, © BIOS Scientific Publishers Limited, 1998.*

Transcription and translation

(i) **HSV**.   HSV, a large DNA virus, replicates mainly in the **nucleus** of the cell, although of course protein synthesis and post-translational modification take place in the cytoplasm. The genome encodes for dozens of virus specific proteins, many with **enzymic** activity (e.g. **thymidine kinase, DNA polymerase**) – such proteins are usually **non-structural** (i.e. will not finish up in the virion). Others are structural proteins and will form the capsid, envelope and tegument (the structure between the capsid and the envelope).

Viruses need proteins at different times and in different concentrations throughout their growth cycles and hence the virus shows transcriptional control. Depending on the timing of their expression in the virus replication cycle, HSV genes are classified as either **immediate early**, **early** or **late** (α, β or γ). The mRNA produced encodes proteins that have control functions, switching on subsequent genes. All proteins, of course, are formed on cytoplasmic host-cell ribosomes and remain in the cytosol or are directed to the endoplasmic reticulum where they undergo the **post-translational** events (e.g. **glycosylation, phosphorylation**) that give them their final identity. Eventually, these proteins find their way (specifically directed, i.e. **chaperoned**) back to the nucleus for assembly.

The double-stranded viral DNA is synthesized by the viral **DNA polymerase** in association with a number of **DNA-binding proteins**. This enzyme has formed the target for a number of **antiviral drugs** as it is significantly different to the host-cell DNA polymerase (see *Fig. 3*).

(ii) **Poliovirus**. Poliovirus is much simpler in its replicative procedures and control, indeed it lacks fine control! The poliovirus genome RNA strand also acts as mRNA (termed a **positive**-sense RNA virus) and is immediately translated into one long **polyprotein** which is subsequently cleaved into a number of structural and non-structural poliovirus proteins. Included are the structural proteins VP1 and VP3 and the **precursor** protein VP0. Non-structural proteins include a **protease** and **RNA polymerase**. Replication takes place in

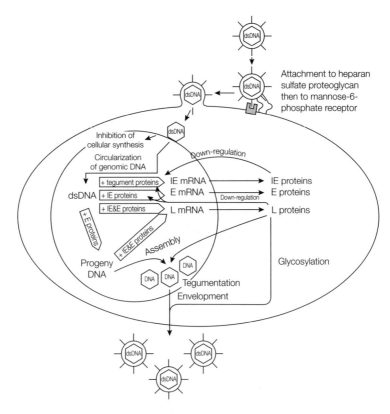

Fig. 3. Herpes simplex virus replication (a dsDNA virus) showing regulation of and by immediate early (IE), early (E) and late (L) proteins. From Harper, D., Molecular Virology, 2nd edn, © BIOS Scientific Publishers Limited, 1998.

the **cytoplasm**, indeed enucleate cells will support poliovirus replication. The ss RNA, under the direction of the viral **RNA polymerase** and cellular factors, replicates via a series of ds **replicative intermediate** molecules which act as template for the synthesis of new positive strands. These are then destined to act as mRNA for further rounds of protein synthesis, or become genomes in newly formed progeny virions (see *Fig. 4*).

(iii) **HIV**.   HIV is a member of the unique family of viruses (**Retroviridae**) that carry a **reverse transcriptase enzyme** in their particles, the enzyme catalyzing the formation of DNA from a RNA template. Thus, HIV goes through a series of events (see *Fig. 5*), which terminates in the formation of a ds DNA circularized molecule, formed from an initial input of HIV ss RNA. This molecule, under the direction of a virally encoded integrase, is inserted into the host DNA as a **provirus**. DNA synthesis is now under the control of the cell – when a daughter cell is produced the provirus is reproduced at the same time. The transcription of viral mRNA is under viral control (from the long terminal repeat (LTR) region of its genome) and a series of **mRNA molecules of various sizes** are transcribed. Translated proteins are in some cases **proteolytically cleaved** by virus **protease** into smaller functional proteins. HIV envelope proteins (gp120 and gp45) are further processed in the endoplasmic reticulum before being laid

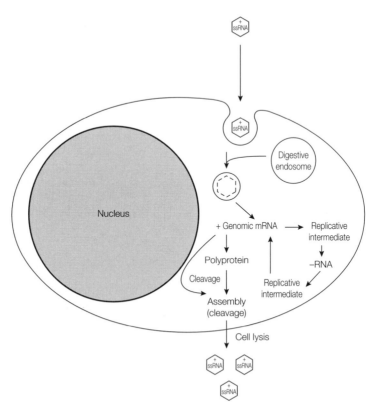

Fig. 4.   Poliovirus replication. From Harper, D., Molecular Virology, *© BIOS Scientific Publishers Limited, 1994.*

down in the plasma membrane of the cell. **Full-length copy RNA** molecules are also transcribed from the provirus DNA, these forming **progeny RNA strands** destined to be encapsidated.

Maturation, assembly and release

As proteins and nucleic acid are synthesized in the infected cell they are channeled to various locations for virion assembly. Capsids assemble in the **nucleus** or the **cytoplasm**. The steps of capsid assembly vary, depending on the complexity of the mature capsid, and for enveloped viruses the site of envelope acquisition is either the **nuclear membrane**, **plasma membrane** and rarely the **endoplasmic reticulum**. Prior to capsids 'budding' through these membranes, virus-specific proteins would have been laid down in the membrane.

(i) **HSV**.   Mature HSV capsids are assembled in the nucleus via a number of **precursor forms**. Assembly is assisted by **scaffolding proteins,** which do not finish up in the mature capsid but facilitate protein–protein and protein–nucleic acid bonding as the sub-units of the capsid come together. The mature capsid (containing the DNA genome) buds through the nuclear membrane to acquire its envelope. Virions are channeled through the ER to the plasma membrane where they are released. Many herpesviruses invade adjacent cells by the process of **cell–cell fusion**. Thus, the plasma membrane of an infected cell fuses with an adjacent normal cell, facilitating the entry of progeny virions which undergo a further replication cycle. In tissue culture this phenomenon can be seen as large areas of **multinucleate fused cells (syncytia)**.

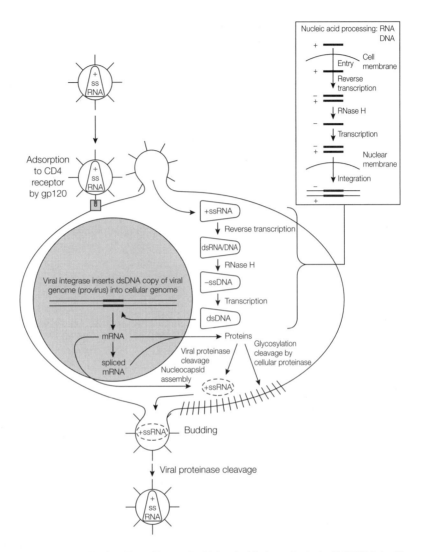

Fig. 5. HIV replication. From Harper, D., Molecular Virology, *2nd edn, © BIOS Scientific Publishers Limited, 1998.*

(ii) **Poliovirus**. Poliovirus is a relatively simple icosahedral capsid with four proteins making up the capsid (VP1, VP2, VP3 and VP4). A further virus-coded protein, VPg, is attached to the ssRNA, serving as a recognition protein. This capsid self-assembles without the need for scaffolding proteins, but does produce an 'immature' capsid form prior to the RNA being inserted into the virion. Thus, the sub-units of the capsid assemble via 'pentamers' to form a nucleic acid-free capsid shell containing VP0, VP1 and VP3. As the RNA is sequestered into the capsid, VP0 is cleaved into VP2 and VP4 and the mature virion is formed. Poliovirus exits the cell by lysis, releasing several hundreds of progeny virions.

(iii) **HIV**. The capsid of HIV assembles in the cytoplasm. Capsid assembly of the icosahedron follows the normal pattern, but the control mechanisms are obviously important here as the mature virion contains within its central core many components: proteins, two identical strands of RNA and the reverse transcriptase, integrase and protease complex of polymers. Budding of HIV is via the plasma membrane.

Many viruses (e.g. influenza, reovirus) have multisegmented genomes. Each segment, during infection, codes for a virus protein essential in replication. In order to be infectious the virion must, therefore, contain a copy of each segment of RNA. Quite how this is achieved remains a mystery!

K8 VIRUS INFECTION

Key Notes

Virus spread

Viruses can gain access to the host through the skin and mucous membranes via the respiratory or gastrointestinal tracts or through sexual contact. Without spread from host to host, viruses will cease to replicate and become extinct. A knowledge of how viruses spread and their routes of entry and exit from the host has allowed a study of the epidemiology of infections, which in turn has helped control the spread.

Clinical results of infection

The outcome of virus infection is dependent on a number of factors (e.g. age and immune status of the host). Infections are either localized at or near the site of entry, or systemic, where the virus spreads from its point of entry to involve one or more target organs. The outcome of a viral infection follows one of several patterns: (a) inapparent infection; (b) disease syndrome, virus eradication and recovery; (c) latency; (d) carrier status; (e) neoplasia; (f) death.

Related topics

Viruses and the immune system (K9)   Antiviral chemotherapy (K11)
Virus vaccines (K10)

Virus spread

In order to persist and evolve in nature, viruses need a large population of susceptible hosts and an efficient means of spread between these hosts. The normal route is termed **horizontal** spread. The most common route of entry and exit of viruses is the **respiratory route**. Following their inhalation, viruses usually infect and replicate in the epithelial cells of the upper and lower respiratory tract (e.g. rhinoviruses, coronaviruses, influenza, para-influenza and respiratory syncytial virus). The viruses produced in these airways exit their host via sneezing and coughing. Many viruses which enter and exit via this route are not 'respiratory viruses' (e.g. chicken pox, measles, German measles). In these cases the virus leaves the respiratory tract to set up infection in other target organs. The **oral–gastrointestinal** route is used mainly by those viruses responsible for gut infections (rotavirus, Norwalk virus and the enteroviruses, including polio and coxsackie viruses). Vast numbers of virus particles can be excreted in fecal material (e.g. in the order of 10^{12} particles g^{-1}), facilitating the easy spread of these viruses in conditions of poor sanitation. Thus, the drinking of fecally contaminated water and consumption of contaminated shellfish or other food prepared by unhygienic food handlers are ways in which these viruses are spread.

Whilst the skin normally provides an impenetrable barrier to virus invasion, infectious viruses can enter following **trauma to the skin**. This may be from the bite of an animal vector (e.g. rabies via an infected canine, yellow fever via an infected mosquito). HIV, hepatitis B and hepatitis C may be transmitted by the injection of blood or blood products either in the form of a blood transfusion, a needle-stick injury, or by intravenous drug abuse.

Sexual transmission of viruses is an important route for the spread of HSV, the papilloma viruses and HIV.

Viruses may also be transmitted **vertically** – that is, from mother to offspring via the placenta, during childbirth, or in breast milk. Examples are rubella virus (German measles) and cytomegalovirus (CMV), acquired by the mother during pregnancy and transmitted to the developing embryo, often leading to severe congenital abnormalities and/or spontaneous abortion. Some viral infections, for example, HSV infections, if acquired *in utero* or during birth can present as an acute disease syndrome in the neonate. In the case of HIV and hepatitis B transmission, the neonate may be born with an asymptomatic infection, the virus persisting in a **carrier state** and developing into disease much later.

Clinical results of infection

The outcome of a viral infection is dependent on a number of factors including **age**, **immune status** and **physiological well-being** of the host. Thus, HSV infection is usually fatal in the neonate but not in the older child. Epstein–Barr virus (EBV) causes a very mild febrile illness in young children but infectious mononucleosis (glandular fever) in teenagers. CMV in a healthy individual may cause a mild febrile illness, but in immunosuppressed individuals can lead to fatal pneumonia. Measles rarely causes severe complications in healthy well nourished children, but kills around 900 000 children per year in 'developing' countries where malnutrition is a problem. Upon infection viruses either remain **localized** or become systemic (*Fig. 1*). The outcome of a virus infection is considered below.

Inapparent (asymptomatic) infection
Many virus infections are **sub-clinical**, there being no apparent outward symptoms of disease. This is virtually always true in the immune host where recovery from a previous infection or vaccination protects the host from virus growth following reinfection by the wild-type virus. However, several viruses (e.g. respiratory and enteroviruses) may not produce clinical symptoms in some non-immune individuals. Thus, polio virus, in 80% of infected individuals, replicates in the epithelial cells of the gastrointestinal tract, is excreted in the feces, but causes no symptoms.

Disease syndrome, virus eradication and recovery
This is the pattern following most viral infections in otherwise healthy individuals – clinical symptoms of various severity (i.e. a **disease syndrome**) followed by virus eradication by the immune system, recovery and often life-long immunity. This is true of most childhood infections, for example, measles, mumps and German measles, and most respiratory diseases, of which there are a high number of viruses responsible. A vast spectrum of other viruses also follow this pattern, including those of hepatitis A virus (infectious hepatitis), rotavirus (gut infections) and coxsackie virus (myocarditis, pericarditis, conjunctivitis).

Figure 2 shows the possible routes of infection by respiratory viruses. One important respiratory pathogen of infants (**respiratory syncytial virus**, RSV) causes severe necrosis of the bronchiolar epithelium, which sloughs off, blocking the small airways. This leads to obstruction of air flows and respiratory disease. Children recovering from acute RSV bronchitis are often left with a weakened

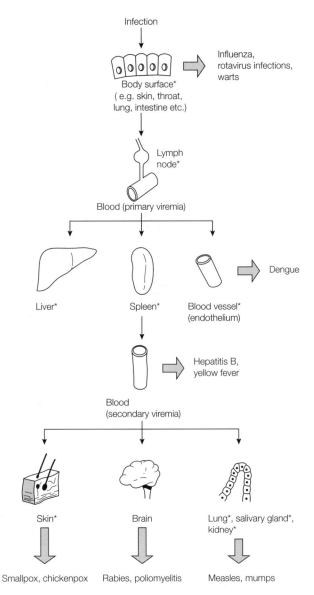

Fig. 1. *Virus spread within the host. Different viruses have different modes of spread within the host. Some viruses remain localized at the site of entry, whereas others may spread to involve other tissues. Routes of infection are shown, together with examples of possible clinical outcomes. *Possible sites of replication; ⇨, sites of shedding. Redrawn from Phillips and Murray (eds),* The Biology of Disease, 1995, *with permission from Blackwell Science Ltd.*

and vulnerable respiratory system, predisposing them to a lifetime of chronic lung disease. Following respiratory infection with measles, the virus replicates in local lymph nodes that drain from the infected tissue. The virus thus enters the blood (**a primary viremia**) where it grows on epithelial surfaces before entering the blood again (**a secondary viremia**). At this point the patient is highly infectious but does not have the distinctive measles rash, which appears about 14 days post-infection (see *Fig. 3* for course of events).

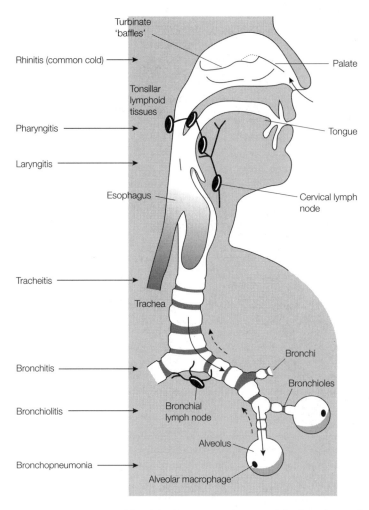

Fig. 2.   Routes of infection in the respiratory tract. Virus infections can produce a variety of respiratory disorders, depending on the area of respiratory tract infected. Dotted arrow, mucociliary flow; solid arrow, air flow. Redrawn from Phillips and Murray (eds), The Biology of Disease, 1995, *with permission from Blackwell Science Ltd.*

Latency

A restricted range of viruses, most being in the **herpesvirus family** (HSV, varicella zoster, Epstein–Barr virus and CMV) are not eradicated from the body following recovery, but instead become **latent** within the host. Virus replication may be initiated some time later (**reactivation**) and cause clinical symptoms which are either similar or different to those observed in primary infection.

HSV can reactivate many times during the life of an individual and produce the typical painful cold sore lesions on the mouth or genitals. The virus lays dormant in the trigeminal or sacral ganglia between periods of reactivation. It is not understood what happens to reactivate the virus at the cellular level, although a number of stimuli including menstruation, exposure to UV light and stress are responsible for initiating it.

A very different clinical syndrome may result from the reactivation of varicella zoster virus. Thus, the primary infection produces chicken pox,

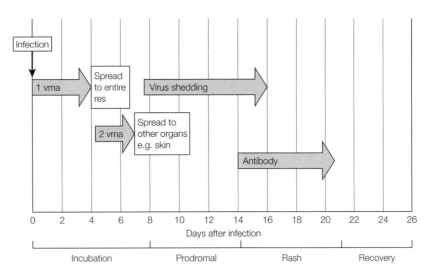

Fig. 3.   The course of clinical measles. Measles is characterized by a primary viremia (1 vma) followed by a second viremia (2 vma). Virus shedding occurs, at which point the patient is highly infectious. A prodromal phase (prior to clinical symptoms) is followed by the typical symptoms and signs of measles, including a rash. The appearance of antibody is usually followed by complete recovery and life-long immunity. res = reticulo endothelial system. Redrawn from Phillips and Murray (eds), The Biology of Disease, *1995, with permission from Blackwell Science Ltd.*

whereas reactivation is associated with the development of **shingles**. Shingles is characterized by a localized area of extremely painful vesicles which clear after 1–2 weeks. However, the patient may suffer from very severe **post-herpetic neuralgia** which can persist for months or years.

Carrier or persistent state
Although a rare event following viral infection, the virus carrier state is induced by **hepatitis B**, **hepatitis C** and **HIV**; 5-10% of individuals infected with hepatitis B virus will carry the infective particles in their blood for months or years. Estimates suggest that 300 million individuals worldwide carry the virus in their blood and body fluids. One reason for this is that the virus is transmitted vertically from mother to offspring. Unfortunately, chronic hepatitis B carriers also have a greatly increased risk of developing **cirrhosis of the liver** and **hepatocellular carcinoma**. Hepatitis C becomes chronic in 80% of infected individuals and persists for many years before signs of liver damage appear.

HIV, following a primary infection, replicates at a low level in **T4 lymphocytes** and other cells. The infected individual becomes **HIV antibody positive** and excretes virus in a range of body fluids. The virus, in most individuals if untreated, will eventually progress to more rapid replication and cause the clinical syndrome known as **AIDS**. Hepatitis B and C and HIV are often referred to as **persistent** viruses.

Neoplastic growth
Introduction of genetic material (**viral oncogenes**) and the rearrangement or switching on of cellular genes (**cell oncogenes**) are events which can be mediated by viruses. In some situations this can contribute to the development of

neoplasia. Feline leukemia virus causes a range of lymphoblastic leukemias in cats, indeed is the second highest cause of death in cats (the highest is road accidents!). Many viruses (e.g. **hepatitis B**, **hepatitis C**, **EBV**, **HSV** and **papilloma viruses)** have been implicated as being co-factors in the development of a range of malignancies.

Death

Whilst many viruses are fatal in distinct circumstances and others in a percentage of victims, some are always fatal. **Rabies**, **HIV** infection leading to **AIDS**, and a range of **neurological conditions** resulting from viral infections (e.g. subacute sclerosing panencephalitis) are examples.

While the incubation period of rabies is 30–90 days, after initial symptoms of the disease are manifest, patients die within 7–12 days. Following a bite, the virus may enter the peripheral nerves and then moves to the spinal cord and brain where it replicates. The virus leaves the cells and spreads to virtually all the tissues of the body including the salivary glands where it is shed in the saliva. The patient develops a variety of abnormalities including **hydrophobia** (aversion to water) **rigidity**, **photophobia** (aversion to light), focal or generalized convulsions and a variety of **autonomic disturbances**. Development of a flaccid paralysis and onset of coma precede death.

K9 VIRUSES AND THE IMMUNE SYSTEM

Key Notes

The immune system	The whole armory of the immune system is harnessed to combat viral infections. Viruses thus react specifically with B cells and T cells. In addition, non-specific responses (e.g. interferon and natural killer cells), have a role in combating virus infection.
Virus antigens	During their replication, viruses code for a range of viral proteins (structural and non-structural) which are foreign to the host. Many induce protective antibody and cytotoxic T lymphocyte (CTL) responses. Others, while not protective, are important in diagnostic tests. The smallest part of a molecule with antigenicity is referred to as an epitope.
Antigen recognition	Antibodies secreted from B lymphocytes (assisted by T4 helper cells) specifically react with antigenic sites on virus particles and may neutralize virus infectivity. CTLs recognize antigens on and lyse infected cells. Antigen–antibody recognition is extremely specific and is determined by the detailed molecular structure of the antigen (epitope) inducing a unique response from a lymphocyte.
Primary response to viruses	Interferon and natural killer cells respond first in combating virus infections. Specific CTL responses take 7–10 days and antibody slightly longer. The secondary response is almost immediate. The primary antibody response is usually an IgM response with smaller concentrations of IgG. The secondary response consists mainly of the more efficient IgG molecules.
Viruses and the immunocompromised	Individuals who are immunosuppressed suffer worse than most from viral infections where the clinical syndrome may be more severe than in an immunocompetent host. HIV causes immunosuppression. In addition, many individuals are born with genetic defects which leave them immuno-suppressed; others are suppressed by medication.
Virus-induced immunopathology	The immune response mounted against an invading virus may itself be responsible for the damage and disease state which follows infection.
Related topics	Virus proteins (K3) Virus vaccines (K10) Virus infection (K8)

The immune system

Virus infections are countered by the host **immune** system. The immune system consists of a wide range of morphologically and antigenically distinct white cells with different functional responses to viral invasion. Simplistically, the major type of white cells are **lymphocytes**, these being **T cells** or **B cells**. T cells (**cytotoxic (CD8)** or **helper (CD4)** cells) pass through the thymus during their maturation. B cells give rise to antibody-producing plasma cells. T4 helper cells help B cells react to highly specific antigenic stimuli (e.g. viral proteins). In addition to these specific responses the host mounts a series of non-specific responses, for example, production of **cytokines** (e.g. **interferon**), **natural killer cells**, **mucociliary responses** and **macrophages**. It is the outcome of the competition between the immune system and the progress of viral replication that determines the outcome of the infection. In **self-limiting** infections viruses are 'cleared' by the immune system. Good humoral immunity requires the induction of high levels of IgG antibodies.

Virus antigens

During the virus-replication cycle, the virus genome is expressed to give a number of proteins which are recognized by the host as being foreign. Many of these proteins induce a **protective response** in the host (important in **vaccine design**). In many viruses these proteins are laid down in the plasma membrane of the infected cells where they are recognized by both T cells and B cells (antibodies). It is usually the viral **structural proteins** (e.g. envelope and capsid proteins) that induce this protective antibody response.

Antigenic recognition

Antibodies, secreted by B cells, recognize whole viral protein antigens, usually in the fluid phase. The specific interaction between the antibody and the antigen has been likened to a 'hand-in-glove' reaction. Antibodies **neutralize** the infectivity of virus particles by agglutination, conformational change, etc. This, of course, reduces dramatically the number of host cells infected by virus. Cytotoxic T cells recognize virus antigens via a **cell-surface heterodimer** (the **T-cell receptor**) in association with the **major histocompatability complex** (MHC-class 1) which is present in almost all nucleated cells with the exception of neurons. Thus, cytotoxic T cells interact with the antigens present on the plasma membrane of infected cells, killing the cell before high concentrations of virus can be produced.

In order for antibody to be secreted by B cells, the specific antigen must be presented to T4 helper cells, this being achieved by, macrophages and dendritic cells (**antigen presenting cells**). In this instance the specific recognition of the antigenic site is made in association with the **MHC-class II** protein (present on selected cells in the body e.g. macrophages and dendritic cells). Primary antigen recognition usually results in the formation of **memory B and T cells** which persist, making antigenic recognition much faster during re-infection.

Primary response to viruses

During a primary infection, the mass of antigen available to stimulate the immune response increases as the organism replicates. Initial responses are made by the host via **interferon** and **natural killer** (NK) cells. Interferons are host cell proteins with antiviral activity. It is α and β interferon that are recognized as having an antiviral effect. Once induced they have an effect against a wide range of viruses. Interferon acts in two stages: **induction** which results in the derepression of the interferon gene and the release of interferon which produces an **antiviral state** in other cells. NK cells exist as large granular lymphocytes in peripheral blood and mediate cell lysis. They are not antigen specific and do not exhibit immunological memory. They appear to have an important role in the control of some virus infections.

The specific T-cell responses peak early (7–10 days) and decline within 3 weeks post-infection. The initial (primary) antibody response usually peaks later than the rise in cytotoxic T lymphocytes (CTL). Antibodies are often barely detectable during the acute stage of infection but increase dramatically 2–3 weeks post-infection. High levels of antibody may linger for several months. Indeed, where virus cannot be isolated from the host, a dramatic increase in **specific antibody** titer from the **acute** to the **convalescent** stage of infection is the main indicator of the cause of the infection. Upon reinfection the **secondary** response is virtually immediate in terms of CTL and antibody response. The mechanisms that combat viral infections are outlined in *Table. 1*.

Viruses and the immuno-compromised

Individuals with natural or artificially induced immunosuppression are at risk from a range of viral infections but particularly those which are responsible for latent infections (e.g. HSV, CMV, chicken pox virus) where the virus reappears to cause disease (**recrudescence**). Thus infants with **severe combined immuno-deficiency** (**SCID**) develop recurrent infections early in life (e.g. rotavirus in the gut, which induces prolonged diarrhea). **Immune suppression** which follows transplant surgery may lead to, for example, generalized shingles (reoccurrence of chicken pox), CMV pneumonia or genital warts. Individuals with little or no antibody production (**hypogammaglobulinemia**) often excrete viruses for many years (e.g. poliovaccines from their gut).

Some viruses induce immunosuppression. Measles is known to slightly suppress T cell responses but the significant culprit is the **human immuno-deficiency virus** (HIV). This virus infects T4 helper cells, predominantly by attachment to the **CD4 receptor** where it persists for several months or years before, usually, progressing to high concentrations of virus which seriously deplete the T4 cell population, thus resulting in serious immunosuppression and susceptibility to a wide range of pathogens (AIDS).

Table 1. Mechanisms to combat virus infection

Stage of infection	Immune response	Mechanism
Early infection (first line of defense)	Interferon, natural killer cells Soluble mucosal surface IgA antibody	Inhibits virus replication Kills virus
Viremia (virus in the blood)	Antibodies, complement Macrophages	Kills virus (neutralizes infectivity) Reduces spread Digests antibody complexes
Target organs	Antibody, complement Cytotoxic T cells, complement Interferon	Lysis of infected cells Kills virus-infected cells, reducing virus replication Inhibits virus replication

Virus-induced
immunopathology
The immune response to viruses can result in damage to the host, either by the formation of immune complexes or by direct damage to infected cells. Complexes can form in body fluids or on cell surfaces. **Chronic immune complex glomerulonephritis** can occur in mice infected neonatally with lymphocytic choriomeningitis virus. In adult mice, direct damage to infected and non-infected brain cells by a T-cell-dependent mechanism is responsible for most of the fatal tissue damage. This is also thought to be true for liver damage in chronic active hepatitis in man.

Viruses may also evoke autoimmunity, probably via **molecular mimicry** (production of an antigen which shares conserved sequences with a host cell protein). As a result, antibodies or T cells are produced which also react with host proteins.

K10 VIRUS VACCINES

Key Notes

Vaccination	Vaccination is the use of a vaccine to stimulate the immune response to protect against challenge by wild-type virus. A good vaccine is designed so as to mimic the host response to infection seen with the wild-type infection and to protect against disease but to have minimal side effects. Vaccines are either live (attenuated) virus or inactivated (killed) virus. Sub-virion protein or DNA vaccines are being developed at present, although most are at an experimental stage.
Live (attenuated) vaccines	Produced by continued passage in tissue culture or by genetic manipulation (e.g. gene deletion). These vaccines are effective in stimulating the full range of immune responses. They are the preferred type of vaccine. Examples are measles, mumps and German measles vaccines.
Inactivated (dead) vaccines	Produced by chemical inactivation of wild-type virulent virus, they are generally less effective than live vaccines but do give significant protection against virus challenge. Examples are influenza and hepatitis A vaccines.
'Sub-virion' vaccines	These are mainly experimental (apart from hepatitis B surface antigen vaccine) and are composed of 'subunits' of the virion. They are used when other vaccines are ineffective or technologically not possible to manufacture.
DNA vaccines	These are totally experimental vaccines and consist of injecting a host with a plasmid containing DNA which encodes antigenic portions of the virus.
Related topics	Virus proteins (K3) Viruses and the immune system Virus infection (K8) (K9)

Vaccination   The early experiments of **Jenner** with cowpox were the basis upon which many of today's vaccination (*vacca* is latin for cow) programs were derived. Jenner introduced the mild **cowpox** virus into some small boys whom he later challenged with **smallpox** virus. The boys were protected against the disease. Cowpox shares a number of antigens with smallpox but does not cause disease in humans. Hence, cowpox replication had induced an immune response which, upon subsequent challenge with smallpox, was sufficient to prevent the smallpox virus undergoing any significant infection. A hybrid cowpox virus, of unknown origin, vaccinia virus, was used in a vaccination program which in 1977 succeeded in eradicating smallpox from the world. Vaccines are basically of two types, live (**attenuated**) or inactivated (**dead**) vaccines. More recently, vaccines which contain selected virus proteins (e.g. hepatitis B vaccine) have been developed. Many, however, are only at the experimental stage of development. Even more recently have been experiments using DNA as a source of vaccination (see later). Vaccines are administered by a number of routes – **intramuscular**, **intradermal**, **subcutaneously**, **intranasal** or **oral**.

Live (attenuated) vaccines

It is generally accepted that live vaccines are the preferred vaccines and examples are shown in *Table 1*. Initially, they were derived by continued passage of virulent wild-type virus in tissue culture, this process selecting for **avirulent viruses** which will replicate *in vivo* but not cause disease. Such viruses could, however, revert back to wild-type phenotype and indeed in some cases during the early years of their use did. This rather empirical approach has now been replaced by the genetic 'manufacture' of viruses with deleted or mutagenized genes. The risk of reversion with these vaccines is minimal. However, many (e.g. Sabin polio vaccine) are the original mutants created by tissue culture passage over 40 years ago.

Live vaccines have distinct advantages over killed. They require small amounts of input virus (which, however, must replicate) they can induce local immunity (i.e. mucosal IgA), they may be given by the natural route of infection (e.g. Sabin polio vaccine which is given orally) and they are usually cheaper. The drawbacks to their use include reversion to virulence, ineffectiveness if not kept live (i.e. refrigerated or freeze-dried), ineffectiveness in individuals with, for example, infections of the gut, where the polio virus vaccine will not grow, contamination with adventitious agents and limited use in immune-suppressed patients. Recent experiments have shown that, for poliovirus type 3, reversion to a virulent form involves only two amino acid changes.

The World Health Organization has an ambitious vaccination program which seeks to get vaccines to most children of the world and to eradicate many viruses in the next decade. Polio has been eradicated from virtually all 'developed' countries, and by using National Immunization Days many millions of people are being made immune to polio and other viruses.

Table 1.   Currently available live attenuated viral vaccines

Vaccine	Vaccine type	Uses
Oral polio	Attenuated trivalent	Routine childhood immunization; mass campaigns
Measles	Attenuated (Schwarz, Moraten, others)	Routine childhood immunization; mass campaigns
Rubella	Attenuated (RA 27/3)	Routine childhood immunization; adolescent girls; susceptible women of childbearing age
Mumps	Attenuated (Urabe or Jeryl Lynn)	Routine childhood immunization
Measles, mumps, rubella (MMR)	Attenuated	Routine childhood immunization (1 or 2 doses)
Varicella	Attenuated (Oka)	Routine childhood immunization (USA); vaccination of susceptible people
Yellow fever	Attenuated (17D)	Routine immunization or mass vaccination in endemic areas; vaccination of travellers to endemic areas
Smallpox	Vaccinia virus	Vaccination of research workers

Inactivated (dead) vaccines

The early polio vaccine (**Salk vaccine**) still used in many countries typifies the approach to dead vaccination; other examples are shown in *Table 2*. High titers of wild-type virus are grown in tissue culture and inactivated chemically by the use of, for example, β-propiolactone or formaldehyde. The killed virus is administered parenterally (subcutaneously or intradermally) in order to stimulate an immune response. With no subsequent virus replication (as seen with live vaccines) the immune response follows that of a 'primary response' to an inert antigen. The necessary high levels of IgG are therefore stimulated only by multiple injections.

Killed vaccines have the advantage of non-reversion to virulence, are not 'inactivated' in the tropics, can be administered to immune-suppressed patients and are usually not affected by 'interference' from other pathogens (important in many developed countries). They are, however, expensive, do not stimulate local immunity, are considered dangerous to manufacture (prior to inactivation) and require multiple doses. The recently derived **hepatitis A vaccine** is a dead vaccine, primarily because it was technically impossible to develop an attenuated virus capable of liver growth but no disease.

It is considered that the use of dead vaccines alone, which often allow some wild-type virus replication at local sites (e.g. polio) will not be sufficient to totally eradicate viruses from the world.

'Sub-virion' vaccines

Attempts to produce these vaccines, often referred to as **sub-unit** vaccines have been prompted by a number of factors. Many viruses do not grow in tissue culture (e.g. hepatitis B); some are considered by some scientists to be too dangerous for use as live or killed vaccines (e.g. HIV); the vaccine may become latent in the body and become reactivated (e.g. HSV); the vaccine may be poorly effective and have side effects (e.g. influenza virus vaccines).

The basis of the approach is to develop a vaccine product which, when injected into the host, will induce protective immunity. The virus protein of choice for such a vaccine is usually a capsid or envelope protein, often the

Table 2. Currently available inactivated viral vaccines

Vaccine	Vaccine type	Uses
Inactivated polio	Killed whole virus	Immunization of immunocompromised people; universal childhood immunization (some developed countries)
Hepatitis B	Purified Hepatitis B surface antigen produced by recombinant yeast, mammalian cells or from plasma	Routine childhood or adolescent immunization: immunization of high-risk adults; post-exposure immunization
Influenza	Killed whole or split virus	Vaccination of high-risk individuals or the elderly
Rabies	Killed whole virus	Post-exposure vaccination; pre-exposure; veterinarians/travellers
Hepatitis A	Killed whole virus	Pre-exposure vaccination: high-risk people and travellers
Japanese B encephalitis	Killed whole virus	Pre-exposure vaccination of travellers to endemic areas; routine or mass vaccination in endemic areas
Tick-borne encephalitis	Killed whole virus	Pre-exposure vaccination of travellers to endemic areas; (?)routine or mass vaccination in endemic areas

protein which binds to the cell receptor (e.g. gp120 of HIV, hemagglutinin of influenza). These proteins can be extracted from the virion by chemical treatment (e.g. influenza hemagglutinin), concentrated from the plasma of infected patients and inactivated (e.g. hepatitis B surface antigen), engineered by recombinant DNA technology (e.g. hepatitis B surface antigen) or synthesized as a peptide (still experimental, but examples include foot and mouth disease virus).

These vaccines may suffer from some of the drawbacks of dead whole virion vaccines in as much as they do not replicate in the host. They are not good inducers of immunity and are linked to immune potentiating chemicals or other molecules linked to presentation. One approach under study at present is to introduce the selected viral gene into a virus vector (e.g. vaccinia virus). The vaccination virus is then used to vaccinate the host, the virus replicating and expressing a number of antigens (including the cloned gene product) which in turn stimulate immunity.

The only subunit vaccine in general use at present is the hepatitis B surface antigen which has been genetically engineered into and expressed by yeast cells. The vaccine is used routinely to induce protection against hepatitis B in members of the medical profession.

DNA vaccines   These appear to be an exciting inovative approach to vaccination. DNA-encoding antigenic portions of viruses (e.g. protective antigens) are inserted into a plasmid which can be injected into the host as 'naked' DNA free of protein or nucleoprotein complexes. Host cells take up the DNA and may express the virally encoded proteins. Such vaccines stimulate excellent cytotoxic T-cell responses. Experimental influenza vaccines in animals have proved effective in challenge experiments.

K11 ANTIVIRAL CHEMOTHERAPY

Key Notes

Historical perspective

Viruses utilize the whole machinery of the cell for their replication and hence most antiviral compounds are also toxic to normal cells. However, recent knowledge of viral gene products has allowed a more selective approach to drug design.

Drug targeting

Our knowledge of the virus growth cycle and those various stages of replication which require virus specific proteins has allowed drug targeting by pharmaceutical companies.

Drug design

Determining the three-dimensional structure of, for example, viral enzymes has allowed drugs to be designed which react specifically with, and inhibit parts of, viral molecules.

Effective concentration 50 and toxicity

Determining the antiviral effectiveness of a compound is achieved usually in tissue culture by assaying its ability to reduce viral infectivity and is expressed as the EC_{50}. Toxicity is assessed and expressed as selective toxicity.

Clinical trials

Phases I–IV of clinical trials assess the pharmacokinetics, pharmacology, metabolism and clinical efficacy of the compound.

Modes of action

Most compounds are nucleoside analogs, others inhibit, for example, proteases, and egress of virus from the cell. Aciclovir, probably the most successful compound to date, is specifically phosphorylated by viral enzymes before being selectively incorporated into the growing virus DNA chain, acting as a chain terminator.

Related topics

Virus replication (K7)   Virus infection (K8)

Historical perspective

As obligate intracellular parasites with very restricted genetic-coding capacity viruses rely heavily on utilizing the metabolic machinery of the cell for their replication. It was therefore considered by many that the concept of **selective toxicity** was an unattainable goal and that interference with viral replication would always bring unacceptable damage to the host. Viruses by definition are resistant to the action of antibiotics.

In the last decade a series of discoveries based on our increasing knowledge of the viral replication cycle and the specific gene products encoded by viruses have resulted in a number of successful antiviral agents being prescribed. However, some still have a level of toxicity which, while acceptable for some diseases (e.g. AIDS), would not be tolerated for less severe diseases.

Chemotherapeutic agents fall into three broad groups. **Virucides** directly inactivate viruses (e.g. detergents, solvents), **antivirals** strive to inhibit viral multiplication but not host-cell metabolism and **immunomodulating agents** attempt to enhance the immune response against viruses (e.g. administration of interleukins). In this chapter **antivirals** will be discussed.

Drug targeting

Viral replication requires the virus to pass through a number of stages which are common to all viruses: **attachment and penetration, uncoating of the nucleic acid, transcription and translation, replication of nucleic acids** and **release of mature progeny**. This is demonstrated in diagrammatic fashion for the growth cycle of HIV in *Fig. 1*.

To date, most antivirals are directed at inhibiting one of the steps shown in *Fig. 1*, although the most common target is interference with nucleic acid metabolism by using many **nucleoside analogs**. Several compounds that interfere with influenza effect the disassembly of particles by blocking a virus protein complex that acts as an ion channel. Many viruses produce a virus-specific protease to process their proteins and this has been targeted in the search to develop an inhibitor of HIV. A series of inhibitors of picorna-viruses act by directly binding to the virion capsid, blocking the interaction between the virion and the receptor on the cell surface that facilitates its entry and disassembly.

Drug design

In the early days, most new antiviral compounds were discovered in an **empirical** fashion. Chemists would produce a wide range of, for example, nucleoside analogs, originally in many cases as anti-cancer drugs, which were tested in tissue culture to determine their antiviral activities. However, with our knowledge of the molecular basis of virus replication, the availability of the entire **nucleotide sequences** of virus genomes and the **three-dimensional protein structure** derived from **X-ray diffraction analysis**, compounds can be designed to interact with specific targets involved in the replicative cycle of viruses. In practice the most useful targets are **virus-induced enzymes** which often have properties different to those of the counterpart enzymes induced by the host cell (e.g. **thymidine kinase, DNA polymerase, reverse transcriptase** and **protease)**. When the enzyme can be crystalized and its three-dimensional structure determined the synthesis of appropriate molecules to interact with particular sites on that enzyme can be determined. However, future antiviral compounds should lack toxicity and should be able to be produced cheaply from available precursors. The chance of a wide range of new products appearing is therefore remote.

Effective concentration 50 and toxicity

Antiviral compounds are usually assessed at an early stage for their ability to interfere with viral growth in tissue culture. The virus is assayed by, for example, $TCID_{50}$ or plaque production with and without the drug. The percentage of reduction in infectivity is plotted against $\log_{10}$ of the drug concentration. The concentration of compound that reduces virus titer by 50% is measured and expressed as the effective dose 50 concentration (ED_{50}) (*Fig. 2*). Tissue culture can also be used to assess the toxicity of compounds, although this is often tested in animals and humans. Animals play a vital role in the study of toxicity and a number of statutory tests are carried out to determine the risks and side effects before new compounds enter clinical trials. The ratio of the 50% toxic concentration to the 50% inhibitor concentration is termed the

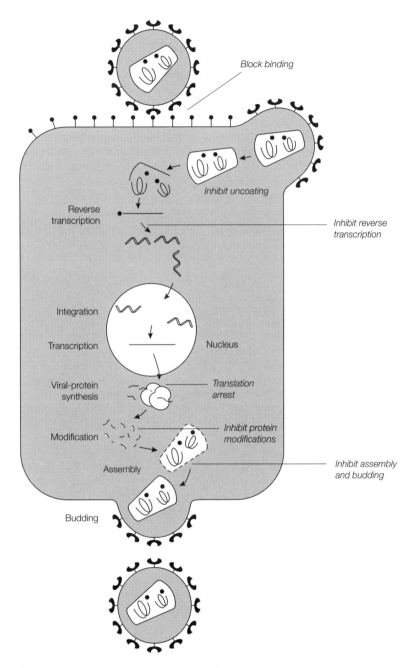

Fig. 1.   *The various stages of a virus replication. Antiviral drugs can be used to inhibit any of these steps. Redrawn from Yarchoan, R. et al.,* Scientific American, *vol. 259, no. 4, p. 90, 1988.*

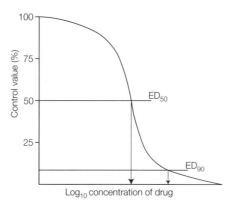

Fig. 2.   Determination of effective dose (ED$_{50}$, ED$_{90}$) for an antiviral drug in tissue culture.

selective index. If this is close to unity the compound is toxic. A high selective toxicity suggests a useful compound. A **benefit/risk** ratio is also considered when using a compound, that is, side effects will be tolerated more if the risk from the disease is high (e.g. AIDS).

Clinical trials   **Phase I** involves administering the drug to healthy human volunteers where studies on the pharmacokinetics, pharmacology and metabolism of the compound are monitored.

In **Phase II** the compound is administered to diseased patients, and data similar to those above are obtained as the metabolism may be different in the diseased patient. Usually <100 patients are used for these trials.

In **Phase III** the drug is usually tested for its clinical efficacy by comparing with **placebos** or existing drugs. The main aim is to determine the **benefit/risk ratio** for the therapeutic course, this requiring 100–1000 patients.

Phase IV studies are usually conducted following marketing approval and increased experience of treating patients, providing more information on safety and efficacy.

Unfortunately, most potential antiviral compounds, although good inhibitors of viral growth in tissue culture, fail to pass successfully through all of the phases of clinical studies.

Modes of action   *Table 1* highlights the modes of action of selected antiviral compounds. Perhaps the most successful antiviral compound is the nucleoside analog **acyclovir** (ACV – now named **aciclovir**) with more than 33 million patients treated. The drug inhibits the replication of HSV, and has been administered prophylactically for over 10 years to individuals, with no ill effects, to suppress recurrences of genital herpes. The compound is related to the natural nucleoside guanosine. To become active the compound must be converted into the triphosphate form by three phosphorylation steps (i.e. ACV mono-, di- and then triphosphate). These steps are not carried out in normal uninfected cells and hence ACV is a poor substrate for cellular enzymes. In contrast, the **HSV thymidine kinase** can convert ACV to ACV-monophosphate (ACV-MP). Cellular enzymes convert ACV-MP to the triphosphate (ACV-TP) where it enters the nucleoside pool and competes with guanosine triphosphate as a substrate for **HSV DNA**

Table 1 Mechanism of action of antiviral drugs

Drug	Mechanism(s) of action	Virus
Aciclovir	Nucleoside analog, inhibits nucleic acid synthesis: active form produced by viral thymidine kinase	HSV, VZV
Ribavirin	Nucleoside analog, inhibits nucleic acid synthesis, possibly by inhibiting viral RNA polymerase	RSV, LFV
Amantadine Rimantadine	Inhibit virus coating, maturation and egress from cell	Influenza
Ganciclovir	Nucleoside analog, blocks nucleic acid synthesis by inhibiting viral thymidine kinase and other enzymes	CMV
Azidothymidine Dideoxyinosine Dideoxycytidine	Nucleoside analogs, inhibit nucleic acid synthesis by inhibition of reverse transcriptase	HIV
Foscarnet	Blocks protein synthesis by inhibition of RNA and DNA	CMV, HSV
Idoxuridine Vidarabine	Nucleoside analog, inhibits DNA synthesis Nucleoside analog, blocks DNA synthesis by inhibiting DNA polymerases	HSV

HSV, herpes simplex; VZV, Varicella–Zoster; RSV, respiratory syncytial; CMV, cytomegalovirus; HIV, human immunodeficiency virus.

polymerase. Cellular DNA polymerases are much less sensitive to inhibition. As the ACV residue is linked to the growing chain of viral DNA it forms a **chain terminator** as there is no 3'-OH group on the ACV sugar moiety to link the next residue to the growing chain of virus DNA.

K12 PLANT VIRUSES

Key Notes

Historical aspects	Plant viruses were discovered over a century ago and have featured greatly in contributing to our knowledge of virus structure (e.g. tobacco mosaic virus, turnip yellow mosaic virus and tomato bushy stunt virus).
Plant viruses	Plant viruses are diverse in size, shape and biochemistry and are present in many virus families which include animal viruses (e.g. rhabdoviridae).
Disease and pathology	Viruses are singly responsible for grave economic loss estimated at being over $70 billion worldwide. They cause necrosis, wilting mosaic formation and other damage, which reduces yields and value of crops, etc.
Transmission, infection and systemic spread	Plant viruses are transmitted mainly by invertebrate animals (e.g. aphids, leaf hoppers) or through infected seeds or 'manually' by contaminated implements. They gain entry by penetrating cuticles of plant cells and need to spread systemically to cause disease (via plasmodesmata).
Control of plant virus disease	Infected plants are virtually impossible to 'cure'. Control is by use of naturally resistant plant varieties or more recently genetically manufactured resistant varieties and by eradication of the transmission vector.
Viroids	These are novel 'virus-like' infectious agents which are composed solely of RNA with a complex tertiary structure (e.g. potato spindle tuber viroid).

Related topics	Virus structure (K1)	Virus nucleic acids (K4)
	Virus taxonomy (K2)	Virus replication (K7)
	Virus proteins (K3)	

Historical aspects Tobacco mosaic virus (TMV) has figured predominantly in early studies on virus structure and replication. The virus, which causes mosaicing of the leaves of the tobacco plant, was first described as being a 'contaguim virus fluidum' in 1898. In 1935 the virus was crystalized and shown to be a 'globular protein'. TMV was the first virus to be observed under the electron microscope, and the first to demonstrate the intrinsic infectivity of extracted RNA and to be assembled *in vitro* from purified preparations of viral RNA and coat protein molecules.

It was studies on turnip yellow mosaic virus and tomato bushy stunt virus that revealed the morphological details of icosahedral viruses.

Plant viruses There are over 1000 plant viruses which have been classified by the ICTV. The virus usually receives its name by a combination of the host plant and the type of disease produced (e.g. **tobacco mosaic**, **turnip yellow mosaic**, **turnip bushy stunt**, **cauliflower mosaic**, **tomato spotted wilt**).

Plant viruses are diverse in their morphology, nucleic acid composition and replication patterns. It is beyond the scope of this book to detail all viruses but *Fig. 1* highlights the various morphologies and groupings of plant viruses. They may be **dsDNA** (Badnavirus, e.g. rice tungrobacilliform virus), **ssDNA** (geminivirus, e.g. maize streak virus), **dsRNA** (reoviruses, e.g. clover wound tumor virus), **RNA viruses with ambi sense RNA** (rhabdoviridae, e.g. potato yellow dwarf virus), **positive-sense ssRNA viruses** with monopartite genomes (e.g. Tobamoviridae tobacco mosaic virus), **bipartite ssRNA** viruses (e.g. Comoviridae cowpamosaic virus) and **tripartite ssRNA** viruses (e.g. Cucumoviridae cucumber mosaic virus).

Diseases and pathology	The impact of viruses globally on the world's plants is enormous and estimates of $70 billion have been quoted as being the monetary loss per year as a result of virus infections. It appears that crops of all types can be affected. One disease, **tristenza**, has wiped out or made non-productive more than 50 million citrus trees worldwide. Swollen shoot disease has destroyed more than 200 million cocoa trees in parts of western Africa. Rice plants, barley, vegetable and field crops all succumb to infections. Many perennial plants are chronically infected with virus. The morphological outcome of a plant infection can be seen as mosaic formation, yellowing, molting or other color disfiguration, stunting, wilting and necrosis. These conditions result from cell and tissue damage which will significantly reduce the yields and commercial value of crops. Thus, photosynthesis, respiration, nutrient availability and hormonal regulation of growth can all be affected. How plant viruses cause such changes remains for the main part a mystery. Plant viruses code for few proteins, but as is the case in many bacteriophage proteins, they are multifunctional and many of their interactions with host proteins, although not part of the virus-replication process, results in pathology.
Transmission, infection and systemic spread	Most plant viruses depend on some kind of 'agent' for their dissemination in the natural setting. Invertebrate animals are the most important of these agents, examples being **aphids**, **leaf hoppers**, **mealybugs**, **whiteflies**, **thrips**, **mites** and **soil nemotodes**. Some viruses are passed directly to progeny plants through infected pollen or seeds and others are present in tubers, bulbs and cuttings. A few viruses can be transmitted as a result of contact with **contaminated implements** (e.g. hoes) or by direct contact between neighboring plants. The stem and leaf surfaces of plants can in many ways be compared to the human skin. As such they do not have receptors for virus attachment and have a protective function. For plant viruses to be infective they must penetrate this barrier and gain access to the metabolic machinery of the plant cell. It is thus through wounds that viruses enter plant tissue, either via a vector or mechanically by manual damage. In the laboratory, infection is achieved by rubbing the leaf with a cloth soaked in a virus suspension, using a fine mesh abrasive method to create the necessary wounds. In effect this procedure mimics the plaque assay described for animal viruses, but is not as accurate! (*Fig. 2*). On entry the virus particle is uncoated and goes through a replication cycle similar to that of animal viruses, this cycle varying from virus to virus. Many plant viruses code for a **movement protein** in addition to enzymes and structural proteins. *Figure 3* represents a diagram of the cycle of TMV. Plant viruses rarely cause significant

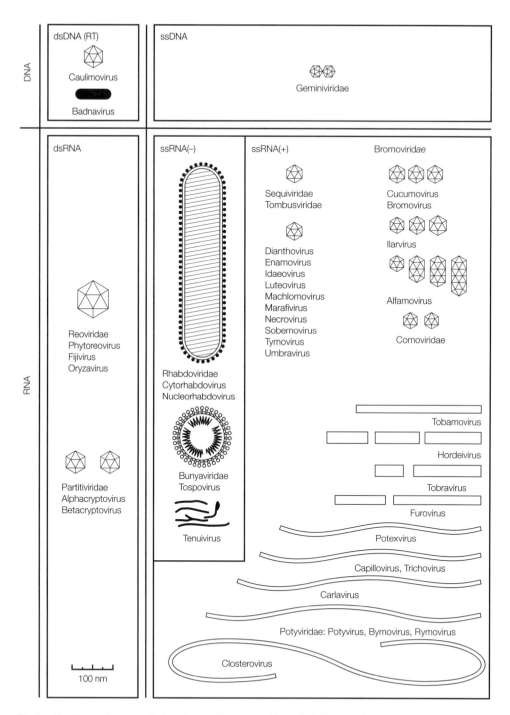

Fig. 1. Families and genera of plant viruses. Reproduced from Sixth Report of the International Committee on Taxonomy of Viruses, *Springer-Verlag.*

Fig. 2. *Focal assay of tobacco mosaic virus on the leaf of a plant. Reproduced from Dimmock and Primrose,* Introduction to Modern Virology, *4th edn, 1994, with permission from Blackwell Science Ltd.*

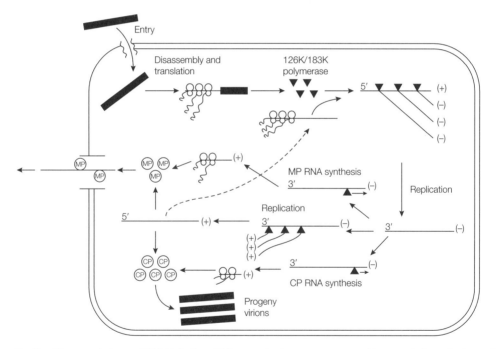

Fig. 3. *Diagram of stages of TMV infection. All the events shown are presumed to occur in the cytoplasm of infected cells. MP, movement protein; CP, coat protein. Redrawn from Fields* et al. (eds), Fundamental Virology, *3rd edn, 1996, with permission from Lippincott-Raven Publishers.*

damage and disease unless they become **systematically** distributed throughout the plant. Failure for this to happen explains why some plants are resistant to particular virus infections. Movement is facilitated by a virus protein which is involved in the transport of virus or virus nucleic acid through the fine pores (**plasmodesmata**) in the cell walls that interconnect plant cells. Movement also occurs through the companion and sieve cells of the phloem, this being facilitated by the viral coat protein.

Control of plant virus diseases

Once infected, it is almost impossible to 'clear up' a virus infection in a plant by use of antiviral agents. Likewise plants do not mount immune responses. Until recently it was a reliance on **horticultural practice** (use of virus-free seeds, eradicating vectors, choosing the time of planting, etc.) that reduced the extent of virus infection. More recently **genetic engineering** has allowed the construction of plants which show **natural resistance** to virus infections. Other scientific measures made use of the fact that previous infections with a non-pathogenic virus appeared to protect the plant from an infection by a more pathogenic strain (but not because of immunity!). This phenomenon has been extended by transforming the plant with the coat protein gene of TMV and then challenging the plant with infectious TMV. The resulting **transgenic** plant was protected from challenge by the wild-type TMV. The phenomenon is referred to as **pathogen-derived resistance** to virus diseases. How and why this phenomenon works is not clearly understood, although in practice the method is receiving a great deal of attention from plant breeders and molecular biologists.

Viroids

Viroids are small, **unencapsidated ssRNA molecules** – the smallest known pathogens of plants. There are some 25 viroids which vary in nucleotide sequence. The first to be examined in detail was **potato spindle tuber viroid** (PSTVd) responsible for significant loss to the potato industry.

The RNA is a covalently closed circle and ranges in size from 246 to 357 nucleotides in length. The RNA has a complex secondary and tertiary structure which gives it a rod-like shape and **resistance to nucleases**. The RNA does not have a characteristic open reading frame and so does not act as mRNA. How this piece of RNA causes disease is largely unknown. It replicates in plant cells with the aid of host-cell enzymes (e.g. RNA polymerase II).

Viroids are spread by plant propagation (e.g. cuttings and tubers) through seeds and by manual mishandling with contaminated implements.

FURTHER READING

There are many comprehensive textbooks of microbiology, but no one book that can satisfy all needs. Different readers subjectively prefer different textbooks and hence we do not feel that it would be particularly helpful to recommend one book over another. Rather we have listed some of the leading books which we know from experience have served their student readers well.

General reading

Madigan, M.T., Martinko, J.M. and Parker, J. (1997) *Brock Biology of Microorganisms*, 8th Edn. Prentice-Hall Inc., Upper Saddle River, NJ.

Prescott, L., Harley, J.P. and Klein, D.A. (1996) *Microbiology*, 3rd Edn. Wm C. Brown Communications Inc., Dubuque, IA.

Singleton, P. and Sainsbury, D. (1987) *Dictionary of Microbiology and Molecular Biology*, 2nd Edn. John Wiley & Sons, New York.

Collier (1997) *Topley & Wilson's Principles of Bacteriology, Virology and Immunity*, 9th Edn. Arnold Publishing, London.

Tortora, G.J., Funke, B.R. and Case, C.L. (1998) *Microbiology: An Introduction*, 6th Edn. The Benjamin-Cummings Publishing Co., Redwood City, CA.

More advanced reading

The following books are recommended for readers who wish to study a particular area in more depth.

Section A   Roberts, D. McL., Sharp, P., Alderson, G. and Collins, M. (Eds) (1996) *Evolution of Microbial Life: Society for General Microbiology 54th Symposium*. Cambridge University Press, Cambridge.

Section B   Dawes, I.W. and Sutherland, I.W. (1992) *Microbial Physiology*, 2nd Edn. Blackwell Scientific Publications, Oxford.

Hames, B.D., Hooper, N.M. and Houghton, J.D. (1997) *Instant Notes in Biochemistry*. BIOS Scientific Publishers Limited, Oxford.

Moat, A.G. and Foster, J.W. (1995) *Microbial Physiology*, 3rd Edn. Wiley-Liss, New York.

Neidhardt, F.C., Ingraham, J.L. and Schaechter, M. (1990) *Physiology of the Bacterial Cell: A Molecular Approach*. Sinauer Associates, New York.

Rhodes, P.M. and Stanbury, P.F. (1997) *Applied Microbial Physiology: A Practical Approach*. Oxford University Press, Oxford.

Schlegel, H.G. (1993) *General Microbiology*, 7th Edn. Cambridge University Press, Cambridge.

Section C   Alberts, B., Bray, D., Lewis., Raff, M., Roberts, K. and Watson, J.D. (1994) *Molecular Biology of the Cell*, 3rd Edn. Garland Publishing, New York.

Beebee, T.J.C. and Burke, J. (1992) *Gene Structure and Transcription: In Focus*. IRL Press, Oxford.

Calladine, C.R. and Drew, H.R. (1997) *Understanding DNA: The Molecule and How It Works*, 2nd Edn. Academic Press, Orlando, FL.

Creighton, T.E. (1993) *Protein Structures and Molecular Properties*, 2nd Edn. W.H. Freeman and Co., New York.

Kornberg, A. and Baker, T. (1991) *DNA Replication*, 2nd Edn. W.H. Freeman and Co., New York.

Lewin, B. (1997) *Genes VI*. Oxford University Press, Oxford.

Lodish, H., Baltimore, D., Berk, A., Zipursky, S.L., Matsudaira, P. and Darnell, J. (1995) *Molecular Cell Biology*, 3rd Edn. Scientific American Books, New York.

Turner, P.C., McLennan, A.G., Bates, A.D. and White, M.R.H. (1997) *Instant Notes in Molecular Biology*. BIOS Scientific Publishers Limited, Oxford.

Smith, C.A. and Wood, E.J. (1991) *Biological Molecules*. Chapman & Hall, London.

Section D

Barrow, G.I. and Feltham, R.K.A. (1993) *Cowan and Steel's Manual for the Identification of Medical Bacteria*, 3rd Edn. Cambridge University Press, Cambridge.

Cappucino, T.G. and Sherman, N. (1996) *Microbiology: A Laboratory Manual*, 4th Edn. The Benjamin-Cummings Publishing Co., Redwood City, CA.

Chan, E.C.S., Pelczar, M.J. and Krieg, N.R. (1993) *Laboratory Exercises in Microbiology*, 6th Edn. McGraw-Hill.

Evans, W.H. and Graham, T.M. (1989) *Membrane Structure and Function*. IRL Press, Oxford.

Isaac, S. and Jennings, D. (1995) *Microbial Culture*. BIOS Scientific Publishers Limited, Oxford.

Logan, N.A. (1994) *Bacterial Systematics*. Blackwell Scientific Publications, Oxford.

Penn, C. (1991) *Handling Laboratory Microorganisms*. Open University Press, Milton Keynes.

Section E

Brown, T.A. (1992) *Genetics: A Molecular Approach*, 2nd Edn. Chapman & Hall, London.

Dale, J. (1998) *Molecular Genetics of Bacteria*, 3rd Edn. Wiley-Interscience, New York.

Freidberg, E.C., Walker, G.C. and Siede, W. (1995) *DNA Repair and Mutagenesis*. American Society for Microbiology, Washington, DC.

Hardy, K. (1986) *Bacterial Plasmids*, 2nd Edn. Van Nostrand Reinhold, New York.

Maloy, S.R., Cronan, J.E. and Freifelder, D. (1994) *Microbial Genetics*, 2nd Edn. Jones & Bartlett Publishers Inc., Portola Valley, CA.

Section F

Janeway, C.A. and Travers, P. (1997) *Immunobiology: The Immune System in Health and Disease*, 3rd Edn. Garland Publishing, New York.

Mims, C., Dimmock, N., Nash, A. and Stephen, J. (1995) *Mims' Pathogenesis of Infectious Disease*, 4th Edn. Academic Press, London.

Russell, A.D. and Chopra, I. (1996) *Understanding Antibacterial Action and Resistance*, 2nd Edn. Ellis Horwood, Hemel Hempstead.

Salyers, A. and Whitt, D.D. (1994) *Bacterial Pathogenesis: A Molecular Approach*. American Society for Microbiology, Washington, DC.

Schaechter, M., Medoff, G. and Eisenstein, B.I. (1993) *Mechanisms of Microbial Disease*. Williams & Wilkins, Baltimore, MD.

Section G

http://phylogeny.arizona.edu/tree/eukaryotes/eukaryotes.html

Hall, J.L. and Hawes, C.R. (1991) *Electron Microscopy of Plant Cells*. Academic Press, London.

Sadava, D. (1993) *Cell Biology: Organelle Structure and Function*. Jones & Bartlett, London.

Smith, C.A. and Wood, E.J. (1996) *Cell Biology*. Chapman & Hall, London.

Section H   http://phylogeny.arizona.edu/tree/eukaryotes/fungi/fungi.html

Alexopoulos, C.J., Mims, C.W. and Blackwell, M. (1996) *Introductory Mycology*, 4th Edn. John Wiley & Sons, Chichester.

Carlile, M.J. and Watkinson, S.C. (1994) *The Fungi*. Academic Press, London.

Deacon, J.W. (1998) *Modern Mycology*. Blackwell Science Ltd, Oxford.

Ingold, C.T. and Hudson, H.J. (1993) *The Biology of Fungi*. Chapman & Hall, London.

Isaacs, S. (1992) *Fungal-Plant Interactions*. Chapman & Hall.

Jennings, D.H. and Lysek, G. (1996) *Fungal Biology*. BIOS Scientific Publishers Limited, Oxford.

Manners, J.G. (1993) *Principles of Plant Pathology*. Cambridge University Press, Cambridge.

Section I   http://phylogeny.arizona.edu/tree/eukaryotes/green-plants.html

Bold, H.C. and Wynne, M.J. (1985) *Introduction to the Algae*. Prentice-Hall, New Jersey.

South, G. and Whittick, A. (1987) *Introduction to Phycology*. Blackwell Scientific Publications, Oxford.

van den Hoek, C. and Mann, D.G. (1996) *Algae, an Introduction to Phycology*. Cambridge University Press, Cambridge.

Section J   Anderson, O.R. (1988) *Comparative Protozoology: Ecology, Physiology, Life History*. Springer-Verlag, New York.

Biagini, G.A., Rutter, A.J., Finlay, B.J. and Lloyd, D. (1998) Lipids and lipid metabolism in the microaerobic free-living diplomonad *Hexamita* sp. *Eur. J. Protistol.* **34**(2), 148–152.

Cox, F.E.G. (ed.) (1993) *Modern Parasitology: A Textbook of Parasitology*. Blackwell Scientific Publications, Oxford.

Gordon, G.LR. and Phillips, M.W. (1998) The role of anaerobic gut fungi in ruminants. *Nutr. Res. Rev.* **11**(1), 133–168.

Green, R.F. and Noakes, D.L.G. (1995) A little bit of sex is as good as a lot. *J. Theoret. Biol.* **174**(1) 87–96.

Ridley, R.G. (1997) Plasmodium: Drug discovery and development – an industrial perspective. *Exper. Parasitol.* **87**(3), 293–304.

Rosenthal, P.J. (1998) Proteases of malaria parasites: New targets for chemotherapy. *Emerg. Infect. Dis.* **4**(1), 49–57.

Stevens, J.R., Tibayrenc, M. (1996) *Trypanosoma Brucei* evolution linkage and the clonality debate. *Parasitol.* **34**(2), 148–152.

Section K   Cann, A. (1993) *Principles of Molecular Virology*. Academic Press, London.

Fields, *et al.* (1996) *Fundamental Virology*, 3rd edn. Lippincott-Raven, Philadelphia.

Harper, D.R. (1998) *Molecular Virology*, 2nd edn. BIOS Scientific Publishers Limited, Oxford.

Primrose and Dimmock (1994) *Introduction to Modern Virology*, 4th edn. Blackwell Science, Oxford.

INDEX